Springer Complexity

Springer Complexity is an interdisciplinary program publishing the best research and academic-level teaching on both fundamental and applied aspects of complex systems – cutting across all traditional disciplines of the natural and life sciences, engineering, economics, medicine, neuroscience, social and computer science.

Complex Systems are systems that comprise many interacting parts with the ability to generate a new quality of macroscopic collective behavior the manifestations of which are the spontaneous formation of distinctive temporal, spatial or functional structures. Models of such systems can be successfully mapped onto quite diverse "real-life" situations like the climate, the coherent emission of light from lasers, chemical reaction-diffusion systems, biological cellular networks, the dynamics of stock markets and of the internet, earthquake statistics and prediction, freeway traffic, the human brain, or the formation of opinions in social systems, to name just some of the popular applications.

Although their scope and methodologies overlap somewhat, one can distinguish the following main concepts and tools: self-organization, nonlinear dynamics, synergetics, turbulence, dynamical systems, catastrophes, instabilities, stochastic processes, chaos, graphs and networks, cellular automata, adaptive systems, genetic algorithms and computational intelligence.

The three major book publication platforms of the Springer Complexity program are the monograph series "Understanding Complex Systems" focusing on the various applications of complexity, the "Springer Series in Synergetics", which is devoted to the quantitative theoretical and methodological foundations, and the "SpringerBriefs in Complexity" which are concise and topical working reports, case-studies, surveys, essays and lecture notes of relevance to the field. In addition to the books in these two core series, the program also incorporates individual titles ranging from textbooks to major reference works.

Springer Series in Synergetics

Founding Editor: H. Haken

The Springer Series in Synergetics was founded by Herman Haken in 1977. Since then, the series has evolved into a substantial reference library for the quantitative, theoretical and methodological foundations of the science of complex systems.

Through many enduring classic texts, such as Haken's *Synergetics* and *Information and Self-Organization*, Gardiner's *Handbook of Stochastic Methods*, Risken's *The Fokker Planck-Equation* or Haake's *Quantum Signatures of Chaos*, the series has made, and continues to make, important contributions to shaping the foundations of the field.

The series publishes monographs and graduate-level textbooks of broad and general interest, with a pronounced emphasis on the physico-mathematical approach.

For further volumes:
http://www.springer.com/series/712

Yoshitsugu Oono

The Nonlinear World

Conceptual Analysis and Phenomenology

Springer

Yoshitsugu Oono
University of Illinois, IL, USA

ISSN 0172-7389
ISBN 978-4-431-54028-1 ISBN 978-4-431-54029-8 (eBook)
DOI 10.1007/978-4-431-54029-8
Springer Tokyo Heidelberg New York Dordrecht London

Library of Congress Control Number: 2012951556

Typesetting: Camera-ready by author

Printed on acid-free paper

Springer is part of Springer Science+Business Media (www.springer.com)

Preface

This book explains a certain way of appreciating "the world filled with nonlinearity." Its core is conceptual analysis and phenomenology, which is backed up by renormalization philosophy. The main target of the book is young people who have just started to appreciate the world seriously. The author wishes the book to be helpful also for those who have been observing the world, but who wish to appreciate it afresh from a different angle.

The most important feature of the nonlinear world is that disparate space–time scales (e.g., macroscopic and microscopic scales) can interfere with each other. Consequently, events of the world directly observable on our own space–time scale are, generally speaking, not closed within themselves. That is, to understand a phenomenon occurring within our human space–time scale, we must often take into account the things happening at space–time scales disparate from ours. It is usually the case, however, that we cannot know what is actually going on at scales far from the ones we can directly observe. The so-called chaos clearly exhibits consequences of this intrusion of the unknowable (at small scales) into the world we experience directly. These intrusions of the unobservable into our directly observable world make the world we wish to comprehend not self-contained. Thus, nonlinearity generates various phenomena that are difficult for us to comprehend. We cannot know everything we need to in order to understand the world in terms of its fundamental laws (even if they exist).

If we wish to understand the world clearly, its description must be precise in terms of unambiguous words. Phenomena we have not understood clearly will demand novel concepts and terminologies to describe them. Trying to characterize the key concepts relevant to a not-yet-comprehended phenomenon may well be the better part of its study. Thus, 'conceptual analysis,' an attempt to make intuitively grasped concepts clear, is crucial. Needless to say, conceptual analysis is always important in any serious study. Therefore, this book begins its main body with the conceptual analysis, following an introductory Chapter 1 that outlines the significance of 'nonlinearity.'

If the unknowable of, e.g., the microscopic world, affects the world we directly experience, it is not always possible to understand what we observe solely in terms of the so-called fundamental principles. We are accustomed to trying to understand a phenomenon in terms of well-established microscopic fundamental laws, but this need not be a suitable strategy to understand a typically nonlinear phenomenon. That is why we need a method to rationalize what we observe without referring to 'fundamental laws.' This is 'phenomenology.' The understanding of the world we wish to attain must not only be clear, but also be general enough. In other words, we are not interested in understanding only a particular system, but wish to reach some level of universal understanding applicable to a class of phenomena. However, the reader will realize that not every phenomenon allows us to attain understanding with some universality; only the phenomenon whose 'good phenomenology' may be constructed allows that. Accordingly, in Chapter 3, after surveying general characteristics of phenomenology and its typical examples, 'renormalization group philosophy' is explained as a means to extract phenomenology, and in Chapter 4, modeling is explained as a means to express the accomplished phenomenological understanding.

Because very different space–time scales entangle and interfere with each other in the nonlinear world, not only chaos but also other well-known peculiar phenomena such as critical phenomena occur. This book only utilizes such phenomena for the sake of illustration and does not aim to explain them in detail. The chief purpose of the book is to advocate a particular stance to appreciate the world. Consequently, standard technical topics such as bifurcation theory will not be discussed in this book at all.

Nonlinearity is indispensable to create a complex system, so numerous books discussing nonlinearity also try to discuss 'complex systems.' Even the present book is no exception, since its last chapter is entitled "Toward Complexity." However, the book is different from other such books in the respect that it clearly explains the distinction between the study of complexity and so-called complex systems studies. It will become clear to the reader that studying complex systems leads us to reflect critically upon the nature of so-called fundamental science. The ultimate goal of genuine complex systems studies must be, from the point of view of this book, to accomplish conceptual analysis of complexity and to construct a phenomenology of complex systems. The last chapter outlines preliminary efforts toward this goal.

The author belongs to a physics department and has always wished to be a student of fundamental physics. A fundamental physicist is a person who pursues 'good ways' to understand an object under study, whatever it may be. Here, 'good ways' imply the 'right frameworks' ('right coordinate systems,' metaphorically) that allow us to understand the object simply and clearly. A fundamental physicist is the kind of person who thinks she is stupid if she cannot find 'good ways.' Biology is so called because it studies biological systems, and chemistry is so called because it studies chemical substances and reactions. Physics is also regarded as a discipline to study the so-called physical world. However, this view is oblivious of the original meaning of *Physica*, which is the study of nature in general. Physics should not restrict its research topics; everything in nature is fair game.

There must be readers who claim that the basic framework explained in the book cannot be general enough for unbiased appreciation of the world; it is only enough for the fundamental physicists' way of appreciating the world. It is the author's firm belief, however, that the way of appreciating the world acquired through fundamental physics is sufficiently universal and is an important part of human culture.

"The way of appreciating things" would be, if perfected, mathematical for objects and phenomena that permit clear description in words. However, being mathematical need not mean being technical. In contrast to the ordinary books devoted to nonlinear science, in this book technical topics will be kept to a minimum. We can appreciate music without much understanding of musical grammar. Without listeners with good taste, no sophisticated musical culture would be sustained, but these listeners need not be familiar with musical grammar. Mathematical and science culture would be similar. Therefore, the book does not emphasize tools and techniques.[1] Some parts may be, however, mathematically rather heavy, but in such cases there are suggestions on how to skip or browse through complicated passages.

The aim of this book is to present a certain "way to think," so inevitably many topics and subjects outside physics must be discussed. The author firmly believes that there should not be the so-called two cultures.[2] Therefore, this book has comments, remarks and footnotes that may urge natural scientists to pay some attention to topics of the humanities.[3] The author wishes to count the people on the humanities side among the readers, but

[1] Some of the more detailed explanations of tools and techniques as well as actual examples will be given in the supporting website as noted below.

[2] ⟪**The two cultures**⟫ "A good many times I have been present at gatherings of people who, by the standards of the traditional culture, are thought highly educated and who have with considerable gusto been expressing their incredulity at the illiteracy of scientists. Once or twice I have been provoked and have asked the company how many of them could describe the Second Law of Thermodynamics. The response was cold: it was also negative. Yet I was asking something which is about the scientific equivalent of: *Have you read a work of Shakespeare's*?" (C. P. Snow, *The Two Cultures and the Scientific Revolution*, The Rede Lecture 1959 (Cambridge University Press, 1961) pp15-16). (However, the reader must be critical about whether Snow's understanding of thermodynamics is sound enough.)

[3] This is an example of such footnotes. This book has many long footnotes. The reader may only browse through such footnotes, regarding them as "boxed notes."
⟪**On scientists, Ortega said**⟫ "That is to say, modern science, the root and symbol of our actual civilization, finds a place for the intellectually commonplace man and allows him to work therein with success. The reason of this lies in what is at the

free use of elementary mathematics in this book probably requires the writing of another book more suitable for a wider audience. However, even if no single formula in this book is comprehended, still a considerable portion of the discourse should be understandable; the reader could browse through only provocative footnotes.

Is this a kind book for students learning natural sciences? The book is for those who wish to reconsider very basic topics. Consequently, it demands some discipline and patience on the readers' side. However, even if early chapters are not understood well, still the remaining parts should make considerable sense. The main ideas of the chapters are related, but can be read almost independently. Chapter 2 is, however, more mathematical than the other chapters, so reading it lightly and going to subsequent chapters might be a wise first reading. Needless to say, Chapter 2 cannot be totally irrelevant to the subsequent chapters. Technical terms and concepts explained in Chapter 2 appear in the subsequent chapters. This chapter also prepares the reader to look at the so-called complex systems studies critically. Reviewing certain parts of Chapter 2 later when needed may be practical.

This book is based on the lecture notes for a special course the author gave at Keio University (Yokohama, Japan) in the summer of 1996 as Toshiba Chair Professor. The course was sponsored by the Toshiba Corporation. The zeroth version was prepared (in English and in Japanese simultaneously) toward the end of the last century, taking into account comments from Shin-ichi Sasa, Ken Sekimoto, Yoichiro Takahashi, Hal Tasaki, and Ichiro Tsuda. Then, a set of lecture notes in English was prepared for a University of Illinois at Urbana-Champaign (UIUC) special physics course. Based on it, a book in Japanese was constructed with the last chapter added, which is based on the commentary portion of the lecture notes called *Integrative Natural History*

---

same time the great advantage and the gravest peril of the new science, and of the civilization directed and represented by it, namely, mechanisation. A fair amount of the things that have to be done in physics or in biology is mechanical work of the mind which can be done by anyone, or almost anyone. For the purpose of innumerable investigations it is possible to divide science into small sections, to enclose oneself in one of these, and to leave out of consideration all the rest. The solidity and exactitude of the methods allow of this temporary but quite real disarticulation of knowledge. The work is done under one of these methods as with a machine, and in order to obtain quite abundant results it is not even necessary to have rigorous notions of their meaning and foundations. In this way the majority of scientists help the general advance of science while shut up in the narrow cell of their laboratory, like the bee in the cell of its hive, or the turnspit in its wheel." [José Ortega y Gasset, *The Revolt of the Masses* (Norton & Co., New York 1932) p110-111]. "The most immediate result of their *unbalanced* specialisation has been that today, when there are more 'scientists' than ever, there are much less 'cultured' men than, for example, about 1750." (p113) "The majority of men of science have given themselves to it through fear of facing life. They are not clear heads; hence their notorious ineptitude in the presence of any concrete situation." (p157)

(for the courses given at UIUC, Keio, and Waseda).[4] On this occasion detailed criticisms/corrections supplied by Hal Tasaki, Hayato Chiba, Masahiko Todoroki, Tohru Tsujishita, and Seiichiro Honjo (in chronological order) and further comments of Hisao Hayakawa, Kuni Kaneko, Yoichiro Takahashi, and Akira Shimizu were incorporated. This present volume is also based on the same source as the Japanese book, but several comments by the editor of the Japanese book, Junsei Kishi, were utilized. The explanation of the proto-renormalization group approach was composed with the help of Yasuhiro Shiwa. Encouraging philosophical discussions with Piet Hut and useful comments of Phil Baldwin are gratefully acknowledged.

At the end of the Foreword to *Homo Ludens*[5] Huizinga wrote in June 1938, "In treating of the general problems of culture one is constantly obliged to undertake predatory incursions into provinces not sufficiently explored by the raider himself. To fill in all the gaps in my knowledge beforehand was out of the question for me. I had to write now, or not at all. And I wanted to write."

So did the author.

Urbana, Illinois, USA, September, 2011 *Yoshitsugu Oono*

[4] A short version of the commentary portion is published in *Physics of Self-Organization Systems* (S. Ishiwata and Y. Matsunaga, editors, World Scientific, 2008).

[5] J. Huizinga, *Homo Ludens, a study of the play-element in culture* (Beacon Press, 1950; based on the German original and the author's translation).

How to Read This Book

The book intertwines general discussions, more technical discussions with formulas, and rather technical explanations. Although some technical topics are relegated to appendices to sections and chapters and fine-lettered notes, still many heavy parts remain in the main text, which the readers seeking an overview may wish to skip. Therefore, 'jumping guidance and instructions' are explicitly written at various strategic positions to facilitate easy browsing. These instructions are in italic in the text.

The reader will realize that most pages have footnotes and some of them are long, even with titles surrounded by ⟪ ⟫. As stated earlier many of the footnotes especially with titles may be read as 'boxed' articles without referring to the main text. Thus, skimming only footnotes may be one way to read this book.

The book has an accompanying support webpage as stated below, where errata as well as additional and augmenting material and updating information may be found. Long materials such as a detailed introductory ε-expansion calculation are posted there. Also useful comments from the readers will be gratefully posted with explicit acknowledgment of the contributors.

The following explanations repeat the key points.

Footnotes
As already noted this book has numerous footnotes, some of which are quite long. Some readers may find them bothersome, but long footnotes, especially with titles, may be read almost independently of the text. Therefore, such footnotes may be regarded as boxed short articles. Some readers may read only these footnotes, since they may be more radical or provocative. This is one way to read this book. The author reads many notes and comments while reading books and is irritated if he has to look for them at the end of chapters and books. This is why remarks and notes are placed as closely as possible to their relevant locations.

Discussions
We will occasionally pose problems and questions as Discussions. Some of them are mere exercises, but some are intended to supply the seeds of discussion. Accordingly, Discussions may have more than one solutions or may have none. The author wishes the reader to pause occasionally to reflect on 'deeper questions.'

Support Pages
A lot of material has been relegated to the Support Pages for this book: `http://www.yoono.org/NonlinearWorldSpringer/Nonlinear_World_Supplements.html`. Additional information, some technical details and updates will be found there. Ideally, the book itself is a *growth kernel of a collection posted here of long and short lecture notes and memos* relevant to any topic discussed in this book.

Contents

Chapter 1
Looking at the Nonlinear World

The purpose of this chapter is to provide an overview of the main topics discussed in this book and to explain why they are selected. The structure of the book is also outlined. The reader might not be able to understand every statement in this chapter, because concepts that are not very familiar may be used without any explanation. The reader has, however, only to understand roughly the salient features of the nonlinear world. Related cultural background is also sketched. The themes in bold face will be explained from the elementary level in subsequent chapters.

The main aim of the book is to explain a particular mode of thinking. Accordingly, to explain the needed technical details is not its main concern. Technical details are given only when they are directly connected to the mode of thinking. However, for the readers who wish to go beyond the level of this book a sort of study guide will be given at appropriate places. Although the book does not pay much attention to required mathematical tools, we must be conscious of the notion that for a given phenomenon there is an appropriate mathematics just as a particular novel requires an appropriate style. For example, we must clearly recognize that to teach classical mechanics without calculus is blasphemous.

The title of this book contains the word 'nonlinear.' Frankly, the author does not wish to single out 'nonlinear something.' Almost everything in the world is nonlinear, so the 'Nonlinear World' is the world itself. If one still wishes to speak specifically about the 'Nonlinear World,' what does one wish to pay special attention to? To answer this question, let us consider the features of 'nonlinear systems.' To this end we first review the characteristics of 'linear systems.'

1.1 Characteristics of linear systems

The term *system* indicates a more or less coherent set of objects which is a part of Nature (e.g., an electronic circuit, a dog, the Earth, a black hole). Etymologically, this word is made up of *syn* (together) and *histanai* (stand), so a system is 'a collection of objects forming a unity by standing together.' The meaning of 'forming a unity by standing together' is, ideally, (1) the objects belonging to the system and all other objects can be distinguished, and (2) the objects belonging to the system interact with one another in a system-specific way and their interactions with the external world may be treated separately.[1]

For a given system S let us assume that its output y for an input x be written as follows with the aid of a map Q (when we say a 'map' mathematically, y must be unique for a given x. For simplicity, we assume this uniqueness):

$$y = Q(x). \tag{1.1}$$

For example, x is an initial condition for a **dynamical system** (roughly, a system whose time evolution rule is given) and y is its state at time $t = 1$.

A system S is called a *linear system*, if the map Q is linear. That is, for arbitrary inputs x_1 and x_2

$$Q(\alpha x_1 + \beta x_2) = \alpha Q(x_1) + \beta Q(x_2) \tag{1.2}$$

holds, where α and β are arbitrary scalar constants. (1.2) is called the *superposition principle*. A system satisfying the superposition principle is a linear system. The key points of this principle are (linear) *scaling invariance*

$$Q(\alpha x) = \alpha Q(x) \tag{1.3}$$

and *additivity*

$$Q(x_1 + x_2) = Q(x_1) + Q(x_2). \tag{1.4}$$

For example, if the extension ΔL of a spring is proportional to the magnitude of the applied force F, i.e., $\Delta L = F/k$ with k being a positive constant (Hooke's constant), the spring is a linear system. If we double the force, the extension certainly doubles. Another well-known example is a static electric field created by a spatial configuration of electric charges; the field is a simple superposition of the fields due separately to individual charges.

---

[1] All things in the world are interacting with one another through inseparable interactions, so one may well conclude that it is impossible to set up a *system* that can be segregated distinctively from the rest of the world. Here, we wish to proceed, assuming the ordinary common-sense understanding of the term *system*. However, we will come back to this question in Chapter 3 (especially toward the end of Section 3.3). According to the consideration there, the interaction between a 'system' and its surrounding external world need not be weak.

Discussion 1.1. The attractive part of the intermolecular force is caused by charge interactions, but it does not satisfy the superposition principle. Discuss why. □

Discussion 1.2. For those who have studied elementary quantum mechanics, it may be a good occasion to review the Hellman-Feynman theorem, and to consider its relation to the preceding discussion topic. □

For simplicity, up to this point, the explanation has proceeded as if whether a system is linear or not is its intrinsic property. The reality is not this simple. Study of a system is possible only through observation, so it is not meaningful to discuss a system without specifying the observables through which we study the system.

For example, the Schrödinger equation (see Note 1.1) is a linear equation, so every isolated system is governed by a linear law. This does not imply, however, that all the properties of an isolated system are linear. Imagine an isolated box containing liquid water and vapor. The system presumably obeys a Schrödinger equation, but nonlinear wave phenomena must be observable in its vapor-liquid interface.[2] For system S described by (1.1), if we observe y^2 instead of y, the story is completely different. In quantum mechanics, the wave function = probability amplitude ψ itself cannot be observed. Even in the linear spring case, the stored potential energy in the spring is not a linear function of F.

In short, it is meaningless to say that a given system is 'intrinsically' linear or not. Therefore, the statement, "A system satisfying the superposition principle is a linear system" is, strictly speaking, meaningless; it should have been stated as "An observable y of a system S exhibits a linear response, if the map Q is linear." In this book, however, for simplicity, we assume that for a given system we have already understood which observables are the ones through which we study the system, and we continue to use an informal expression such as 'a system is linear.'

It is very convenient if a system can be described as a linear system, so the question through what observables we can assume a wide range of linearity to describe the system is an important one. However, general strategy to answer this question does not seem to exist.

Note 1.1. Schrödinger equation.
Suppose there are N point particles in a three-dimensional space. The mass of the j-th particle is denoted as m_j $(j = 1, 2, \cdots, N)$. Let the position vector of the j-th particle be $\boldsymbol{r}_j = (x_j, y_j, z_j)$. If the potential energy of the interacting

[2] As can be seen from the relation between Hamilton's equation of motion and the Liouville equation in classical mechanics, nonlinear ordinary differential equations can always be rewritten as linear partial differential equations. In this book such a method utilizing an infinite dimensional (or functional) space is not considered. The Schrödinger equation reminds us of the Liouville equation, but here we stick to the ordinary interpretation of the equation and regard it as the fundamental equation of motion.

particles is given by $V(\boldsymbol{r}_1, \boldsymbol{r}_2, \cdots, \boldsymbol{r}_N)$, the Schrödinger equation for this N-particle system is

$$i\hbar\frac{\partial\psi}{\partial t} = -\sum_{j=1}^{N}\frac{\hbar^2}{2m_j}\Delta_j\psi + V(\boldsymbol{r}_1, \cdots, \boldsymbol{r}_N)\psi, \tag{1.5}$$

where Δ_j is the Laplacian with respect to $\boldsymbol{r}_j$:

$$\Delta_j \equiv \frac{\partial^2}{\partial x_j^2} + \frac{\partial^2}{\partial y_j^2} + \frac{\partial^2}{\partial z_j^2},$$

$\hbar$ is the Dirac constant (the Planck constant h divided by 2π; $h = 6.626\times 10^{-34}$ Js, $\hbar = 1.0546 \times 10^{-34}$ Js), and $\psi = \psi(t, \boldsymbol{r}_1, \cdots, \boldsymbol{r}_N)$ is the wave function. Let $d\tau_j$ be the volume element centered at $\boldsymbol{r}_j$ (let us use the same symbol to denote its volume). Then, $|\psi(t, \boldsymbol{r}_1, \cdots, \boldsymbol{r}_N)|^2 d\tau_1 \cdots d\tau_N$ is proportional to the probability of the event that, for all $j = 1, \cdots, N$, particle j is found in $d\tau_j$ simultaneously at time t.

Just as Newton's equation of motion cannot be logically derived from more fundamental laws, the Schrödinger equation cannot be derived from a set of more basic laws. Its legitimacy lies in its consistency with the empirical facts. Schrödinger gave various plausibility arguments in the paper(s) in which he proposed this equation (1926[3]), but the agreement of the energy eigenvalues of the hydrogen atom to the empirical results finally convinced him of the correctness of the equation.[4]

Discussion 1.3. In the above it is stated that liquid water and vapor in an isolated box obeys a Schrödinger equation. How can you be sure about this statement, or how can you confirm this statement empirically? □

Let us study a typical example of linear systems. A thin copper rod is thermally isolated except for its ends, that is, heat exchange does not occur except at the ends. At the ends we may impose various boundary conditions. Its temperature distribution (field) at time t (numerically in seconds) is written as $T(t, x)$, where the temperature may be measured in °C and x is the position coordinate (numerically in meters). Let us assume that the origin of the x-coordinate is at the left end in Fig. 1.1. If the initial distribution of the

[3] [1926: Byrd reached the North Pole by plane.] This year Schrödinger published four papers "Quantization as an eigenvalue problem I-IV" and completed singlehandedly in a single stroke the elementary quantum mechanics that we are taught now. All the papers (translated into English) are found in *Collected Papers on Wave Mechanics* (American Mathematical Society; 3rd Revised edition (2003)). The English titles of the papers in this collection is "Quantization as a problem of proper values I-IV." A slightly advanced knowledge of analytical mechanics is required to digest the papers.

[4] There are numerous textbooks of quantum mechanics. As an introduction D. J. Griffiths, *Introduction to Quantum Mechanics* (Prentice Hall, Upper Saddle River, NJ, 1995) is excellent. After studying the basics (and thicker standard textbooks such as by J. J. Sakurai), it is recommended to read at least the first half of P. A. M. Dirac, *Principles of Quantum Mechanics* (Oxford University Press, 1982; paperback). The author could understand (the spirit of) quantum mechanics thanks to this book, when he was a beginning student (of organic chemistry). The third volume of the Feynman Lectures (R. P. Feynman, R. B. Leighton and M. Sands, *The Feynman Lectures on Physics III* (Addison-Wesley, 1971)) provides an excellent introduction.

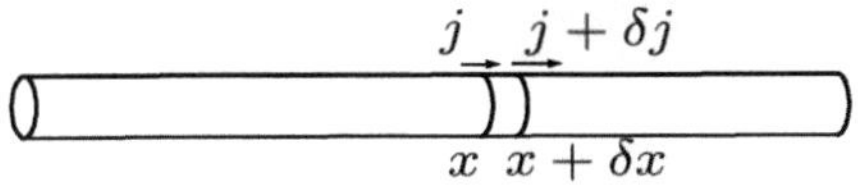

Fig. 1.1 If we consider the thermal energy balance for the slice between x and $x+\delta x$ of the rod, we obtain the diffusion equation. Here, j is the heat flux through the cross section at x from left to right, and $j + \delta j$ is analogous at $x + \delta x$.

temperature is not too wild, the time evolution of the temperature field obeys the *diffusion equation*

$$\frac{\partial T}{\partial t} = \kappa \frac{\partial^2 T}{\partial x^2}, \tag{1.6}$$

where κ is a positive constant called the *thermal diffusion constant.*

Discussion 1.4. What happens if the temperature distribution changes wildly from point to point? What do we mean by 'wild change' of the temperature? □

Let us review the derivation of (1.6). Let j be the *heat flux* (amount of thermal energy passing through the unit cross-sectional area of the rod per unit time in the positive coordinate direction). In the present case, since thermal energy does not transform into any other form of energy, we can obtain the following relation from the thermal energy balance (see Fig. 1.1; the symbol o in the following equation generally denotes a higher-order smaller quantity in an appropriate limit: $\lim_{\delta x \to 0} o[\delta x]/\delta x = 0$ in the present case):

$$\delta h = [j(x,t) - j(x + \delta x, t)] S \delta t = \left[-\frac{\partial j}{\partial x} \delta x + o[\delta x] \right] S \delta t, \tag{1.7}$$

where δh is the thermal energy[5] increase in the small slice of width δx during the small time interval δt, and S is the cross section of the rod. Let C be the heat capacity per unit volume (i.e., the specific heat) of the rod, and let δT be the temperature change of the small volume under consideration during δt. Its volume is $S\delta x$, so $\delta h = (CS\delta x)\delta T$.

We assume a linear law (*Fourier's law*) between the heat flux and the temperature gradient:

$$j(x) = -K \frac{\partial T}{\partial x}, \tag{1.8}$$

where K is called the *thermal conductivity* and must be positive due to the second law of thermodynamics (due to *Clausius' principle*). From (1.7) and (1.8) we obtain

---

[5] In Fourier's day, the nature of heat was not well understood. It was regarded as a kind of substance (caloric theory). Just as mass is conserved, it was thought, so is 'calorique.' This notion is not so bad if thermal energy does not transform into any other forms of energy.

$$C\delta T = \left[\frac{\partial}{\partial x}\left(K\frac{\partial T}{\partial x}\right) + o[1]\right]\delta t. \tag{1.9}$$

That is,

$$C\frac{\partial T}{\partial t} = \frac{\partial}{\partial x}\left(K\frac{\partial T}{\partial x}\right). \tag{1.10}$$

(1.10) reduces to the diffusion equation (1.6) with $\kappa \equiv K/C$, if neither C nor K depends on temperature (thanks to the uniformity of the rod, these quantities do not explicitly depend on x). The *differential operator*[6] appearing in the partial differential equation (1.6) satisfies the superposition principle (1.2), so the diffusion equation describes a linear system.

An important lesson we should learn from the derivation of the diffusion equation just reviewed is that we cannot derive linear equations unless we assume some sort of simplification/idealization (e.g., C does not depend on temperature). Almost everything in the world is nonlinear.

Note that we have assumed that T is a sufficiently smooth function of space and time. Since we set up the partial differential equation assuming this smoothness, we cannot justify our argument up to this point, unless we can demonstrate that the solution to (1.6) exists and is unique and sufficiently smooth. Natural scientists including physicists have great confidence in their own intuition, and also trust mathematics almost unconditionally, so they never feel any necessity of checking such basic properties of the solutions. Consequently, even if mathematicians prove the unique existence of the solution and demonstrate its smoothness (real analyticity[7] as a function of spatial coordinates), scientists are never impressed nor do they recognize the significance of the mathematicians' work. Mathematicians ask grave questions: Is the scientists' intuition correct, and if so, can our mathematics correctly capture it? In the case of the diffusion equation, everything goes well and scientists' belief seems vindicated. However, sometimes considerably counterintuitive things may happen (due to the idealization required, e.g., to formulate a linear theory) as can be seen from Discussion 1.5 below, so we should not look down on mathematicians as strange creatures talking about the self-evident. Still, mathematicians should take special care when they handle (e.g., educate) scientists, because scientists' intuition is very often correct.

---

[6] ⟪**Differential operator**⟫ Let f be a function of an independent variable y (which may be, e.g., a vector). If the value of $M(f)$ at y depends only on y, and the values of f and its derivatives at y, then M is called a differential operator. $\partial/\partial t$ and $\partial/\partial x(K\partial/\partial x)$ in (1.10) and three-dimensional Laplacian $\Delta \equiv \partial^2/\partial x^2 + \partial^2/\partial y^2 + \partial^2/\partial z^2$ are typical linear differential operators. Therefore, the Schrödinger equation is a linear equation. [Mathematically precisely speaking, operators cannot be defined by their 'forms' alone. Their domains (the set of functions upon which they can act) must be explicitly stated to have complete definitions.]

[7] As a function of x it may be expanded into a Taylor series. This is not a formal expansion; the convergence of the series is implied.

Discussion 1.5. A space between two infinitely large parallel planes is filled with a uniform thermal medium which is initially everywhere at 0°C. The reader may idealize that the thermal diffusion constant is a constant. Let us assume that the two planes are always kept at 0°C. What can be said about the temperature at a point (say, in front of the reader) of the medium in the future? □

That linear systems can generally be obtained only after idealization (or that they have only finite domains of validity) may be understood directly from the scaling invariance (1.3). If T satisfies (1.6), then T multiplied with any number also satisfies (1.6). For example, $10^{10}T$ must be accepted as a feasible solution. However, no one would believe this is possible for a copper wire (cf. the surface temperature of the sun is 6,000 K, to be sure).

If the absolute value (the value itself)[8] of a certain quantity is physically meaningful, and if the equation (or the law) governing the quantity is linear, it does not express a precise natural law. Linear laws are only 'local laws' with respect to the magnitude of the quantity. For example, in (1.6) the numerical value of T itself has the meaning of the temperature measured in Celsius, so it holds good only in a limited range of T values. In the example of the spring we discussed before, everyone knows that the spring is deformed plastically if the magnitude of the force F is too large. The fundamental equations of classical electromagnetism, Maxwell's equations (in the vacuum), cannot describe a fundamental law for the same reason; it is a surprise that no one in the past seems to have criticized the equations in this way. However, the equations can be rigorously linear for quantities whose absolute values do not have any physical meaning. There is nothing fundamentally wrong for the Schrödinger equation to be rigorously linear, because the absolute value of the wave function does not have any physical meaning.[9]

The diffusion equation was written down for the first time by Fourier (1768-1830) around 1805.[10] He was then the Prefect of Isère (Grenoble is the capital of this département), and played an important role in cultural history, beginning an epoch in Egyptology through authoring the academic report of Napoleon's expedition (1798-1801), and through encouraging Champollion (1790-1832), who would decipher hieroglyphs (published in 1824), and preserving him from conscription.[11] Actually, as we will see soon, he played a much bigger cultural role in history and has probably a much deeper influence

[8] 'Absolute value' here is used in contrast to 'relative value,' and does not mean $|x|$ for x. This is also a standard parlance among scientists.

[9] For the same reason, the Liouville equation that governs the density distribution function of the ensemble of a classical mechanical system is rigorously linear.

[10] [1805: F. von Schiller (b. 1759) died; the battle of Trafalgar took place]

[11] T. W. Körner, *Fourier Analysis* (Cambridge University Press, 1988) Chapter 92 "Who was Fourier?"; I. Grattan-Guinness (in collaboration with J. R. Ravetz), *Joseph Fourier, 1768-1830: a survey of his life and work, based on a critical edition of his monograph on the propagation of heat, presented to the Institut de France in 1807* (MIT Press, 1972).

on the world today than his famous contemporaries Göthe and Schiller.[12] However, the ordinary history timetables never mention Fourier. This exemplifies some grave defects inherent in our culture (Appendix 1A to this chapter is intended to compensate for this to some extent).

Fourier's most important cultural contribution is to have provided the growth nucleus of modern mathematics through his provocative assertion that "any function can be expanded in (if we use the modern terminology) a *Fourier series*" in conjunction to a general method to solve the diffusion equation. Fourier asserted[13] that the solution T to (1.6) may always be written in the following *Fourier series*: with time-dependent functions $a_n(t)$ and $b_n(t)$

$$T(t,x) = \frac{1}{2}a_0(t) + \sum_{n=1}^{\infty}\{a_n(t)\cos nkx + b_n(t)\sin nkx\}. \tag{1.11}$$

Here, we have assumed that the length of the copper rod is L and $k \equiv 2\pi/L$. That is, he asserted that the solution can be written as a superposition of all the standing waves that can exist on the rod. Furthermore, Fourier proposed a way to compute the coefficients:

$$a_n(t) = \frac{2}{L}\int_0^L T(t,x)\cos nkx\,\mathrm{d}x, \quad b_n(t) = \frac{2}{L}\int_0^L T(t,x)\sin nkx\,\mathrm{d}x. \tag{1.12}$$

These formulas can be checked easily with the aid of the orthogonality of trigonometric functions:

$$\frac{2}{L}\int_0^L \mathrm{d}x\,\cos nkx\cos mkx = \delta_{nm}, \ \frac{2}{L}\int_0^L \mathrm{d}x\,\sin nkx\sin mkx = \delta_{nm}, \tag{1.13}$$

if we may exchange the order of summation and integration (for $n = m = 0$, use $(1/L)\int_0^L \mathrm{d}x = 1$). Fourier gave this level of proofs; he was not advocating mere guesses. If we put (1.11) into (1.6) and if we may exchange summation and differentiation, we obtain:

[12] Familiar results of those influences include the music compression technology MP3 and picture compression technology JPEG, whose core is Fourier expansion (and information theory). Fourier was also the first to point out the greenhouse effect of gases in the atmosphere (in 1824).

[13] All the subsequent formal calculations can be justified. The key is that the convergence of the Fourier series is very quick; the essence is that contribution from the spatially rapidly changing modes (large n modes) is not large. This should be physically no surprise. The diffusion equation is an equation describing the process that the distribution of temperature or material becomes blurred, so small-scale structures in the initial distribution should be washed away as time proceeds. Therefore, the solution must be smooth and its Fourier series should converge quickly.

$$\frac{1}{2}\dot{a}_0(t) + \sum_{n=1}^{\infty}\{[\dot{a}_n(t) + \kappa n^2 k^2 a_n(t)]\cos nkx \\ + [\dot{b}_n(t) + \kappa n^2 k^2 b_n(t)]\sin nkx\} = 0, \tag{1.14}$$

where $\dot{}$ denotes the time derivative.

If we may exchange the order of infinite summation and integration, with the aid of the orthogonality of trigonometric functions (1.13), we may separate the time evolution of the distinct 'modes' (i.e., terms described by individual trigonometric functions). That is, the time evolution of the amplitude of each mode is closed within each mode. For example,

$$\frac{da_n(t)}{dt} = -\kappa n^2 k^2 a_n(t) \tag{1.15}$$

and neither a_m $(m \neq n)$ nor b_m show up in this equation. Thus, the original partial differential equation has been decomposed into a bunch of ordinary differential equations describing non-interfering individual modes. That no modes interfere with each other implies (in the present example, clearly) that no different spatial scales interfere with each other.[14]

The idea that various functions can be described in terms of Fourier series as (1.11) was for the first time proposed by Daniel Bernoulli (1700-1782) to solve the one-dimensional wave equation (the equation of motion of a string) about 50 years before Fourier.[15] This sounds like a natural idea, if we imagine various vibrational modes of the string. However, no one contemporaneous to D. Bernoulli accepted the idea as a general method, and soon the study was forgotten. Even after 50 years the situation was not very different, and the leading mathematician at that time, Lagrange (1736-1813), did not accept

---

[14] ⟪**Sturm-Liouville operator**⟫ If the rod is not spatially uniform in the above heat-conduction example (i.e., C or K depends on x), we cannot decompose the temperature field into non-interfering modes with the aid of trigonometric functions (we may decompose it into the sum of trigonometric functions, but the amplitudes cannot be separated as (1.15)). Even in this case, the theory of generalized Fourier expansion allows us to decompose the temperature field into non-interfering modes. If the rod is not spatially uniform, the equation becomes $\partial_t T = C^{-1}\partial_x(K\partial_x T)$. The differential operator $C^{-1}\partial_x K \partial_x$ is generally called the *Sturm-Liouville operator* and its eigenfunction system $\{\varphi_n(x)\}$ makes a complete orthogonal system. That is, 'any' function $T(t,x)$ may be expanded as $T(t,x) = \sum_n a_n(t)\varphi_n(x)$, where each $a_n(t)$ evolves in time without interfering with other amplitudes. Generally speaking, the spatial variation of $\varphi_n(x)$ becomes greater as n becomes larger. In this sense, disparate spatial scales do not interfere in the generalized Fourier expansion, either.
⟪**Remark on more general differential operators**⟫ Not all the linear operators allow orthogonal bases consisting of their eigenfunctions. Then, the generalized Fourier expansion does not work and various pathological things may happen. See L. N. Trefethen and M. Embree, *Spectra and pseudospectra: the behavior of nonnormal matrices and operators* (Princeton UP, 2005).

[15] For the history of partial differential equations see H. Brezis and F. Browder, "Partial differential equations in the 20th century," Adv. Math. **135**, 76 (1998).

the idea. However, Cauchy (1789-1857), who is regarded as the founder of modern analysis, became serious about the foundation of analysis (at least partially) because of Fourier's provocative assertion: "any function may be written as a sum of trigonometric functions."

We may even say that a sizable portion of modern mathematics was created to understand this statement correctly and to justify it. Fourier's contribution to mathematics was a far bigger contribution to human culture than his contribution to Egyptology, which was not small at all. If the reader wishes to know the sequel to this story, go to Appendix 1A at the end of this chapter, where we have a glimpse of the history of mathematics while reviewing some elementary mathematical facts.

1.2 Characteristics of nonlinear systems

A nonlinear system is a system that is 'not a linear system.'[16] We may say that, since the meaning of 'linear' was clarified in the preceding section, the meaning of 'nonlinear' is now clear, but this is a bit pathetic. Since we have recognized that the almost everything in the world is nonlinear, don't we wish to characterize the term 'nonlinear' positively instead by negation of something else?

For a linear system we may decouple disparate scales, so in order to understand linear phenomena occurring near our human scale, we have only to consider what is happening near our space-time scale. For a 'non'-linear system the superposition principle (1.2) does not hold, so Fourier expansion is not very useful to obtain (exact) solutions. This is because we cannot remove interferences among different modes (i.e., *mode coupling* occurs). Consequently, we cannot necessarily ignore scales removed from the scale of our immediate interest. Doesn't the existence of such *scale interferences* positively characterize nonlinear systems?

Roughly speaking, if a mode observed on our scale[17] is globally stable but

---

[16] As stated in the earlier part of the preceding section, the term 'nonlinear system' is a simplified statement that a certain aspect of the system we are interested in has a description that does not satisfy the superposition principle. However, we will use informal expressions just as in the linear cases.

[17] ⟪**Our scale**⟫ 'Our scale' sounds vague, but this indicates the length scale that is not much smaller than 1 μm (10^{-6} m) nor not much larger than a few km (10^3 m). We did not realize for a long time that we are made of cells, so the length scales smaller than a few tens of μm are safely regarded as disparate length scales from our length scale. The opposite limit of our scale range cannot be much larger than 100 m (the distance we can throw a stone). That is, our everyday scale is within the range of $10^{\pm 4}$ cm.

Isn't such an anthropocentric point of view unscientific? Not quite, because we human beings are products of Nature. What determines the size of *Homo sapiens*?

locally linearly unstable as $\dot{x} = x$, then we may observe *chaos*. Chaos is a phenomenon in which a deterministic system (a system whose behavior in the past up to now completely determines its future behavior) exhibits behavior that is virtually impossible to predict. Chaos exhibits unpredictability despite determinism simply because its time evolution enormously expands the distance between adjacent points in its phase space (the space in which its trajectories lie). What happens is that small-scale events influence global behavior; it is nothing but scale-interference itself. Events in very small scales are in principle (i.e., not due to the technological limitations today) impossible for us to know.[18] Therefore, it is often the case that even a very simple deterministic system is not 'closed' within 'the knowable.' The significance of chaos is that it readily exhibits this fact. We must clearly recognize that the macroscopic world is not closed within what we can know, even when we ignore quantum effects.[19]

Discussion 1.6. For an ordinary differential equation, no two trajectories cross each other due to the uniqueness of the solution to its initial value

---

Large animals, or, more generally, the so-called *megaeukaryotes*, are often constructed with repetitive units such as segments. The size of the repeating unit is at least one order larger than the cell size. Consequently, the size of 'advanced' organisms must be at least 2-3 orders as large as the cell size. The size of the eukaryotic cell is about 10 μm, so we cannot be much smaller than 1mm (the smallest frog is about 10 mm; the smallest invertebrate is about 0.5 mm). If we demand that the organism must have eyes with considerable resolving power, the body size easily becomes 1-2 orders larger, so 1 m is a natural scale. Thus, the crucial question is why the eukaryotic cell is about 10 μm. There is no precise argument about the size of the cell. Still, we could argue that it should be at least 2 orders above the protein size, 1 nm, which is almost determined by physics. If we accept the universality of biological systems to use informational macromolecules (e.g., DNA), then its total volume corresponds to that of a sphere of diameter $\sim$1 μm (the representative size of bacterial cells). Therefore, even a crude argument can tell us that we cannot be smaller than 1mm, and are likely to be of order 1 m.

[18] ⟪**'in principle'**⟫ Here, 'in principle impossible' implies that 'impossible not only at present but in any future'. Suppose we wish to specify the initial position of a particle with a precision of 10^{-60} m. The uncertainty principle tells us that confining the particle within this spatial range requires an outrageously deep potential well. It is not certain whether quantum mechanics still holds at such a small scale, but, in any case, realizing such precision must be impossible. Such outrageously accurate specification of initial conditions is required, if we wish to predict the particle position reasonably accurately a few minutes in the future under a time evolution that doubles the small spatial displacement in every second (this is not an extreme case for chaotic motion). Besides, to this end, we must extremely carefully exclude any influence of the outside world (e.g., the gravitational effect due to the mass distribution change in the sun caused by convection) on the system. May we move our hands near the system?

[19] However, the author does not necessarily assert that chaos is a very important phenomenon. Chaos is not required for such unpredictability to occur, although it is quite interesting that very simple systems can exhibit various unexpected phenomena. The real significance of chaos lies in its pedagogical value as a conceptual counterexample illustrating that very simple systems can already exhibit that the world observed around our scale is not self-contained.

problem. Thus, in a two-dimensional space trajectories can be arranged as cooked noodles on a plate without crossing, so not a very complicated configuration of trajectories is realizable. Therefore, we can intuitively understand that we need at least three variable systems to observe chaos. Still, it is interesting to know what the 'maximally strange' behavior is in two-dimensional spaces. See the work by Denjoy.[20] □

If the mode on the scale we can observe is stable but not linearly stable (for example, as $\dot{x} = -x^3$ [21]) *critical phenomena* may occur.[22] A critical phenomenon is one in which numerous linearly almost neutral modes on various scales couple with each other to build up large fluctuations. Small-scale modes change rapidly and look like noise from larger scales. These modes interact with larger-scale modes and drive the latter. These larger-scale modes are linearly neutral, so their motions do not decay quickly. Thus, the effects of the smaller-scale modes build up and the outcome drives further larger-scale modes. They are also hard to decay and the effects driven by smaller scale modes accumulate. Consequently, due to this cascade very large space-time scale fluctuations are realized. This is a critical phenomenon.

In the cases of both chaos and critical phenomena, the modes on the space-time scales quite removed from our own cannot be observed nor controlled, so we may regard them as *noise* (or noise sources). 'Noise' is a term to describe, roughly, unpredictable and uncontrollable perturbations coming from outside the scales that we can directly observe (usually from smaller scales we cannot observe). In contrast to the usage of this word as in 'what he says is a mere noise,' there is no connotation in this book that what we call noise is unimportant or trivial. Rather, noise is regarded (almost with awe) as an information source beyond our reach ('The Almighty Noise').

The effect of noise on a stable linear system is not significant. For example, consider the following linear *Langevin equation*:[23]

$$\dot{x} = -x + \nu, \tag{1.16}$$

---

[20] See I. Tamura, *Topology of Foliations: An Introduction* (Translations of Mathematical Monographs) (AMS, 2006) Section 1.6.

[21] Here, $x = 0$ is stable, but if linearized around this point, the system reduces to $\dot{x} = 0$. We may express this as 'linearly neutral.'

[22] **⟪No critical phenomena in chemical reaction systems⟫** For this to occur, we need many 'linearly neutral' modes distributed over wide-length scales. As has been stressed by Y. Kuramoto, if the smallest scale of the neutral modes is not very small (e.g., 0.1 mm) as, e.g., in elementary chemical reactions in pattern formation, no sufficient building up of fluctuations occurs (i.e., the cascade is too short). Consequently, in practice, we observe no critical phenomena in chemical reaction systems.

[23] A differential equation with a stochastic driving term (= noise term) is generally called a Langevin equation. In physics, this name covers a much wider class of 'noise-driven' equations than stochastic differential equations in mathematics.

where ν is the noise with average zero. If we denote the ensemble average (which is obtained by preparing numerous systems obeying the same equation, and by simply averaging their results) as $\langle x \rangle$, we obtain

$$\langle \dot{x} \rangle = -\langle x \rangle. \tag{1.17}$$

That is, noise has no effect on the ensemble average. This is another expression of the statement that, although the world we directly observe is not self-contained, if it were linear, we could be oblivious of the open-ended nature of the world observed on our own scale. Such a simple conclusion cannot generally be true for nonlinear systems, because $\langle f(x) \rangle \neq f(\langle x \rangle)$ for a general function f. Noise can cause truly new problems only in nonlinear systems. The effect of noise could accumulate and would have macroscopic effects. Thus, critical phenomena occur.

As critical phenomena and chaos illustrate, the world we observe on our own scale is not self-contained; we could generally call the influence of the scales removed from ours 'noise.' If a noise sequence (noise signal as a function of time) is **random** (we will discuss randomness at length towards the end of the next chapter), it can contain any melody as a segment (thus, 'noise' is almighty). Therefore, noise could be used as a source to create a novel order in the system. On the other hand, however, noise could have a destructive influence as everybody has experienced, so there must be mechanisms to suppress the effect of noise as well as to exploit its creative influence.

An extremely effective way to cope with noise and to produce stable structures against noise is the use of the law of large numbers.[24] To utilize this law, a space-time scale significantly larger than the microscopic scale is required (in order to sum noises up over a large space-time scale). In particular,

---

[24] ⟪**The law of large numbers**⟫ Roughly speaking, the law of large numbers formalizes the observation that the sum of many ($N \gg 1$) mutually unrelated fluctuating variables $x_1, x_2, \cdots, x_N$ behaves as $\sum_{j=1}^{N} x_j = N\langle x \rangle + o[N]$, where $\langle x \rangle$ is the average of $\{x_j\}$. $o[N]$ denotes the term that fluctuates randomly, but the notation expresses that its magnitude is much smaller than N, to which the systematic part is proportional (see Section 3.10). If we collect numerous fluctuating variables, a large non-fluctuating portion emerges.

More precisely, the (weak) law of large numbers can be stated as follows: Let $\{X_i\}$ be a set of iid (= independently and identically distributed) stochastic variables. Whatever $\varepsilon > 0$ we choose,

$$\lim_{N\to\infty} P\left(\left|\frac{1}{N}\sum_{k=1}^{N} X_k - \langle X_1 \rangle\right| > \varepsilon\right) = 0$$

holds, if $\langle |X_1| \rangle < \infty$ (i.e., if the distribution is not too broad) [here, X_1 is used as the representative of the stochastic variables, since they are iid], where P denotes the probability of the event bracketed with parentheses, and $\langle \ \rangle$ denotes the expectation value.

slowly changing macroscopic (at least mesoscopic[25]) degrees of freedom are required. A general way to produce such degrees is the lowering of the system symmetry (that is, the symmetry of the state becomes lower than that of the laws governing the system). The macroscopic variable x thus created typically obeys a differential equation such as $\dot{x} = x - x^3$. $x = 0$ is unstable, and $x = +1$ or -1 is the destination state after a long time for this equation. The equation is invariant (symmetric) under $x \leftrightarrow -x$, but the destination is one of ± 1, so the long time fate does not reflect the symmetry of the equation itself. This is (spontaneous) *symmetry breaking*[26] (lowering of the symmetry). The equation itself cannot determine which of $x = \pm 1$ is the future state; this is determined solely by the initial condition. Small noise around $x = 0$ (a symmetric state) in the initial condition can select the destination; in this case, noise + nonlinearity can create 1 bit of information on our scale. Consequently, the macroscopic degree of freedom x as a memory variable allows the initial noise to become an information source.

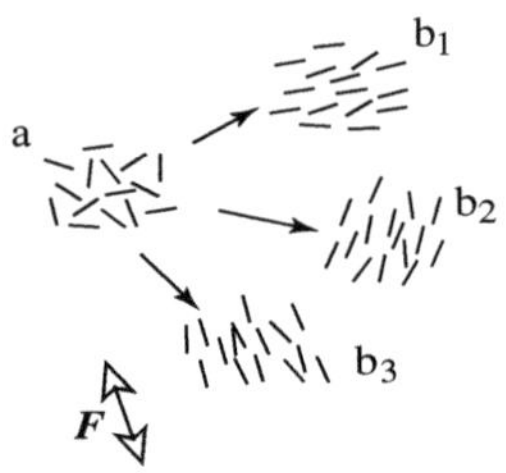

Fig. 1.2 Symmetry breaking (isotropic liquid-nematic liquid crystal phase transition). **a** is the high-temperature disordered state (isotropic liquid phase) and **b**'s exemplify the low-temperature ordered phase (nematic phase). If there is no external field $\boldsymbol{F}$, the direction of the director (small bars) cannot be chosen (a special configuration, e.g., b_3 cannot selectively emerge). However, this macroscopic degree of freedom due to self-organization can be macroscopically fixed (controlled) by an external field, by a boundary condition, etc. Consequently, it can be used to store information.

Typical examples of 'x' are order parameters of ferromagnets and liquid crystals. Below a certain temperature, spins in a macroscopic magnet align

[25] 'Mesoscopic scales' indicates the scales intermediate between really microscopic (molecular) and really macroscopic (our) scales. In practice, mesoscopic quantities exhibit small fluctuations around some systematic behaviors.

[26] ⟪**Symmetry breaking**⟫ Symmetry breaking can occur for various reasons, but those without a microscopic materialistic basis are usually vulnerable to external perturbations. Therefore, (equilibrium) phase transitions due to microscopic structural change and macromolecular conformation changes due to allostery or complex formation are the most important reasons for symmetry breaking in the cell. The word 'spontaneous' is often added, because microscopic entities, though they themselves behave symmetrically under thermal noise, interact with each other and 'spontaneously' cause mutual constraints, resulting in a global structure whose symmetry is lowered by the mutual constraints. Even in the case of a switch-like conformation change of proteins, they cannot determine whether the on or the off state is the destination; if a switch determines its state by itself, it is useless as a switch.

(the ferromagnetic phase emerges). In a liquid crystal, slender molecules align to make the nematic phase (Fig. 1.2). In both cases, the direction of the alignment is determined not by the fundamental law, but by certain contingent factors such as a weak external field (stray field) or fluctuations before the symmetry-breaking process occurs in earnest. Ferromagnetism is the physical principle behind the magnetic memory. Thus, symmetry-broken phases can amplify noise and fix it as memory, resulting in an apparent creation of macroscopic information.

Complex systems emerge through a cascade of selection steps of macroscopic (or mesoscopic) variables created by symmetry-breaking processes. Although, in the above, symmetry-breaking due to phase transition is illustrated as an example, this could happen in a single or a few macromolecular systems as well. We will come back to this topic in the last chapter. The spontaneous symmetry breaking itself is an intrinsic property of the system (equations of motion, interactions in the system, etc.) and is an ordinary research topic of physics and materials science. Complex phenomena can happen only beyond symmetry breaking, because they concern what 'sectors' realizable by symmetry breaking are actually chosen. The ordering process due to symmetry breaking is often called a self-organization process. Thus, self-organization processes are often required for complex phenomena to occur, but we must not forget the obvious but crucial fact that the self-organization process itself cannot select its own destination (as illustrated in Fig. 1.2). That is, the essence of complex phenomena and complex systems is how they specify the outcomes of self-organization processes. These processes must be steered; how they are steered characterizes complex systems.

A complex phenomenon is, so to speak, a painting that uses symmetry breaking (and accompanying self-organization) as artist's supplies. Thus, the fundamental *Physica* of complex phenomena and complex systems is the study of the universality exhibited in the paintings painted with the materials prepared by symmetry-breaking processes.[27]

[27] The effect of noise discussed here may be regarded as the scale interference due to its effect on the system that exhibits discontinuous behaviors (which cannot be linear). Multistability, hysteresis, discontinuity, etc., are the relevant standard topics, but they will not be discussed in this book. For these topics, see, for example, Z. Yoshida, *Nonlinear Science: The Challenge of Complex Systems* (Springer, 2010) (this book, however, discusses only paints and not paintings. That is, despite its subtitle, the book does not discuss complex systems.)

1.3 Intrinsically nonlinear systems

Even though we say we will discuss nonlinear systems, not even one soliton[28] has appeared. None will appear in the subsequent pages of this book, either.

Imagine a crumpled sheet of paper. If the sheet was flat before it was crumpled, there is a way to make it flat again. To make it flat might require a lot of trouble and perhaps even some intelligence, but, in any case, we need not tear it. However, if it had a curvature before being crumpled, it is absolutely impossible to make it flat without tearing it. As Gauss realized, there is a fundamental distinction between the two cases. There must be something analogous to this distinction for nonlinear systems: apparently nonlinear systems ('crumpled linear systems') and intrinsically nonlinear systems.[29] Integrable systems are the systems that can be transformed into linear systems; they are, so to speak, disguised linear systems. To see through the disguise may be difficult, and successful theories are beautiful, but such disguised systems lack scale interference. Therefore, integrable systems are not considered in this book. There are other reasons to do so (needless to say, other reasons than that the author is quite ignorant about integrable systems).

The prestige in being integrable is not very important when we consider the noise response of systems. For integrable systems trajectories are arranged in an orderly fashion in the phase space, but the phase space is generally transformed by a nontrivial transformation just as a sheet of paper is crumpled. Consequently, trajectories separated far from each other before crumpling may be brought within a close distance unexpectedly, so small noises can cause havoc to integrability very easily. That is, noise itself causes scale interference. Then, the long-time behavior of integrable systems under noise and that of intrinsically nonlinear systems should not be very different. Furthermore, it is unlikely that there is an **algorithm** (a definite procedure that produces the solution to a given problem after a finite number of well-characterized steps) that can discriminate intrinsic and non-intrinsic nonlinear systems.[30] Then, when we start to study a new system, it is natural to assume that it is intrinsically nonlinear.

Are not the majority of nonlinear systems intrinsically nonlinear? Even if the answer is affirmative, it does not immediately imply the importance of intrinsically nonlinear systems, and does not diminish the importance of inte-

---

[28] A solitary wave packet that behaves as if it were a particle is called a soliton. 'Behaves as a particle' implies not only it does not diffuse away as an ordinary wave packet, but also behaves like individual particles when they collide with each other.

[29] Here, as already stated before, we do not consider methods that change the dimensionality of the system drastically. As was already mentioned, any nonlinear system can be mapped to a linear system on a function space. See, for example, I. P. Cornfeld, S. V. Fomin and Ya. G. Sinai, *Ergodic Theory* (Springer, 1982) Part III.

[30] However, the so-called *Painlevé test* may not be ignored.

grable systems. However, characteristic phenomena discussed in the preceding section are never observed in integrable systems that lack scale interference. Also as discussed just above in the presence of noise, integrability is not very important. Therefore, this book does not discuss integrable systems.

1.4 What do we mean by 'appreciating the world'?

Thus far, 'nonlinearity' has been explained.

In this book, making efforts to find the 'essence' of various natural phenomena is understood as the meaning of 'appreciating the world.' It is synonymous with fundamental physics, if the word 'physics' is properly understood. Physics is not a discipline characterized by its subjects of study in contradistinction to chemistry or geology. Physics is not a discipline characterized by 'what we study,' but by 'how we study', or by the stance taken toward Nature. Subjects can be anything.

It is not easy to characterize the meaning of the word 'essence,'[31] but, simply put, what scientists feel intuitively as the 'essence' of a phenomenon is the (abstract) features that allow them to reconstruct the characteristic features of the phenomenon. Many details may be required to reconstruct the phenomenon actually, but its main features can be recovered/understood "if we only know....." 'Essence' is abstract by its very nature, so its expression often becomes mathematical. It is said that to explain is to replace visible complexity with invisible simplicity.[32]

The discipline that attempts to understand the abstract essence of things and matters is, in other words, the "effort to appreciate the world system-

---

[31] "The intrinsic nature or indispensable quality of something that determines its character." according to the *Oxford American Dictionary.*

[32] ⟨⟨**Replacing visible complexity with invisible simplicity**⟩⟩ It is quoted in R. Thom, "The qualitative and the quantitative in natural science," Kagaku **48**, 296 (1978) as J. Perrin's words. This assertion sounds even biologically natural. 'Abstract concepts' may be sensed directly by organisms lacking sophisticated nervous and sensory systems. For example, take 'brightness.' Our ancestors of one billion years ago should have sensed this without any sophisticated eyes, so we may say they lived in a highly abstract world. Therefore, it is reasonable to assert that abstract understanding is much more primitive and fundamental than we usually suppose. See footnote 32 of Chapter 3.

However, the above interpretation by R. Thom seems to be a 'correct misinterpretation' of the original statement by J. Perrin in *Atoms* (`http://www.archive.org/details/atomsjean00perrrich`). Perrin meant something else; he meant to understand something complicated we can actually observe in terms of something simple we cannot see, but this something unseeable is not an abstract concept but merely the concrete objects too small to be seen (in his time). The relevant original quotation (in English) is: "To divine in this way the existence and properties of objects that

atically," is the fundamental natural science. Its non-empirical part is the fundamental mathematical science, but the author does not wish to distinguish mathematics and physics meticulously in this book.

It may well be the case that there is no 'essence' for any thing, or even if there is, it may not be wise to pay too much attention to it. For example, the assertion is rather familiar that biology and 'essentialism' are not in conformity with each other.[33] However, our human insight may not be so powerful to assert *a priori* that 'essence' does not exist without pursuing it seriously. If the pursuit of essence is in principle impossible, the student of the fundamental science must demonstrate its impossibility. Fundamental physicists always think at the onset that explanation of a natural phenomenon is simply complicated due to their stupidity or due to their lack of a proper way of looking at the phenomenon, rather than due to its intrinsically complicated nature. To always take this stance of the fundamental physicist when we observe the world is the meaning of 'to appreciate the world' in this book.

There is of course a sober opinion that complicated things are complicated irrespective of the way of looking at them. This may be true, but a student subscribing to this opinion is not a fundamental scientist. What if no appreciation of the world in this sense is possible? Then, we would have demonstrated an important characteristic of the world.

Since being nonlinear seems to be a universal property of almost all phenomena in the universe, appreciating the 'nonlinear world' is simply equivalent to appreciating the 'world.' Then, why do we mention 'nonlinear' in particular? The essential feature of nonlinearity is scale interference. Because of this we cannot exclude the 'unknowable.' The closed understanding of this world within what we can (in principle) know becomes impossible. Since the world we observe directly is not self-contained, unexpected novel phenomena do exist. Their description will require new concepts not very familiar to us, and the description of phenomena without assuming the complete knowledge of the system. Thus, this book focuses on conceptual analysis to prepare for

---

still lie outside our ken, *to explain the complications of the visible in terms of invisible simplicity*, is the function of the intuitive intelligence which, thanks to men such as Dalton and Boltzmann, has given us the doctrine of Atoms. This book aims at giving an exposition of that doctrine." (p. vii, translated by D. L. Hammick). Here, the italics are in the original.

[33] For example, see E. Mayr, *Towards a New Philosophy of Biology* (Belknap-Harvard, 1988). However, if physicists read this book, they would feel that an inferiority complex to physical sciences is behind the pages. D. J. Futuyma, "Wherefore and whither the naturalist?," American Naturalist **151**, 1 (1998) is a wonderful article that frankly admits the existence of this feeling of inferiority, and still emphasizes the importance of being a 'naturalist.' The author's effort, *Integrative Natural History*, already mentioned in the Preface, tries to proceed along this line through integrating the scales from molecules to the macroworld of organisms, and disciplines from field biology to mathematics.

description of new phenomena and on the way to rationalize what we observe without a firm basis of the so-called fundamental laws, i.e., phenomenology. This is the reason why the word 'nonlinear' appears explicitly in the title of this book.

The usual usage of terms such as 'nonlinear science' or 'nonlinear physics' often has a connotation that the discipline studies nonlinear phenomena that are not discussed in the 'standard physics subfields' (such as many-body theory, statistical physics, etc.). However, 'to appreciate the nonlinear world' must be 'to have an overview of general phenomena due to nonlinearity, including those covered by these standard fields' (although the author does not mean that he can do so). The author does not mean to establish a new discipline. He never has any intention to create an alternative science that is not reductionistic. The students of nonlinear physics must learn the standard material represented by, e.g., the Landau-Lifshitz volumes and basic mathematics more than physicists of a particular subfield.[34] If one wishes to understand organisms as a fundamental scientist, s/he should learn elementary biology after her/his consolidation of elementary mathematics and physics skills.

Discussion 1.7. What is learning? What do we mean by the usefulness of learning? When the author discusses natural science, he does not have any of its applications in the usual sense in his mind. Important applications of natural science lie not in engineering or medicine, but in changing the way that we human beings look at the world. Engineering and technology have changed our way of understanding the world profoundly, but we must recognize the more direct influence of science as well. □

1.5 The structure of this book

This section is a sketch of the contents of this book.

The major theme of Chapter 2 is conceptual analysis: to describe clearly and explicitly what we feel we have understood intuitively. We need clarity to exchange ideas properly with other people. Clarity is required by any serious studies. Chapter 2 will make solid efforts to analyze the concept 'chaos' in order to illustrate conceptual analysis. To this end we cannot avoid the question: what is **randomness**? Inevitably, we will have to pay due attention to the **theory of computation** (what is an algorithm?) that may be close to the foundation of mathematics. The reader may say that such indirect topics

[34] "Even in the future when physics departs from materials science and expands its realm, I believe the wealth we can obtain from the accumulated treasure is significant. Ignoring this and pursuing fashionable topics will produce only shallow results." (Kyozi Kawasaki, *Nonequilibrium and Phase Transitions* (Asakura Shoten, 2000 [in Japanese]).

are likely to be useless to appreciate the real world. However, we will learn that very abstract concepts may well be the most primitive even biologically.

Chapter 2 aims to illustrate what we mean by pursuing clarity. The author acutely recognizes the opinion that worrying too much about the definitions of concepts is unwise for scientists. Any creative research starts from a state of confusion and never with clear definitions. There is a mound of complicated arguments about what a biological 'species' is. It is a difficult concept, and some people even regard it as a pseudo-concept lacking any clear definition. If one had worried too much about its definition, it would have been impossible to write *The Origin of Species.*[35]

However, that we cannot give any explicit definition of a concept implies (as long as it is not an undefinable fundamental concept) either that our understanding of the concept has some lacunae or that no clear definition is possible (i.e., it is a pseudo-concept). Even if the latter is the case, the impossibility should be clearly recognized only after attempting seriously to give its clear definition. Therefore, to illustrate conceptual analysis must not be meaningless even in practical researches. If we wish to exchange ideas between distinct individuals while avoiding any word game, mathematics is the clearest way the human being is allowed to use.[36] To use mathematics is to describe something outside mathematics using mathematics. Therefore, the concept that is to be expressed mathematically must be explicit at least to some extent previously. To this end we need conceptual analysis. In any case, our research objective is not narrowly restricted, so concepts we will need cannot be expected to be given beforehand. Thus, conceptual analysis is a major tool to appreciate the world.[37]

It is natural to expect that nonlinearity makes the world complicated due to scale interference. We cannot easily know (or often cannot in principle

---

[35] Let us listen to J. T. Bonner: "But as to the question of the desirability of setting formal definitions in biology, I have to say that I'm often against it. The reason is that it prematurely puts rigidity into your argument. An example would be the term "species". I don't care what anybody says, there has been no definitive definition of this term and I am not sure [it] would be helpful to find one either at present or even in the future." ("Interview with J. T. Bonner," BioEssays **25**, 727 (2003)) A recent popular account of 'species' may be E. Marris, "The species and the specious," Nature **446**, 250 (2007) (Linnaeus' 300th birthday special issue).

[36] "It is still true that mathematics is man's most extensive and most profound effort to achieve precise and effective thinking and what it accomplishes measures the capacity of the human mind. It represents the upper limit of what we can hope to attain in all rational domains." (M. Kline, *Mathematics: the loss of certainty* (Oxford University Press, 1982), Chapter XIV Whither Mathematics? p326).

[37] "I believe the fundamental scientific attitude is in logical conceptual analysis. I learned this most from Galileo Galilei. ··· The analysis of the concepts being used today must have the priority. Although it is surely important to respect the empirical data, only with the latter we cannot go any further beyond the Greek science." (M. Taketani, *Contemporary Theoretical Problems* (Iwanami Shoten, 1968) p1-2 [in Japanese]).

know) what is going on on the scale removed from our own scale. Consequently, through scale interference the things we cannot know affect the phenomena on our scale and make the world we can directly observe open-ended. Thus, the world could be indefinitely complicated and even lawless, but in reality the world does not look that disordered. It looks even comprehensible to some extent.

Isn't it an amazing empirical fact that the world full of noise and nonlinearity does not give us surprise at every moment? To exploit this observation is the aim of Chapter 3.

Understanding the world must not require understanding its every detail. If understanding our surroundings demands their detailed understanding, we would wake up every morning into an extremely fresh world. However, "the thing that hath been, it is that which shall be; and that which is done is that which shall be done: and there is no new thing under the sun." We do not feel every day is really novel, because the parts of our environment which we feel are vital do not change very much day by day. They are rather stable (even invariant). Put precisely, if we dissect our environment into the unimportant trivial part and the more vital part, and if the latter does not cause any surprise, we feel we have understood (or at least we have gotten accustomed to) our environment.

We cannot avoid the question of what to regard as trivial. Let us dissect a phenomenon into its stable part (the part which is stable against changing details) and the rest that is sensitive to details. To understand the former clearly (hopefully, to understand its essence) is called the **phenomenological understanding** of the phenomenon.[38] The way we understand our environment is an example of phenomenological understanding. That we can live in this world with its phenomenological understanding is a nontrivial empirical fact. This must be thanks to a certain property of the world that undeniably exists externally and objectively to us, that is, thanks to a certain lawfulness of the external world. Konrad Lorenz (1903-1989) wrote in his *Behind the Mirror: A Search for a Natural History of Human Knowledge*[39] roughly as follows: The fish body with a fluid dynamically admiring shape is the reflection of a property of liquid water that its long-time behavior is governed by fluid dynamics. However, that the long-time behavior of liquid water obeys fluid dynamics is not because fishes swim in it. Liquid water existed long before fishes emerged, and even without fishes the law of fluid dynamics continues to hold for liquid water. The relationship between our brains and the world

---

[38] 'Detail' and 'trivial part' above sound the same, so how can we know what is a detail without knowing what is trivial? This is a good question, but in practice we know some aspects of events on very small scales are clearly irrelevant to our scale. Thus, we know, at least sometimes, small-scale details are indeed 'details' we may ignore.

[39] Mariner Books, 1978 in English; the original in 1973. Incidentally, for dog lovers, *Man Meets Dog* (1950; Routledge, 2002) is highly recommended. This also discusses cats.

must be understood in a similar fashion. That our brains are made to recognize some kind of laws is because the law-like relationships really exist in the world independent of our existence.

Chapter 3 explains the idea of renormalization group theory as a theoretical and conceptual framework to extract phenomenology. Its crux is to extract 'stable' features from the models supposedly describing the phenomenon we wish to understand. Here, whether a property is stable or not may be seen by 'shaking' the system (e.g., by perturbation). If some parts stay unaffected even by shaking, we can have a phenomenological theory of the phenomenon in the precise sense of the word. To have a nontrivial phenomenology the stable parts must not be trivial. A phenomenon that exhibits stable features (and that are not trivial in the ordinary sense) is said to be **renormalizable.**[40] If a phenomenon is renormalizable and has a definite phenomenological description, we say the phenomenon has a renormalizable structure. What is explained above may be summarized as: there are lots of phenomena with renormalizable or approximately renormalizable structures in the world. This fact also allows us to feel that the world is made of various subsystems that are more or less independent of each other. This is the secret of the world that allows us to understand it. In other words, this is the secret of the world that could produce intelligent organisms like us.

Incidentally, isn't phenomenological understanding regarded as inferior to understanding based on microscopic fundamental theories? Weinberg asked in his famous book, *Dreams of a Final Theory* (Pantheon Books, 1992) why a piece of chalk is white, and stated that its fundamental understanding is possible through going back to atomic structure and eventually to elementary particle theory. Then, why is not only a piece of chalk but also flour white? In this case, what is the significance of going back to atomic structures to answer the question?

There are many phenomena the pursuit of whose materialistic basis does not contribute significantly to their understanding. *A fortiori*, understanding of phenomenology is often unavailable through the pursuit of its materialistic or microscopic basis. Take thermodynamics as an example. History teaches us that this phenomenological framework transcends mechanics. It was thermodynamics that supplied important clues to quantum mechanics.[41] The reader may say statistical mechanics justifies thermodynamics, but this opinion is based on a total misunderstanding of what is basic, because the thermodynamic variables cannot be chosen without referring to thermodynamic laws; they are chosen so that the laws are satisfied. However, the second law of thermodynamics in its general form can never be demonstrated by mechan-

---

40 Our usage of the term 'renormalizable' may be a bit of shock to high-energy physicists who introduced this terminology, but in this book the term is used in this very broad sense.

41 Section 35 of P. T. Landsberg- *Thermodynamics with quantum statistical illustrations* (Interscience Publishers, New York, 1961) is a summary of thermodynamics paving the way to quantum mechanics.

ics.[42]

The reason why the opinion that microscopic approaches are more fundamental than phenomenological approaches sounds natural is that we never think 'there is no water without waves.'[43] Since we believe there is no phenomenon without the materials supporting it (the principle of no ghost, so to speak), we tend to think that studying materials thoroughly is the key to obtaining a deep understanding of the phenomenon. However, it is also self-evident that there is no (liquid) water that cannot support waves. If there is no matter, there is no phenomenon, and if there is no phenomenon, there is no matter. Therefore, when we try to understand nature, the point of view centered around matter and that centered around phenomena are both equally fundamental.[44]

As the reader has already sensed, this book uses the word 'phenomenological' as an affirmative and positive adjective in contrast to the usual usage in conventional science (see Section 3.1). When we appreciate paintings, we do not usually appreciate materials. What Weinberg emphasizes is often materials instead of the painting itself. That we can appreciate paintings even through their black and white photos implies that there is something in the painting that transcends the materials used, an obvious fact. Needless to say, to appreciate a painting is to appreciate something expressed by the painting transcending the materials used.

Discussion 1.8. What is the (in)significance of elementary particle theory in the above context? □

Phenomenological understanding is often detached from materials, so it has to be mathematical. In fact, to pursue phenomenological understanding is to explore a minimal mathematical structure behind a set of phenomena. If we have a reasonable mathematically expressed model of these phenomena,

---

[42] **⟪Can mechanics explain thermodynamics?⟫** Those who are familiar with modern equilibrium statistical mechanics may counter that Jarzynski's equality [C. Jarzynski, "Nonequilibrium equality for free energy differences," Phys. Rev. Lett. **78**, 2690 (1997)] or Lenard's theorem [A. Lenard, "Thermodynamical proof of the Gibbs formula for elementary quantum systems," J. Statist. Phys. **19**, 575 (1978)] implies Planck's principle: an adiabatic cyclic process cannot decrease the system energy. However, a thermodynamically 'adiabatic condition' is not the same as 'isolation,' so there is no way to describe a thermodynamic adiabatic condition in its generality in terms of mechanics. Thus, mechanics cannot demonstrate the second law.

A more fundamental problem of statistical mechanics may be: Statistical mechanics assumes that all the particles obey mechanics simultaneously, but there is no way to demonstrate this. That is, if you are fastidious, you must admit that this assumption is a metaphysical assumption. (May we introduce beliefs into natural science? Yes, we may, if the result is indifferent to the belief.)

[43] This is based on a poem of a Japanese Zen monk of the 15th century, Ikkyu Sojun.

[44] Or, equally biased. According to the philosophy of Avicenna (980-1037), who systematized Islamic scholastic philosophy, matter and form are both substance, "because form is based on matter that contains it and exists inside matter, but matter that corresponds to the base cannot exist without form that exist inside it."(from T. Izutsu, *History of Islamic Philosophy* (Chuokoron, 1975 [in Japanese]).

it is likely that a renormalization approach can extract the phenomenological theory common or universal to them. However, Nature does not give us models. We must devise them. Accordingly, Chapter 4 illustrates modeling of natural phenomena with the aid of phase transition dynamics as an example.

There are two major usages of models. One is as a tool to describe phenomena. The other is as a tool to study the consistency of ideas. The second usage is common in mathematical logic and theory of computation as discussed in Chapter 2. As a tool of describing a phenomenon a minimal model that reproduces a given phenomenology with a 'minimal structure' indicates the culmination of our understanding of the set of phenomena described by the phenomenology. Therefore, constructing models is a fundamental task.

The goodness of description of a phenomenon may be judged, for example, by the extent of quantitative success of the corresponding model in describing the relevant observables. However, we cannot easily say that a model is successful because it successfully describes the target phenomenon. The successful description of the phenomenon is necessary, but it is not a sufficient condition for a good model. Sufficient conditions may not be explicitly itemized, but something more than successful description is required. What should we require? This is the problem of the goodness of the model. This will also be discussed in Chapter 4.

As already discussed in Section 1.2, a complex system is possible by specification of numerous indeterminate variables created by, e.g., spontaneous symmetry breakings, so it is important that these variables should not spontaneously assume particular values. That is, lack (or incapability) of self-organization is a key for a complex system. The reader might think that physicists have been studying 'complex systems,' but almost all such systems have been created by self-organization. In this sense, the so-called 'complex systems studies' have been the study of pseudocomplex systems.[45] Examples of complex systems no one would object to are eukaryotic biological systems.[46] To understand the mathematics of, e.g., their organization will allow us to reach some aspects of the essence of complex systems; eukaryote physics may be the core of complex systems studies. We cannot establish such a discipline now, needless to say, but the last chapter will critically reflect on the

---

[45] Social and economic phenomena are also often considered from the self-organization point of view. However, even for bee and wasp societies, a crucial point is that they are not collections of simple points, but collections of organisms that are regulated by brains interacting with self-organizing feedback loops. See J. J. Boomsma and N. R. Franks, "Social insects: from selfish genes to self organization and beyond," Trends Ecol. Evol., **21**, 303 (2006). That is why, for example, neuroeconomics becomes increasingly important.

[46] All the organisms on the earth are classified into three domains, Bacteria, Archaea and Eukaryota, although Archaea may not be a natural group. We are a member of one (Opisthokonta) of the six kingdoms of Eukaryota including all fungi and animals. See Note 5.2.

nature of fundamental physics (as a representative of fundamental sciences) and review some possible starting points to explore true complex systems. The goal of genuine complex systems studies must be conceptual analysis of complexity and the construction of phenomenology of complex systems. Hopefully, Chapters 2-4 supply at least a part of the needed preparation for this ultimate goal in the future.

As the reader must have realized, the core ideas of the chapters are related, but the adopted main illustrations are widely different from chapter to chapter. Even if some chapters may not be understood well, this should not greatly hamper understanding the rest. It is even possible to browse through topics; instructions as to how to skip some parts are often explicitly given. Reading only some footnotes may be one way to read this book, as already noted in the Preface.

Note 1.2. "Explaining the world"
Elementary particle theory is a fundamental branch of science just as thermodynamics and animal taxonomy are. However, the ordinary (rather shallow) opinion is that elementary particle theory is fundamental because it is the study of the foundation of the world or, in other words, because it is the starting point of all the explanatory arrows if we follow Weinberg.

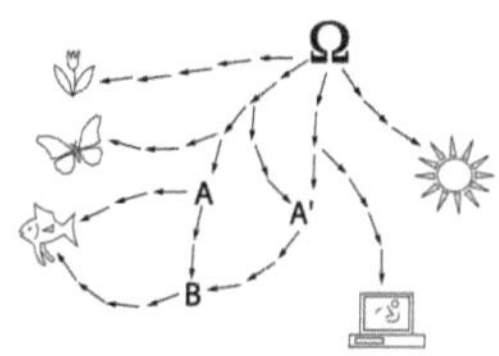

Fig. 1.3 Weinberg's view of explaining the world. All the explanation arrows are embedded in the cascade emanating from the ultimate Ω.

In Fig. 1.3, an arrow of explanation is drawn from A to B as A $\rightarrow$ B, when A explains B. If we recall the example of a piece of chalk and flour, indeed it may be A $\rightarrow$ B, but A' $\rightarrow$ B is also true, so for B A need not be a prerequisite. It might well be the case that explanations for A and A' both are embedded in the flow of arrows emanating ultimately from some Ω. However, those who still claim A is the explanation of B must have some prejudice (some faith) or must be logically sloppy. Weinberg interprets thermodynamics as patterns of reasoning appearing repeatedly in this flow of explanation. However, since the principle of equal probability cannot be demonstrated by mechanics without a certain fundamental assumption, it is clear that the pattern recognized by thermodynamics is not due to Ω. That is, the explanation of the world cannot be complete with Ω alone. Even if it is granted that all the arrows emanate ultimately from Ω, the network of the arrows is not governed by Ω. This implies that Ω does not determine what is explained by what.[47]

[47] The reader may have realized immediately that the correct aspects of Weinberg's idea may be category-theoretically formulated. Such a trivial statement is not very enlightening, but it might be a good occasion to learn a category-theoretic approach to the foundation of mathematics. See S. MacLane, *Mathematics Form and Function* (Springer, 1986) Chapter XI for an excellent introduction. This book is recommended for a general audience.

Appendix 1A: From the seed sown by Fourier

As we have seen in Section 1.1, Fourier's assertion, "Every function may be expanded in a trigonometric series," is an ultimate use of the superposition principle.[48] The heart of linear mathematical science is this idea of Fourier's. The idea was so fundamental that efforts to make it mathematically respectable have produced the framework of the contemporary mathematics. Let us briefly review the subsequent development initiated by Fourier's work. The reader might say this review considerably departs from natural science, but the history of mathematics is packed with excellent examples of conceptual analysis.

In Fourier's day even the concept of function was not definite. Slightly before that time, d'Alembert (1717-1783) and Euler (1707-1783) argued about the definition of functions. D'Alembert regarded only those expressible explicitly with formulas as functions. In contrast, Euler asserted anything whose graph could be drawn by hand was a function. After many twists and turns,[49] the concept of a function as a map was eventually accepted: f is a function defined in the domain D, if $f(x)$ is uniquely given for any x in D.

A difficulty due to this very general notion of function is the interpretation of integrals we find in the calculation of the Fourier coefficient, e.g.,

$$a_n = \frac{2}{L}\int_0^L f(x)\cos\frac{2n\pi x}{L}\mathrm{d}x. \tag{1.18}$$

The integration was often regarded as the inverse operation of differentiation to compute primitive functions. Then, one must be at a loss as to how to interpret such an integral as (1.18). Riemann (1826-1866) defined his integral in his thesis to clarify the meaning of the integrals needed to compute Fourier coefficients (1853[50]). That is, the modern theory of integration was created to formulate Fourier's idea precisely.

The next natural question is the relation between function f and its Fourier

---

[48] T. W. Körner, *Fourier Analysis* (Cambridge University Press, 1988) is a superb introduction to Fourier analysis (more generally, real analysis). Its Section 7 exemplifies the importance of linearity well.

[49] ⟪**History of functions**⟫ There is a summary of the history of function concept in Section 6.5 of S. Stahl, *Real Analysis—a historical approach* (Wiley, 1999). According to Stahl Fourier had a considerably modern notion of functions, and this seems to be due to his study of trigonometric series.

According to M. Takahashi (*The beginning of infinite analysis* (Chikuma, 2009) [in Japanese] pp108-118), Euler had the notion of function as a map, when he said arbitrarily drawn curves define functions; he simply did not have an adequate means to express his idea. Euler conceived various notions of functions during 1745-48. These years should be noted as the key period when the concept of functions was born.

[50] [1853: Taiping Tianguo soldiers took Nanjing, Livingstone started his African expedition, Wagner started to compose *Der Rheingold.* Boole's (1815-1864) *Laws of Thoughts* was published the next year.]

coefficients a_n, b_n. The Fourier coefficients of f are said to be determined by the formula such as (1.18), but how precisely are they determined by f? When Cantor (1845-1918) got his first (respectable) academic position in Halle,[51] Heine (of Heine-Borel's covering theorem; 1821-1881), who first recognized the importance of uniform convergence, proposed this problem to Cantor.

Discussion 1.9. State the definitions of the following convergence concepts and give examples: absolute convergence, conditional convergence, uniform convergence, maximal convergence. □

First, Cantor computed Fourier coefficients of continuous functions according to (1.18) and demonstrated that if all the coefficients are zero, then indeed $f = 0$ (that is, for continuous functions the function and its Fourier coefficients have a one-to-one correspondence).[52] Cantor did not stop here and discovered that Fourier coefficients may be determined uniquely even if the value of f is not known at infinitely many points.[53] Then, at how many points do we not need the values of the function? While thinking about this question, he realized the distinction between the infinity of integers (*countable* or denumerable infinity) and the infinity of reals (*uncountable* infinity) (December, 1873[54]) (See A1 below).

In 1891[55] Cantor found an alternative proof of this fact (*diagonal argument*) (see A2). This proof tells us that there is an indefinitely 'larger' infinity than a given infinity (transfinite number). Take an arbitrary set A (this is anachronistic, but let us use the modern terminology) and write the totality of its subsets as 2^A. Cantor proved that 2^A is a 'higher' infinity than (has a larger *power* than) A. $\boldsymbol{R} = 2^{\boldsymbol{N}}$ holds ($\boldsymbol{R}$ is a standard symbol for the totality of real; $\boldsymbol{N}$ is the standard symbol for the totality of non-negative integers).

---

[51] ⟪**Biography of Cantor**⟫ J. W. Dauben, *Georg Cantor, His mathematics and philosophy of the infinite* (Princeton, 1979) is excellent. His first position was in a girls' high school in Berlin. It is not written in this biography, but Cantor was a thesis committee member of Husserl, who was to be an assistant to Weierstrass, and it is said that they continued their good relationship subsequently (B. Smith and D. W. Smith, Introduction to *The Cambridge Companion to Husserl* (Cambridge UP 1995) p4). Husserl's thesis (1882) was about variational calculus, which was related to the problem more generally treated in Zermelo's thesis (1894). Incidentally, Husserl and Zermelo are buried in the same cemetery.

[52] It was the 19-year-old Fejér (1880-1959) who actually reconstructed the continuous function from its Fourier coefficients (1899). Incidentally, Fejér was the adviser of J. von Neumann, P. Erdös, G. Polya, C. Lanczos and others.

[53] Cantor's progress is recorded in the exchange of letters between him and Dedekind [*Briefwechsel Cantor-Dedekind* (Actualites scientifiques et industrielles, 518) edited by E. Noether and J. Cavailles (Hermann & C., 1937). The author has not read this book, but learned about it from the expositions by K. Shiga in *Have Fun with Mathematics* vol. 1, p96 (Nihon Hyoronsha, 1997) [in Japanese].

[54] [1873: The Great Depression began in Europe, Livingstone died; Bruckner completed his Third Symphony (Ver. 1), van der Waals proposed his equation of state.]

[55] [1891: Edison applied for a patent for the radio, Java man was discovered (E. Dubois), T. Hardy wrote *Tess of the d'Urbervilles.*]

Cantor then introduced the concept of the set to furnish a solid foundation to his theory of transfinite numbers (1895[56]).[57] However, his notion of the set was ambiguous, and various paradoxes resulted from it.[58] Thus, set theory had to be specified more precisely and Zermelo (1871-1953) initiated axiomatic set theory.[59] There are various axiomatic systems of sets. The question which best describes the world could be a question of natural science (empirical problem).[60]

As we have seen, even the axiomatic set theory that is the foundation of mathematics may be said to have been created almost under the direct impact of Fourier's idea. The most important lesson we must learn is that purely mathematical foundational researches and concrete problems in the

---

[56] [1895: The Sino-Japanese war ended; an automobile with a gasoline engine was invented, wireless transmission became possible, and X-ray was discovered; Lorentz (1853-1928) completed his electrodynamics; P. Curie (1859-1906) discovered Curie's law for magnetism; Mahler's Second Symphony premiered.]

[57] "Beiträge zur Begrundung der transfiniten Mengenlehre," Math. Ann. **46**, 481 (1895) [*Contributions to the Founding of the Theory of Transfinite Numbers* (Dover, 1955)].

[58] See, for example, M. Kline, *Mathematics, the loss of certainty* (Oxford, 1980), Chapter IX. For the so-called Russell's paradox, see A3.

[59] ⟨⟨**Zermelo and the axiom of choice**⟩⟩ G. H. Moore, *Zermelo's Axiom of Choice, its origins, development, and influence* (Springer, 1982) is a very readable account (perhaps more captivating than average novels). H.-D. Ebbinghaus, *Ernst Zermelo, an approach to his life and work* (Springer, 2008) is the definitive biography. Zermelo is known to physicists as a person who pointed out a logical difficulty of Boltzmann's explanation of irreversibility. Boltzmann's reply had a tint of mocking just as the author feels sometimes while talking with theoretical physicists ("it is impossible to talk with those who lack common-sense..."). See `http://plato.stanford.edu/entries/statphys-Boltzmann/`. However, Zermelo was a man of polemic, so both may be to blame. In those days, he was the assistant to Planck and was critical of atomism together with his boss. His most important work is, without doubt, the construction of axiomatic set theory, but he never lost his interest in statistical mechanics. It was he who translated Gibbs' famous book into German (1908). We may now state that Zermelo destroyed the ergodic-theoretical foundation of equilibrium statistical mechanics.

Whatever Zermelo's direct motivation may have been, his aim was to extract principles of set theory sufficiently restrictive "to exclude all the paradoxes" and sufficiently powerful "to keep all the valuable conclusions." This is nothing but the essence of successful modeling. Indeed, to create an axiomatic system for the mathematics that can be used in physics is equivalent to modeling the (mathematical structure of the) world, or to making a 'minimal model' of the mathematical structure of the world. Now, the standard axioms is called ZFC (Zermelo-Fraenkel + Axiom of Choice). The system Zermelo proposed in 1908 included Axiom of Choice but was without Axiom of Replacement (due to Fraenkel and Skolem).

[60] Which axiomatic system to adopt is not a question of logic. See P. Maddy, *Realism in Mathematics* (Oxford, 1990) and *Naturalism in Mathematics* (Oxford University Press, 1997). However, J. Hintikka, *The Principles of Mathematics Revisited* (Cambridge University Press, 1996) is critical (p167). See, for the 'transcendental part,' J. Mycielski, "A systems of axioms of set theory for the rationalists," Notices AMS **53**, 206 (2006).

actual world are directly connected. Fourier's work is nothing but the best example of applied mathematics.

As explained above, the concept of integration was refined by Riemann. We may understand integrating $f(x)$ as the computation of the area between the graph of the function and the x-axis. There are functions which are not Riemann-integrable but the area of the shape surrounded by their graphs and the x-axis is intuitively obvious. A typical example is the Dirichlet (1805-1859)[61] function $D(x)$. The function is defined as follows: if x is rational, $D(x) = 1$, otherwise 0; $D(x)$ is the indicator of rational numbers $\boldsymbol{Q}$ (this is also the standard symbol). The set of rational numbers is a countable set (the number of elements in the set = the cardinality of the set is at most countable infinite), so, as can be seen from Cantor's argument in A1 below, its total length is zero (we say the set is *measure zero*). Therefore, intuitively,

$$\int_0^1 D(x)\,\mathrm{d}x = 0. \tag{1.19}$$

$D(x)$ is not Riemann integrable, but we know the length of the set $\{x : D(x) = 1\}$, so this integral must have a definite value (0). That is, the concept of integration can be further refined. Lebesgue (1875-1941) clarified the concept of area and volume in his thesis (again!), and established the Lebesgue measure theory (1902[62]). His motivation was also the Fourier series. Since probability theory is based on Lebesgue measure theory, it is outlined in Appendix 2.4A.[63]

Thus, the contemporary mathematics, especially a sizable chunk of real analysis, started with the study of Fourier series, which is the study of linear systems. Does the same mathematical framework really furnish an ideal framework to understand nonlinear systems?

A1 Reals are not countable.
If all the real numbers in $[0, 1]$ may be counted (countable), that is, if there is one-to-one correspondence between the positive integers and all the real numbers in $[0, 1]$, then we can write $[0, 1] = \{\omega_1, \omega_2, \cdots, \omega_n, \cdots\}$. Let ε be a positive number and introduce the small closed interval U_n whose length is $\varepsilon/2^n$ centered around ω_n (i.e., $U_n = [\omega_n - \varepsilon/2^{n+1}, \omega_n + \varepsilon/2^{n+1}]$). The totality of these closed intervals covers any ω_n at least once for any small positive

61 Young Dirichlet was influenced by Fourier and attached importance to partial differential equations. Since Riemann admired Dirichlet, his complex function theory was based on the Cauchy-Riemann equation. See D. Laugwitz, *Bernhard Riemann 1826-1866: Wendepunkte in der Auffassung der Mathematik* (Vita Mathematica) (Springer, 1995).

62 [1902: The Anglo-Japanese Alliance; J. W. Gibbs (1839-1903) published his statistical mechanics book, Mahler's Third and Sibelius' Second symphonies premiered.]

63 A. N. Kolmogorov and S. V. Fomin, *Introductory Real Analysis* (translated and edited by R. A. Silverman, Dover, 1975) is highly recommended.

ε. Therefore, $\cup_n U_n \supset [0,1]$. The total length of U_n $(n = 1, 2, \cdots)$ cannot be larger than $\varepsilon/2 + \varepsilon/2^2 + \varepsilon/2^3 + \cdots = \varepsilon$, but ε may be as small as we wish. Therefore, $\cup_n U_n \supset [0,1]$ cannot be true. We must conclude that the infinity of the set of reals is much much bigger than that of integers.

The above argument immediately tells us that any countable set is measure zero (= can be covered by a set whose total length is smaller than any positive number). □

A2 Diagonal argument
To specify a subset B of A is equivalent to specifying the indicator of B: χ is the indicator of B, if $x \in B$ implies $\chi(x) = 1$, otherwise 0. Assume that there is a one-to-one correspondence between A and the totality 2^A of its subsets. That is, we assume that for any subset of A is a unique corresponding element of A and vice versa. This correspondence may be expressed in terms of a two-variable function $\varphi(\cdot, z)$ which gives for each $z \in A$ the indicator of the subset (denoted by A_z) of A corresponding to z (i.e., if $x \in A_z$, then $\varphi(x, z) = 1$, if $x \notin A_z$, then $\varphi(x, z) = 0$). Consider $g(x) = 1 - \varphi(x, x)$ as a function of $x \in A$. For each x, this takes 0 or 1, so it is a respectable indicator. Therefore, it must specify a certain subset of A. However, there is no z satisfying $g(x) = \varphi(x, z)$, because for any z_0 $g(z_0) \neq \varphi(z_0, z_0)$. Thus, our assumption is not correct: there is no one-to-one correspondence between A and 2^A. Obviously, 2^A has a larger power than A.[64] □

A3 The Russell paradox
It is known that Zermelo already knew and recognized the seriousness of this paradox a few years before Russell (1872-1970) (in Göttingen Hilbert and his colleagues knew it), so Fraenkel proposed calling it the Zermelo-Russell paradox.[65] Let us write a collection of the sets Z satisfying a condition $P(Z)$ as $\{Z : P(Z)\}$.

If we naively regard a set as a collection of things, then we can classify sets into two groups: the sets containing themselves (Z satisfying the condition $Z \in Z$, for example a collection of concepts is a concept) and the rest ($Z \notin Z$). Let us define $A = \{Z : Z \notin Z\}$. Is this a meaningful set? If we assume A to be a set such that $A \notin A$ (respectively, $A \in A$), then by extensionality $A \in A$ (respectively, $A \notin A$), a contradiction. □

To avoid this paradox (due to allowing the existence of any set $\{x : P(x)\}$ corresponding to any logical formula $P(x)$) the standard axiomatic system (the Zermelo-Fraenkel system) characterizes sets with a set of weaker axioms.

It is often said that self-referencing is a cause of paradoxes, but not all the

---

[64] Notice that starting with $A_0 \equiv A$, we can construct $A_{n+1} = 2^{A_n}$, a set with indefinitely large power.

[65] Russell communicated this paradox to Frege. J. van Heijenoort, *From Frege to Gödel, a source book in mathematical logic, 1879-1931* (Harvard University Press, 1967) reproduces their original exchange and contains Russell's moving letter to van Heijenoort of November 23, 1962.

self-referencing statements are paradoxical (e.g., 'this statement is true.'). However, in the standard axiomatic system the Foundation Axiom (FA) is postulated to allow transfinite induction. This axiom forbids infinitely nested structures. That is, it forbids a set being made of elements, which are regarded as sets individually and made of another elements, which are $\cdots$, and demands that this regression stops after a finite number of steps (so $a \in a$ is forbidden). Moss wrote as follows:[66] the picture of the mathematical universe as generated in stages from an empty set (or even from atoms) is related to the view that the physical world is built from individual particles, or that the social world is composed primarily of independent individuals. This connection is the real cultural significance of FA in mathematics. It connects us with a deeply rooted atomistic paradigm that occurs throughout science. Conversely, to deny the iterative conception is to challenge "common sense."

Even mathematics is strongly restricted by social factors. However, the reader should not fall into the relativistic pitfall of concluding that mathematics is an arbitrary system. Culture and society do not exist apart from human beings. They are made by living human beings. Living human beings are strongly constrained by phylogeny and are fundamentally regulated by the structure of the world. Ignoring this fundamental fact may be called '*humanistic fallacy*' (in contrast to 'naturalistic fallacy'[67]). □

References

Bonner J T (2003) Interview with J. T. Bonner. BioEssays 25:727-733

---

[66] L. S. Moss, Bull. Amer. Math. Soc. **20**, 216 (1989), which is a review of J. Barwise and J. Etchemendy, *The Liar* (Oxford UP, 1989).

[67] "We first started with severing human beings from Nature and establishing our supreme reign over Nature. We believed in this way we could erase the most undeniable among the human characteristics that we are living existences. And to become blind to this common characteristic of living beings, all kinds of corrupt practices multiplied." (C. Levi-Strauss, *Fondateur des Science de l'Homme (Jean-Jacque Rousseau)* (Editions de la Baconniére, 1962). To claim that there is a naturalistic fallacy as asserted by G. E. Moore (1873-1958) is a typical humanistic fallacy.
⟪**Naturalistic fallacy**⟫ (Here the popular understanding of this concept is adopted.) The name is coined by Moore for the fallacy to understand or to characterize ethical concepts in term of naturalistic concepts such as 'pleasant, 'more evolved, 'desired, etc. Although 'the good' cannot be defined, there is no doubt about the fact that its core is biologically supported or is a biological phenomenon.

The naturalistic fallacy is related to, and often confused with, the is-ought problem. As a result, the term is sometimes used loosely to describe arguments that claim to draw ethical conclusions from natural facts (this is from `http://en.wikipedia.org/wiki/Naturalistic_fallacy`). Moore rigorously distinguished 'is' and 'ought' and asserted that the latter cannot be deduced from the former. In other words, Moore clearly pointed out the irreducibility of ethical rules (according to Iwanami *Dictionary of Philosophy and Thoughts* (Iwanami, 1998) edited by W. Hiromatsu). However, such distinction is a useless distinction from the evolutionary epistemology point of view (or from the point of view of the people who wish to live). As we will see later (Section 4.2, for example), even our judgement of what exists depends on our value system, which is biologically grounded.

Boomsma JJ, Franks NR (2006) Social insects: from selfish genes to self-organization and beyond. Trends Ecol Evol 21:303-308
Brezis H, Browder F (1998) Partial differential equations in the 20th century. Adv Math 135:76-144
Cantor G (1895) Beiträge zur Begrundung der transfiniten Mengenlehre. Math Ann 46:481-512 [Cantor G (1955) Contributions to the founding of the theory of transfinite Numbers. Dover]
Cornfeld IP, Fomin SV, Sinai YaG (1982) Ergodic theory. Springer
Dauben JW (1979) Georg Cantor, his mathematics and philosophy of the infinite. Princeton University Press
Dirac PAM (1982) Principles of quantum mechanics. Oxford University Press
Ebbinghaus H-D (2008) Ernst Zermelo, an approach to his life and work. Springer
Feynman, RP, Leighton RB, Sands M (1971) The Feynman lectures on physics III. Addison-Wesley
Futuyma DJ (1998) Wherefore and whither the naturalist? Am Nat 151:1-5
Grattan-Guinness, I (in collaboration with Ravetz JR) (1972) Joseph Fourier, 1768-1830: a survey of his life and work, based on a critical edition of his monograph on the propagation of heat, presented to the Institut de France in 1807. MIT Press
Griffiths DJ (1995) Introduction to quantum mechanics. Prentice Hall
Hintikka J (1996) The Principles of mathematics revisited. Cambridge University Press
Izutsu T (1975) History of Islamic philosophy. Chuokoron
Jarzynski C (1997) Nonequilibrium equality for free energy differences. Phys Rev Lett 78:2690-2693
Kawasaki K (2000) Nonequilibrium and phase transitions. Asakura Shoten
Kline M (1982) Mathematics: the loss of certainty. Oxford University Press
Kolmogorov AN, Fomin SV (1975) Introductory real analysis (translated and edited by Silverman RA). Dover
Körner TW (1988) Fourier analysis. Cambridge University Press
Landsberg PT (1961) Thermodynamics with quantum statistical illustrations. Interscience Publishers
Laugwitz D (1995) Bernhard Riemann 1826-1866: Wendepunkte in der Auffassung der Mathematik (Vita Mathematica). Springer
Lenard A (1978) Thermodynamical proof of the Gibbs formula for elementary quantum systems. J Statist Phys 19:575-586
Levi-Strauss C (1962) Fondateur des science de l'homme (Jean-Jacque Rousseau). Editions de la Baconniére
Lorenz K (1950) Man meets dog. Routledge
Lorenz K (1978) Behind the mirror: a search for a natural history of human knowledge. Mariner Books (original in 1973)
MacLane S (1986) Mathematics form and function. Springer
Maddy P (1990) Realism in mathematics. Oxford University Press
Maddy P (1997) Naturalism in mathematics. Oxford University Press
Marris E (2007) The species and the specious. Nature 446:250-253
Mayr E (1988) Towards a new philosophy of biology. Belknap-Harvard
Moore GH (1982) Zermelo's axiom of choice, its origins, development, and influence. Springer
Moss LS (1989) Bull Amer Math Soc 20:216-225, which is a review of Barwise J, Etchemendy J (1989) The Liar. Oxford University Press
Mycielski J (2006) A systems of axioms of set theory for the rationalists. Notices Am Math Soc 53:206-213
Schrödinger E (2003) Collected papers on wave mechanics (3rd Revised ed-

ition). American Mathematical Society
Smith B, Smith DW (1955) Introduction: The Cambridge companion to Husserl. Cambridge University Press
Stahl S (1999) Real analysis — a historical approach. Wiley
Takahashi M (2009) The beginning of infinite analysis. Chikuma
Taketani M (1968) Contemporary theoretical problems. Iwanami
Tamura I (2006) Topology of foliations: an introduction (Translations of Mathematical Monographs). American Mathematical Society
Trefethen LN, Embree M (2005) Spectra and pseudospectra: the behavior of nonnormal matrices and operators. Princeton University Press
van Heijenoort J (1967) From Frege to Gödel, a source book in mathematical logic, 1879-1931. Harvard University Press
Weinberg S (1992) Dream of a final theory, the scientist's search for the ultimate laws of nature. Vintage
Yoshida Z (2010) Nonlinear science: the challenge of complex systems. Springer

Chapter 2
Conceptual Analysis

—Prerequisite for Clear Argument—

It is usually the case that no science research begins with clear definitions of key concepts, but it occurs frequently that to try to clarify relevant concepts is a productive strategy. Since 'definition' is mere 'rephrasing' (according to a formal statement by Frege (1848-1925)[1]), it is logically superfluous. It is true that a formal logical system is indifferent to whether a concept is important (significant) or not, so what concepts should be defined is not a problem within a formal system. For natural scientists logic and mathematics are tools to understand the real world. Therefore, it is crucial to make efforts to map the real world precisely to the formal world (mathematics). At this juncture 'to understand the world' and 'to map it precisely to the formal system' are in a chicken-and-egg relation, so "we cannot start any research from precise definitions." Still, to map the world precisely to the world of mathematics is a part of the effort to understand the world. We should not forget that efforts to give unambiguous definitions of, e.g., 'continuity' or 'volume' have been very rewarding. This is because even for mathematics its driving force is not its formal logical aspect but its relation to the real world.

In this chapter, *conceptual analysis*—pursuit of clear expression of things we feel we know intuitively—is illustrated with the aid of 'chaos.'[2] As was outlined in the preceding chapter, if a system is locally unstable, 'what we cannot know previously' can be amplified and the information we already

---

[1] G. Frege, *On the Foundations of Geometry and Formal Theories of Arithmetic* (translated and edited by E. H. Kluge, Yale University Press, New Haven, 1971).

[2] ⟪**Abusing chaos**⟫ The importance of conceptual analysis is not restricted to the study of nonlinear systems. However, it should be instructive to illustrate conceptual analysis with the analysis of 'chaos,' because there is a wrong trend to regard nonlinearity, especially chaos as radically contradicting the logical and rational intellectual tradition (the tradition of the Enlightenment). Incidentally, there is an interesting reference that summarizes abuses of chaos: C. Matheson and E. Kirchhoff, "Chaos and literature," Philosophy and Literature **21**, 28 (1997).

have turns out to be grossly insufficient for the prediction of its future behavior. Thus, 'chaos,' an apparent random behavior of a deterministic system, emerges.

The reader without previous knowledge of chaos should be able to understand this chapter. Still, if the reader knows at least vaguely what chaos is like, the chapter should become easier to understand. Therefore, at the end of Section 2.1 is Appendix 2.1A illustrating the key points with the aid of a simple example. If the reader feels this appendix is enough as an illustration of 'chaos,' she could *jump from the end of this preamble to Section 2.2.*[3]

There are various ways to explain or describe things. For example, we can use a parable, some examples, or some concepts whose interpretation is assumed to be shared with the audience. The last method seems to be common in the so-called humanities, but this method has a lot of problems. It is hard to remove vagueness from our language (natural language). Niels Bohr is quoted as having said, "We never know what a word means exactly, and the meaning of our words depends on the way we join them together into a sentence, on the circumstances under which we formulate them, and on countless subsidiary factors."[4] This is all too natural, because natural language tries to describe our real lives. However, there have been efforts to polish a distilled natural language to be a precise tool of discourse. The quintessence of such efforts is 'Mathematish,' the way to describe in terms of mathematics. This is the only way available to us to speak as clearly as we can: Mathematics "represents the upper limit of what we can hope to attain in all rational domains."[5] Consequently, to try to talk clearly about something without any aid of actual examples must be to develop its mathematical formulation. To furnish a tool of deduction is an important function of formal logic, but to supply a tool of precise description is also an important function of a formal system.[6] To use this tool we need refined analysis of concepts. Disambiguation of concepts is a part of the core of mathematics.

The author, however, does not assert that concepts that defy mathematical formulation are unimportant. What we can talk about clearly is only a small portion of what we can think about clearly.[7] There are many things we can

---

[3] ⟪**Scope and limitation of this chapter**⟫ The roadmap of this chapter is posted at the end of this preamble. Although the main purpose of this chapter is *not* to explain chaos, basic concepts required to understand chaos (such as Ruelle's inequality) are explained to a considerable extent. The chapter discusses very basic topics of chaos only. For those who are interested in some practical aspects of chaos, the following book may be of use: H. Nagashima and Y. Baba, *Introduction to Chaos, physics and mathematics of chaotic phenomena* (Institute of Physics Publishing, Bristol, 1999).

[4] W. Heisenberg, *Physics and Beyond* (translated by A. J. Pomerans, Harper & Row, 1971) 11, p134.

[5] M. Kline, *Mathematics; the loss of certainty* (Oxford University Press, 1982) p326.

[6] As stressed by J. Hintikka, *The Principles of Mathematics Revisited* (Cambridge University Press, 1996), p9.

[7] "It is sometimes said that animals do not talk because they lack the mental capacity. And this means: "they do not think, and that is why they do not talk." But—they

think of, but cannot speak of. 'Speaking clearly' is required to communicate ideas with minimal misunderstanding among people who cannot be expected to share cultural backgrounds. Science is a communal activity built upon shared information and ideas. We need not be clear or logical when we are creative, because we need not have discussions with other people. However, science requires that created results must be 'expressed clearly' as much as possible.[8]

The source of ideas for theorization is intuition. The source of intuition is the structure of the world built into our body (thanks to the 'blood and tears' of our ancestors). Conceptual analysis is to pursue mathematical formulation of things maximally consistent with our intuition, but it is not an easy task to distill intuitive images to more or less precise descriptions, because what we human beings know is much broader than what we can actually speak clearly of. Furthermore, we must not forget that our intuition evolves, and also that intuition without careful introspection may not be very reliable. The crux of conceptual analysis seems to be intimate conversation between intuitive understanding and mathematical understanding.

How can we, then, perform conceptual analysis? First of all, (1) we must have a rough idea of 'what we wish to describe clearly.' This would be grasped intuitively through various experiences. Therefore, we must scrutinize typical examples. Next, (2) we try to describe what we have grasped intuitively in the ordinary language. The next step is to reflect on whether we really know the meanings of the words used in the obtained description. After this step, (3) we may compile the crucial points of 'what we wish to describe clearly' into a working definition. Then, (4) we must check how consistent the (working) definition is with our intuition. We must also think of whether it is a good concept or not. Obviously, we must think about what we wish to mean by 'a good concept.'

This chapter has the following structure. In Section 2.1, we 'watch' a typical 'chaotic' system. This step corresponds to (1) above. After defining dynamical systems in Section 2.2, in Section 2.3 we define (mathematically formulate) 'chaos' using an intuitively obviously random process. This corresponds to step (3) above. Here, we bypass (2) (an intuitive description of 'chaos' with the aid of the word 'randomness') with the aid of an actual typical example, and then wish to return there later.

---

simply do not talk. Or to put it better: they do not use language—if we except the most primitive forms of language, —Commanding, questioning, storytelling, chatting, are as much a part of our natural history as walking, eating, drinking, playing." [L. Wittgenstein, *Philosophical Investigations* (Blackwell, 2001; original 1953) 25 on p11^e].

[8] However, we should not forget the possibility that even if 'objects' may not be describable unambiguously, discourses relying on the actual objects whose identities are agreed upon are clear. Taxonomy may be understood as a discipline to try to supply this sort of clarity to our discourses.

How closely connected is the obtained definition of chaos to its intuitive understanding? Thus, (4) above is the next topic. This is related to information and entropy. After preparatory sections about measure and entropy (Section 2.4), information (Section 2.5) and measure-theoretical dynamical systems (Section 2.6), we will discuss how to quantify the 'extent of chaos' in Section 2.7 (Appendix 2.7A is an introduction to the Kolmogorov-Sinai entropy), and then confirm that this quantification is consistent with our intuitive feeling; dynamical behaviors judged as chaos with this quantitative measure coincide with chaos characterized in Section 2.3. Appendix 2.4A is an introductory explanation of measure and probability, exhibiting good examples of conceptual analysis.

(2) above remains: Do we really know the meaning of the word (concept) 'random(ness)' that we wish to use? How can we formalize 'randomness' that is introduced intuitively? As a natural approach to this question, algorithmic randomness is introduced in Section 2.8. To describe this concept precisely we start thinking about computation in Section 2.9. This supplies another classic example of conceptual analysis. The needed tool, the Turing machine, is introduced in Section 2.10. With these preparations, in Section 2.11 we can define algorithmic randomness unambiguously. Using all the theoretical machinery introduced up to this point, in Section 2.12, we will see quantitative vindication of the claim that dynamical systems whose trajectories are algorithmically random are chaotic dynamical systems. This is our ultimate analysis result of 'chaos.' We must still think about how satisfactory the result is, so Section 2.13 contains some comments on the definition of randomness, etc.

As the astute reader has realized, the chapter follows the progress in Russia chronologically 'backward,' starting from chaos and going to entropy and algorithmic randomness.[9] The order follows just the author's personal 'ontogeny.'

Various concepts that appear in this chapter are frequently encountered in the so-called 'complex-systems studies,' so this chapter prepares the reader to look at 'complex systems studies' critically. The last section suggests that the popular view or the popular characterization of complex systems is seriously flawed (according to the natural idea that typical eukaryotes are the typical complex systems).

---

[9] Y. G. Sinai, "Chaos Theory Yesterday, Today and Tomorrow," J. Stat. Phys. **138**, 2 (2010). "My personal experience shows that people in the West consider the development of Chaos Theory differently from their Russian colleagues, mathematicians and physicists." [This seems a great euphemism.] "Many people share the point of view that the beginning of chaos theory dates back to 1959 when Kolmogorov's paper ["New metric invariant of transitive dynamical systems and automorphisms of Lebesgue spaces," Dokl. Acad. Nauk SSSR **119**, 861 (1958)] on the entropy of dynamical system appeared."

2.1 Starting with typical examples—chaos as an example

In natural sciences the motivation to analyze a concept often comes from observation of a concrete phenomenon. The chapter intends to outline a practice of conceptual analysis with the aid of 'chaos' as an illustrative device, so let us scrutinize 'chaotic' behaviors exhibited by Keisuke Ito's 'great earthquake model,' which is a coupled-relaxation-oscillator system.[10] The reader may well say it is not a real phenomenon. However, a very similar system can be realized in the lab. In the following, it will eventually be shown that the model is closely related to the famous Lorenz model. On the way from Ito's model to the Lorenz model various ways to observe and describe a dynamical system will be illustrated and the reason for the peculiar behavior of the great earthquake model will become obvious. 'Chaotic behaviors' are common in simple systems. The reader will really feel surprised if such behaviors do not exist in reality. The subsequent content of this section was regarded, when it was first reported, as a sort of geometrical acrobat, so some concentration may be required to read the figures. However, the outline of the story can be glimpsed from Fig. 2.9 and Fig. 2.10. If the reader thinks that the appendix to this section, Appendix 2.1A, is already enough, *she can simply glance through this section and can go to Section 2.2.*

The motivation for Ito's great earthquake model is as follows. Island arcs such as the Japanese Archipelago consists of blocks spaced by faults. Each block is being pulled down by a subducting ocean plate and accumulates strain energy (Fig. 2.1). When this energy reaches a threshold, the block breaks (a great earthquake occurs) and the stored strain energy is quickly reset to its lowest value; the model chooses this value as the energy origin. If there is no interaction among blocks, earthquakes occur periodically (Fig. 2.2). This is probably the simplest model of the relaxation oscillator. If there

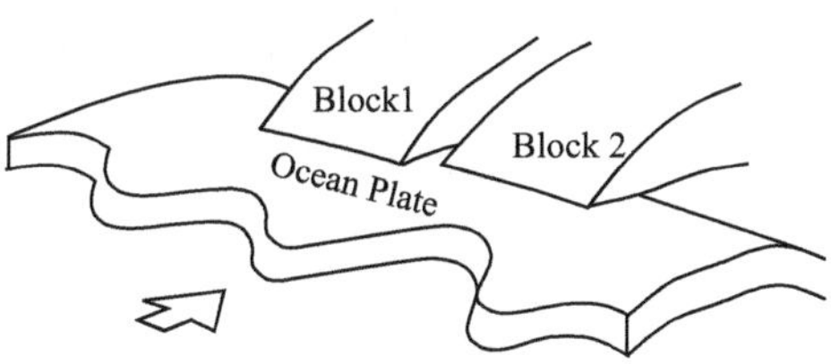

Fig. 2.1 The great earthquake model. The subducting ocean plate pulls two blocks down. The blocks accumulate strain energy and at some point they break apart from the plate and earthquakes occur. Then, the blocks return to the original positions.

---

[10] K. Ito, Y. Oono, H. Yamazaki and K. Hirakawa, "Chaotic behavior in great earthquakes—coupled relaxation oscillator model, billiard model and electronic circuit model—," J. Phys. Soc. Jpn. **49**, 43 (1980).

are two or more blocks, the earthquake occurring in one block must affect the surrounding blocks. Ito thought that the earthquake produces cracks in the neighboring blocks that decelerate the accumulation rate of the strain energy.

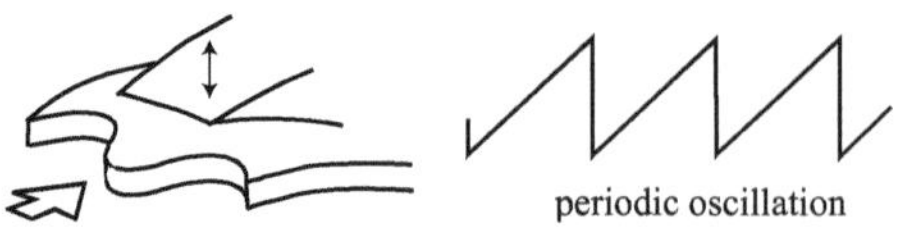

Fig. 2.2 If there is only one block, it is a simple model of the relaxation oscillator.

The simplest nontrivial case is the two-block case. The following rules of time evolution can model the idea depicted above. Let u_i be the strain energy stored in block i ($i = 1$ or 2).

(1) The rate of increase of the strain energy is initially b, which is assumed to be a positive number.

(2) If u_i reaches 1 an earthquake occurs in block i. Subsequently,

 (2a) u_i is reset to 0, and the rate of increase of strain energy is also reset to b; if the rate is already with this value, it is maintained as b.

 (2b) The rate of increase of the strain energy in the other block without the occurrence of the earthquake becomes b^{-1}; if the rate is already with this value, it is maintained as b^{-1}.

A typical behavior under the above rule is illustrated in Fig. 2.3. If the rate of increase of the strain energy $b > 1$, we see apparently complicated behavior. The time evolution of the two blocks can be illustrated as a trajectory on a square $[0, 1] \times [0, 1]$, if we plot u_1 on the horizontal axis and u_2 at the same moment on the vertical axis (Fig. 2.4).

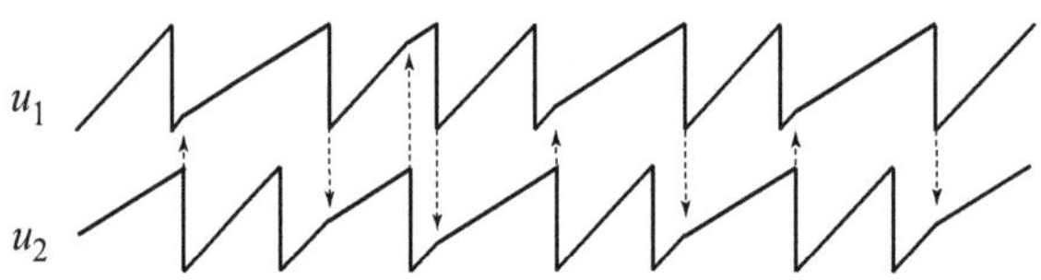

Fig. 2.3 A typical behavior of stored energies u_1 and u_2 in two coupled blocks 1 and 2, respectively. This is the case of mutual hindrance $b > 1$. The earthquake in one block decelerates the energy increase in the other block. Vertical arrows denote hindrances. (Except for the beginning stage of the behavior, the rates of energy increase of the blocks never coincide.)

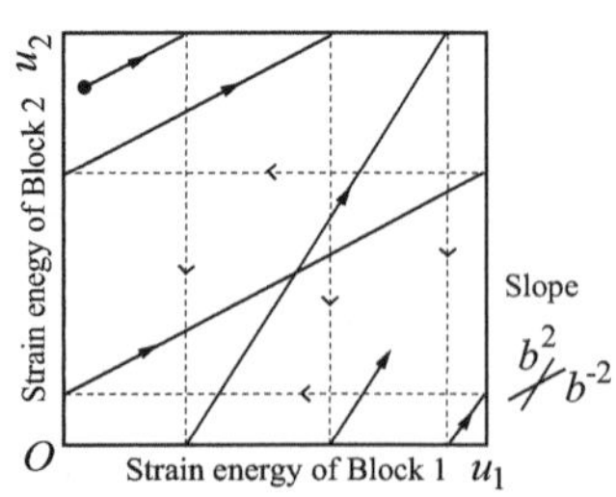

Fig. 2.4 The typical time dependence of the stored energy in the blocks may be depicted as a trajectory on $[0, 1] \times [0, 1]$. Horizontal jumps correspond to the earthquakes in block 1 and the vertical jumps those in block 2.

The rules mentioned above are illustrated in Fig. 2.5(i) on the square $[0,1] \times [0,1]$ as the rules for the trajectory of the two-dimensional vector (u_1, u_2). We can write the rules in formulas, but no new insight is obtained by doing so. For example, a portion of the trajectory with a slope less than 1 on the square implies that the energy increase rate in block 1 is larger than that in block 2. If we glue the left and the right edges of the square and then glue the top and the bottom edges, we obtain a 2-torus[11] (the surface of a donut). The strain energy changes in these two blocks may be described by a motion of a point on the torus.

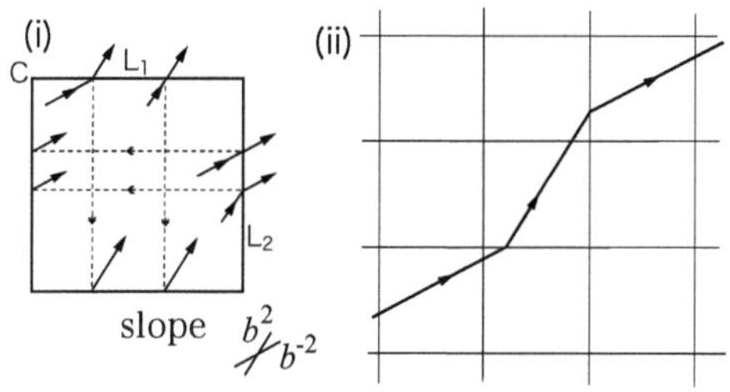

Fig. 2.5 The rules for the motion. **(i)** is the extension rule: after a jump how the slope changes is specified. If the point reaches L_1 (the top edge; an earthquake in block 2), it jumps down to the point following the broken line, and then starts running with slope b^2 (because u_2 increases at rate b and u_1 at rate b^{-1}); if the point reaches L_2 (the right edge), it jumps to the left according to the broken line and starts running with slope b^{-2} ($b > 1$ assumed). If the opposite edges of the square are glued, we can make a torus, and we obtain a continuous trajectory on the torus (as seen in Fig. 2.4). **(ii)** exhibits a typical trajectory on the so-called universal covering space that is made by tessellating the squares instead of forming a torus. The crossings with the vertical lines correspond to earthquakes in block 1 and those with the horizontal lines to those in block 2.

Instead of gluing the opposite edges of the square to make the torus, we may tessellate many copies of the square to make the so-called universal covering space.[12] In this space (i.e., instead of returning to itself, moving on to the next tiles) we can clearly see what is going on (Fig. 2.5(ii)). In Fig. 2.6 three trajectories starting from closely located initial points are depicted.[13] Earthquakes occur in block 1 (respectively, block 2) when the trajectory

---

[11] = two-dimensional torus; generally, 'n-object' implies n-dimensional object. For example, 2-sphere is the ordinary sphere, which is the surface of a 3-ball.

[12] To obtain elementary knowledge of topology I. M. Singer and J. A. Thorpe, *Lecture Notes on Elementary Topology and Geometry* (Undergraduate Texts in Mathematics) (Springer 1976; original 1967) is the best.

[13] The distance between the two adjacent points increases as $e^{t\lambda}$ as a function of time on the average, where λ is called the Lyapunov exponent (a slightly more detailed explanation is in Appendix 2.7B). In the present example, it can be computed as λ'/τ, where λ' is the exponent defined for the corresponding discrete system in Fig. 2.13 and τ is the average sojourn time on a single tile. To estimate λ' may not be easy, but it is obviously positive as can be trivially seen from the reduced map in Fig. 2.14.

crosses the vertical (respectively, horizontal) lines. If $b > 1$, two initially close trajectories come apart exponentially in time, but if $b < 1$, they eventually

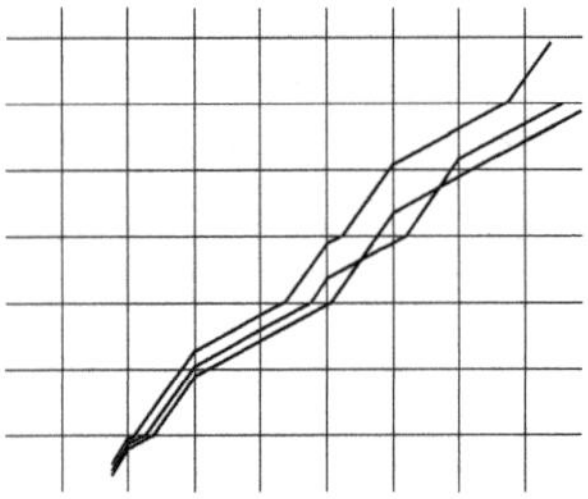

Fig. 2.6 Trajectories on the universal covering space. Locally, trajectories depart from each other exponentially in time. (Around the lattice points something complicated may happen, but this is a global problem.) (To prove that the trajectories separate from each other exponentially on the average, the easiest way is to use the one-dimensional map illustrated in Fig. 2.14.)

fall into a periodic orbit. The great earthquake model is with $b > 1$, so according to this model long-term prediction of earthquakes is impossible. Whether the model is plausible geophysically is not our concern. The crucial points here are that we have probably the simplest model of mutually hindering coupled relaxation oscillators, and that its behavior is not periodic, but is such that small deviations in initial conditions can have a large effect eventually.

Discussion 2.1. For $b < 1$ sketch the figures corresponding to Figs. 2.3-2.6 (although the answer is already given in the above). If the reader realizes that $b \to b^{-1}$ is a time reversal operation, the answer is easy to guess. □

Let us investigate why such complicated behaviors are possible. If the reader has already understood the mechanism, *she can jump from here to around Fig. 2.9.* The following story tells the reader that Ito's great earthquake model is equivalent to the so-called Lorenz template. It is somewhat like a geometrical acrobat, so details need not be followed closely; the key idea is in Fig. 2.9.

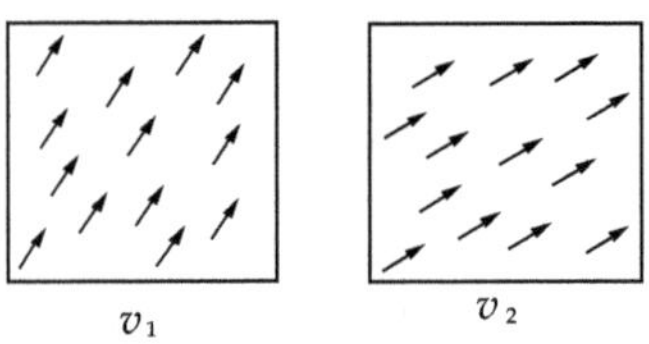

Fig. 2.7 In Fig. 2.4 at a particular point on the torus the trajectory runs in one of the two directions depending on the history up to the point. That is, the trajectory runs on the torus according to one of the two vector fields, v_1 or v_2, depending on the history.

If Fig. 2.4 is understood as a description of a two-dimensional motion, we must conclude that the trajectory follows the vector field v_1 in Fig. 2.7 at one time, and then v_2 at another. The history up to the point determines on which flow (vector field) the trajectory follows. A two-valued vector field dependent on history is not a very convenient object (a flow field that dras-

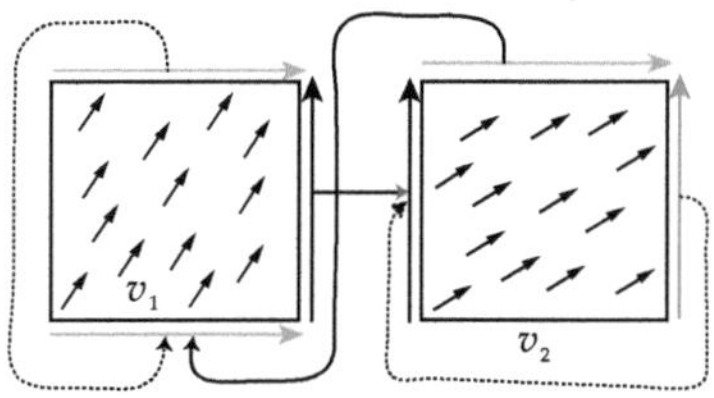

Fig. 2.8 The univalent vector field independent of the history on the connected two copies of the original torus. The two vector fields prepared in Fig. 2.7 are connected according to the rules consistent with the trajectory connection rules in Fig. 2.5(i). How to glue the tori is shown by straight arrows with large arrowheads (the directions denote correct orientations to glue).

tically changes according the history of the object being flowed is rather unusual). In order to introduce a univalent (i.e., unique time-independent) vector field (i.e., to describe the trajectory as an integral curve of a single-valued time-independent vector field), we prepare two copies of the torus, on one of which is v_1 and on the other v_2, and then connect these two tori according to the connection rule consistent with Fig. 2.5(i) (Fig. 2.8).

As can be seen from Fig. 2.5 or Fig. 2.6, something special happens when a crossover between two tori occurs. To understand what happens there, it is convenient to glue only the edges connected by broken arrow curves in Fig. 2.8 as A to B in Fig. 2.9 instead of completing the two tori. The trajectories coming into the connection edge from the left in Fig. 2.8 (along the short horizontal arrow) have smaller angles with the connection edge than the ones going out to the right as clearly illustrated in Fig. 2.9C, so we see the trajectories are spread and then are inserted into the trajectories on the right cylinder. That is, the event happening at the connection is akin to card shuffling. Thus, we understand intuitively why the system 'produces randomness' despite its deterministic nature.

It is clear that the apparently random behavior observed in Fig. 2.3 is caused by the intrinsic nature of this *deterministic dynamical system* (= a system whose behavior is uniquely determined if its past is known) and that external noise has nothing to do with it. If we knew the trajectory precisely without any error, we would not lose any information even if there is an expansive tendency of the trajectory bundle. However, we can never know very small scales. This unknowable is amplified by the expansion of the trajectory bundle and then is fixed into the large-scale behavior of the system by the insertion occurring at the connection to the other cylinder, as illustrated in Fig. 2.9C. Consequently, we feel the behavior of this system to be random.[14]

---

[14] As can be seen from this example, nonlinearity is not needed to expand small scales. Nonlinearity is, however, usually needed to contain the system within a finite

If we deform the system slightly further, we can map it to a system that must be familiar to chaos aficionados. If we squish the cylinders in Fig. 2.9B on

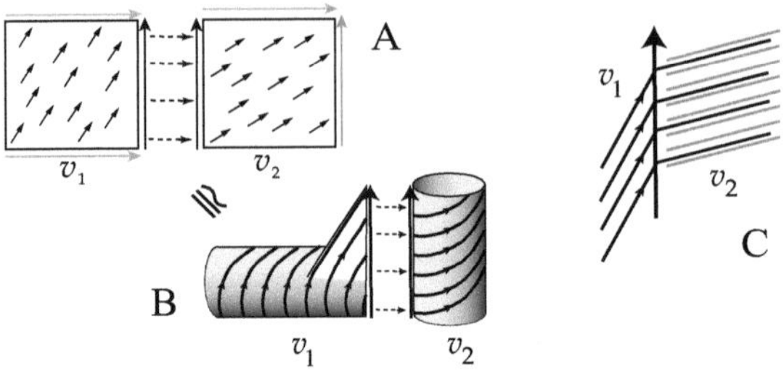

Fig. 2.9 Let us pay attention to the connection from v_1 to v_2 only (the broken arrows in A, which is almost a copy of Fig. 2.8). Gluing only one pair of edges in A, we obtain B. The 'screw' part of B corresponds to the inside of the squares in A, where the trajectories run in parallel. In order to make easy to observe the crossover from the left to the right square (or torus), a cut is introduced in the left cylinder that does not affect the parallel trajectories and the edge to be glued is made straight. What happens at the connection between v_1 and v_2 is equivalent to spreading out the trajectory spacings and then inserting trajectories into the right cylinder as illustrated in C; it reminds us of shuffling cards.

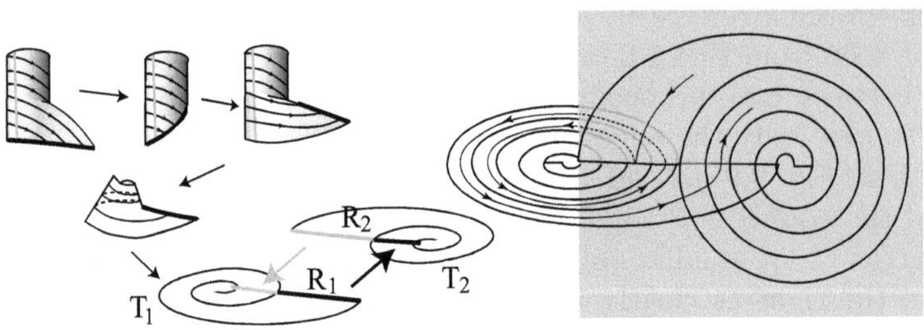

Fig. 2.10 Further topological acrobatics. The portions where the cylinders are connected are the same in Fig. 2.9B, but the ways to extend the trajectory ends are different. After running the spiral portion, trajectories are inserted into the other spiral. If we dovetail these two sheets with spiral trajectories, we obtain the rightmost figure. T_1, and T_2 correspond to the tori and R_1 and R_2 are connecting edges, where insertion and fixation of the expansion outcomes occur. The difference from Fig. 2.9B is only that the spiral on the cylinders becomes that on the disks, so, similarly as before, we can see how expanded microscopic details influence decisively the world around our scale.

which the motion is 'spiral,' we obtain Fig. 2.10. The figure reminds us of the famous *Lorenz model* (Fig. 2.11). The Lorenz model is defined by the

---

range despite local linear expansion. Therefore, if the phase space is intrinsically compact, then we might be able to say nonlinearity is not absolutely necessary for chaos. Actually, a typical chaotic system is given by linear maps from a torus onto itself. The decisive paper on this topic is R. L. Adler and B. Weiss, "Similarity of automorphisms of the torus," Memoir. Am. Math. Soc. **98** (1970).

following set of differential equations:[15]

$$\dot{x} = -\sigma x + \sigma y, \tag{2.1}$$

$$\dot{y} = -xz + rx - y, \tag{2.2}$$

$$\dot{z} = xy - bz, \tag{2.3}$$

where σ, b, and r are constants. A typical choice is $\sigma = 10$, $b = 8/3$, $r = 20$. The behavior of the system under this choice is seen in Fig. 2.11. This system is a differential equation system, so its trajectories cannot cross each other (due to the uniqueness of the solution). Consequently, the two disk-like structures have some thickness. If we ignore this thickness (projecting perpendicular to the disk surface), we obtain the branched manifold model of the Lorenz model (called the *Lorenz template*, Fig. 2.12). This is equivalent to Ito's great earthquake model.[16]

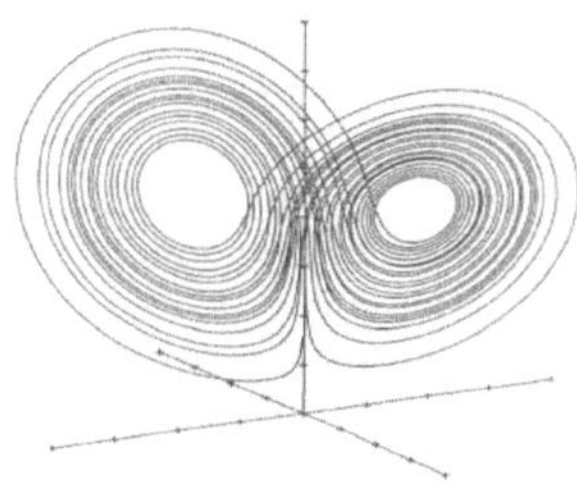

Fig. 2.11 Solution curves of the Lorenz system with $\sigma = 10$, $b = 8/3$, $r = 20$ (*Mathematica* was used). The disk-like structures on both sides do not form a single disk but a structure like mille-feuille; numerous sheets lie side by side.

Fig. 2.12 The Lorenz template (from the American Mathematical Society home page: `http://www.ams.org/featurecolumn/archive/lorenz.html`)

[15] E. N. Lorenz, "Deterministic nonperiodic flow," J. Atmospheric Sci. **20**, 130 (1963). The model was obtained by keeping only three long-wavelength modes in the Fourier expanded form of the Bousinesque equation describing the Bénard convection. It was also Lorenz who recognized that such a bold truncation could still exhibit some salient behaviors of the original system.

[16] It is homeomorphic. R. F. Williams, *The universal templates of Ghrist*, Bull. Amer. Math. Soc. **35**, 145 (1998) is a good review of the research with the aid of templates. See also `http://www.ams.org/featurecolumn/archive/lorenz.html`. For a mathematical proof that the Lorenz system is indeed chaotic, see the following expository article: M. Viana, "What's new on Lorenz strange attractor," Math. Intelligencer **22**(3), 6 (2000). E. W. Weisstein, "Lorenz Attractor," From MathWorld—A Wolfram Web Resource, `http://mathworld.wolfram.com/LorenzAttractor.html` gives a good succinct summary.

The above earthquake model exhibits complicated, even random-looking motion of the particle riding on a vector field (flow vector field), but the 'flow' itself is surprisingly simple. This implies that even if flows are simple (laminar) without any random element, still the location of the particle flowing with it can be hard to predict. Such a phenomenon is called *Lagrangian turbulence.*[17]

Discussion 2.2. We coupled two relaxation oscillators to produce 'chaotic behaviors,' but we can also consider a periodically perturbed single relaxation oscillator: when a pulse arrives, the oscillator changes its energy increase rate to b^{-1} (if already with this value, keeps it); after an 'earthquake' due to u reaching the threshold (unity) this rate is reset to b. Here, we assume $b > 1$. Show that this model exhibits 'chaotic behavior.' (That is, a nonlinear oscillator + periodic perturbation can already produce 'strange behaviors.') □

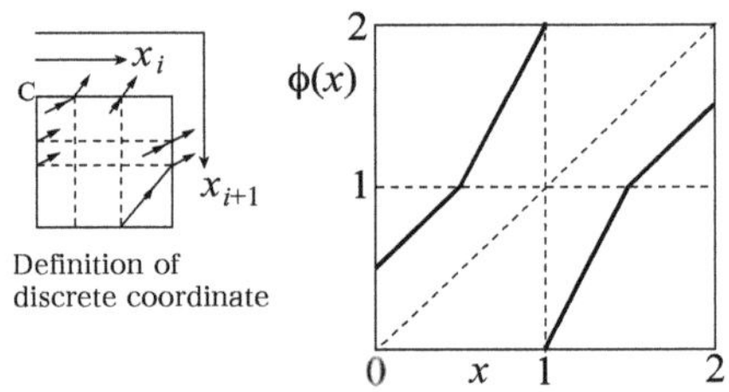

Fig. 2.13 If the successive lattice crossing positions of a trajectory are described in terms of the distance measured along the edges of the square from the corner C, we obtain a one-dimensional map $[0, 2] \to [0, 2]$: $x_{i+1} = \phi(x_i)$.

The above earthquake model can be understood as a one-dimensional map whose example is used in Appendix 2.1A to illustrate chaos. A trajectory in the universal covering space is piecewise linear as we have seen in Fig. 2.6, so we must be able to record a trajectory only by recording its breaking points. We have only to pay attention to the points where the trajectory crosses the lattice lines in the universal covering space, and convert the continuous time system to a discrete time system (however, it is not a simple discretization such as observing a system periodically with stroboscopic light). More explicitly, in Fig. 2.5(i) (copied in Fig. 2.13) if we record the distance x measured along the edges from the corner C when the trajectory crosses L_1 or L_2, we can map a trajectory to a sequence $\{x_i\}$. This correspondence is one to one.[18] That is, from $\{x_i\}$ we can reconstruct the original continuous trajectory, because we know the speed of the point along the trajectory. (The actual detail explained below is not important, so *the reader can jump from the end of the next paragraph to the next section.*)

If the sequence $\{x_i\}$ can be reconstructed by a certain recursive rule, the

[17] H. Aref, "Chaotic advection in a Stokes flow," Phys. Fluids **29**, 3515 (1986) pointed out such a phenomenon can happen in Stokes' flows, where the Reynolds number is zero.

[18] Here, we ignore the 'loose ends' of the trajectories between the lattice crossings.

system may be described more simply. Here, a 'recursive rule' means a rule that can give the next time state in terms of the current state (corresponding to the equation of motion in classical mechanics). Actually, this sequence $\{x_i\}$ is determined as a solution to an initial value problem of a (nonlinear) difference equation

$$x_{n+1} = \phi(x_n), \tag{2.4}$$

where $\phi : [0,2] \to [0,2]$ is given by the graph in Fig. 2.13 Right (how to read this diagram is explained in Fig. 2.14 or Fig. 2.15A in Appendix 2.1A).

This discrete time model is equivalent to the (2-block) earthquake model defined above (as we have just seen, the continuous time model can be reconstructed from the discrete model). This graph (or the map ϕ defined by it) is the recursive rule. We have dissected the discrete point sequences into a recursive rule and initial conditions. In this example, even if we change the initial condition, we do not observe qualitatively different behaviors.[19] Hence, finding a recursive rule (corresponding to the fundamental laws) is the essential part of understanding this system.[20]

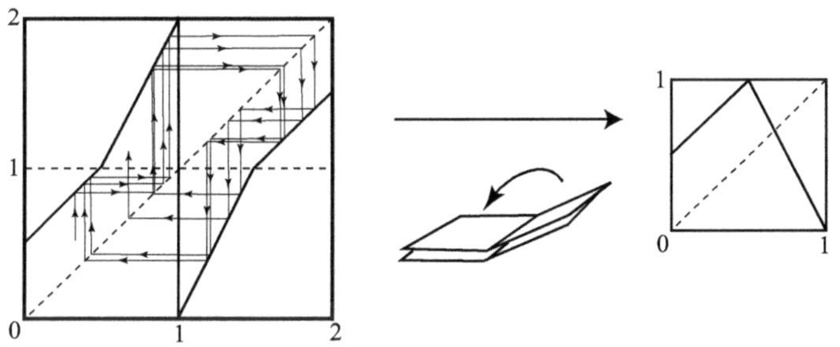

Fig. 2.14 A reduced discrete map may be constructed by 'folding' the original map. Such unimodal piecewise linear maps are mathematically thoroughly understood. On the left the way to chase a discrete history is illustrated.

This discrete model can be, as explained in Fig. 2.14, folded into a unimodal piecewise linear map from $[0,1]$ into itself.[21] Folding implies identifying two points in $[0,2]$. Therefore, the original behavior cannot be reconstructed from the reduced system. However, the correspondence is simple, so we can learn various things about the original system from the simplified system. For example, if the behavior of the simplified system is complicated, the behavior

[19] Actually, there are infinitely many initial conditions that give periodic orbits, but they are measure zero (for measure zero see **A1** of Appendix 1A at the end of Chapter 1).

[20] As we will see in Chapter 5, many interesting systems lack such a property.

[21] Unimodal piecewise linear maps are thoroughly studied in the following papers: Sh. Ito, S. Tanaka and H. Nakada, "On unimodal linear transformations and chaos I, II," Tokyo J. Math. **2**, 221, 241 (1979).

of the original system cannot be simple. That is, we may 'bound from below' the complicatedness of the original system.

Note 2.1. History of the study of chaotic systems
The accomplishment of Lorenz (1963[22]) was to propose a method to demonstrate that apparently irregular signals are not simple noise, and to demonstrate convincingly by this method that his computational result was not merely due to numerical errors.[23] This was an epoch-making work showing that chaotic behaviors existed in a system more or less related to an actual natural phenomenon. The work was publicized by McLaughlin and Martin,[24] becoming well known among physicists. Behind their work was also the proposal of dynamical-system-theoretical interpretation of turbulence (proposal of strange attractors) by Ruelle and Takens (1971).[25]

The possibility of apparently random solutions for simple deterministic dynamical systems was pointed out by Poincaré in conjunction with the three-body problem, and even before him, Maxwell already knew the possibility. In the 20th century Levinson's work on the periodically perturbed nonlinear oscillator (1949)[26] and the work by Cartwright and Littlewood[27] that motivated Levinson's work were pioneering. To understand their work, Smale introduced the horseshoe dynamical system and then the Axiom A system, and made various deep conjectures about the generic properties of dynamical systems.[28]

In Russia fundamental studies on dynamical systems were vigorously pursued since Lyapunov (1857-1918), and many important results were obtained by 1970. For example, the thermodynamic formalism for dynamical systems that connect statistical mechanics and the theory of dynamical systems was initiated by Sinai in 1967.[29] The characterization of chaos in terms of symbolic

---

[22] [1963: Assassination of President Kennedy, Freedom March on Washington, Beetles fever began]

[23] See also E. N. Lorenz, "The problem of deducing the climate from the governing equation," Tellus **16**, 1 (1964).

[24] J. B. McLaughlin and P. C. Martin, "Transition to turbulence in a statically stressed fluid system," Phys. Rev. A**12**, 186 (1975). [1975: the Vietnam War ended, Asilomar Conference on genetic engineering]

[25] D. Ruelle and F. Takens, "On the nature of turbulence," Commun. Math. Phys. **20**, 167 (1971). See also *ibid.* **23**, 343 (1971). See Ruelle's comment on this work in D. Ruelle, *Chance and Chaos* (Princeton University Press, 1993). [1971: Kissinger-Zhou Enlai secret talks]

[26] N. Levinson, "A second order differential equation with singular solutions," Ann. Math. **50**, 127 (1949). [1949: People's Republic of China, Russian atomic bombs, G. Orwell's *1984*]

[27] M. L. Cartwright and J. E. Littlewood, "On non-linear differential equations of the second order: I the equation $\ddot{y} + k(1 - y^2)\dot{y} + y = \lambda \cos(\lambda t + a)$, k large," J. London Math. Soc. **20**, 180 (1945). As to Cartwright see S. McMurran and J. Tattersall, "Mary Cartwright (1900-1998)," Notices AMS. **46**, 214 (1999).

[28] S. Smale, "Differentiable dynamical systems," Bull. Amer. Math. Soc. **73**, 747 (1967), which probably physicists could understand.

[29] R. Bowen, *Equilibrium states and the ergodic theory of Anosov diffeomorphism*, Lecture Notes in Math. **470** (Springer, 1975) is pedagogical (for the author). [1967: Gulf of Tonkin Incident, President Johnson's 'Great Society Program']

dynamical systems was proposed by Alekseev in the 1960s.[30] Smale's work was welcomed in Russia. In the US Ornstein[31] proved a decisive theorem on the classification of Bernoulli systems along the line of the Russian program initiated by Kolmogorov on the entropic characterization of dynamical systems (see footnote 9). These were, in retrospect, all fundamental contributions to understanding chaos. In sum, by around 1980 conceptually important works had mostly been finished. The works of, especially, non-mathematicians in the West after 1980 was on the more applied side and in popularization of chaos. To this end computers (esp., graphics) were indispensable. This is in sharp contrast with the fundamental conceptual works, for which computers were irrelevant.

Appendix 2.1A An example of simple but genuine chaos

Here, although artificial, a simple example of chaos is illustrated. The roadmap of this chapter in the light of this example is also given. If the reader understands this example, *she will probably be able to do without the example in the main body of this section.*

Time evolution exhibiting chaos: A map T from $[0, 1]$ onto itself is defined as

$$Tx = 2x \mod 1. \tag{2.5}$$

Such a system is generally called a *dynamical system* (see Section 2.2). If we use the notation $\{r\}$ to denote the fractional part of a real number r, we may write $Tx = \{2x\}$ (i.e., multiply 2 and then remove the integer part). Important points are summarized in Fig. 2.15.

History is determined by initial condition: If an initial condition is given, an indefinitely long sequence $\{x_n\}$ may be constructed as $x_1 = Tx_0$, $x_2 = Tx_1 = T^2x_0, \cdots$; the future is perfectly determined by x_0. How to chase this trajectory graphically is illustrated in Fig. 2.15A.

Chaos directly connects the unknown world to our world: Fig. 2.15B illustrates how chaos expands the world of microscopic scales and connects it to the world we can directly observe. In this example, the microscopic world is doubled every time T is applied. Therefore, although the system we consider at present is a deterministic system, we lose our predictive power quickly. Roughly speaking, for large n x_n becomes indistinguishable from a random number.

Coding trajectories or correspondence to number sequence: To see the random nature of chaotic trajectories more explicitly, their discrete coding $s_0, s_1, \cdots$ is introduced in terms of symbols s_n that take 0 or 1. It is a rather obvious transformation in this case; the interval $[0, 1]$ is divided into two intervals, $[0, 1/2)$ and $[1/2, 1]$, and we call them $[0]$ and $[1]$, respectively. Generally, $[s_0s_1 \cdots s_n]$ is defined as a set $\{x : T^kx \in [s_k] \text{ for } k = 0, \cdots, n\}$, where s_k are 0 or 1. In Fig. 2.15C,[32] we see that if we apply T once to the interval, say, $[011]$, it is mapped onto $[11]$. One more application of T to it gives $[1]$ (it is easy to

---

30 V. M. Alekseev, "Quasirandom dynamical systems I, II, III," Math. USSR Sbornik **5**, 73 (1968); **6**, 505 (1968); **7**, 1 (1969).

31 D. S. Ornstein, "Bernoulli shifts with the same entropy are isomorphic," Adv. Math. **4**, 337 (1970); D. S. Ornstein, *Ergodic Theory, Randomness, and Dynamical Systems* (Yale University Press, 1974); I. P. Cornfeld, S. V. Fomin, and Ya. G. Sinai, *Ergodic Theory* (Springer, 1982).

32 The reader might worry about to which interval the boundary points belong, but in this example, careful assignment of symbols is not important.

see this, if we chase the end points as explained in Fig. 2.15A). That is, [011] is a bundle of trajectories for which x_0 is in [0], x_1 is in [1] and x_2 is in [1] (such a bundle of trajectories is called a *cylinder set*. See Section 2.4).

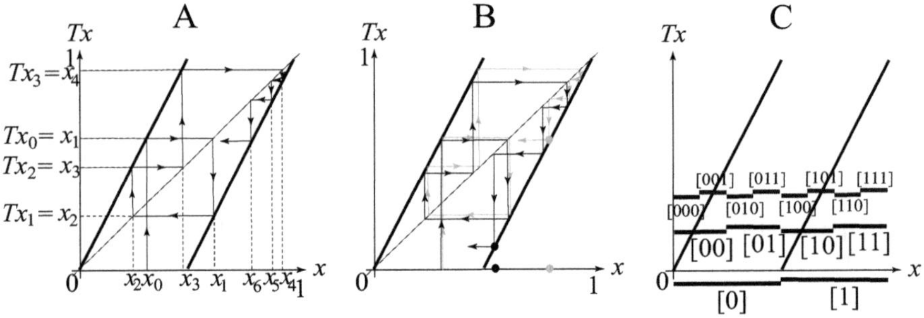

Fig. 2.15 A simple map $Tx = \{2x\}$ that produces chaos.
A: How to chase history graphically: A method to chase the trajectory which is determined by the initial condition x_0 on the graph is illustrated. The broken line diagonal denotes $y = x$ where the values on the horizontal and vertical axes coincide. Oblique thick parallel lines denote the graph of $y = Tx$. If an initial condition x_0 is given on the horizontal axis, look vertically upward to find the point on the graph of T. Its vertical coordinate is x_1. To find x_2, we must find x_1 on the horizontal axis, and then $Tx_1 = x_2$ may be obtained just as before. To this end, with the aid of the diagonal, we can fold the vertical axis onto the horizontal axis and locate x_1 on the latter. Therefore, if we chase the vertical or horizontal lines with an arrow, we can successively find $x_2, x_3, \cdots$.
B: Exponential separation of nearby trajectories: Doubling of the gap is shown due to each application of T between the gray and black trajectories that are initially very close. The positions after 6 applications of T are denoted by the gray and black small disks on the horizontal axis.
C: Coding of trajectory: How to convert a trajectory into a 01 symbol sequence is illustrated. The interval $[0, 1]$ is divided into two and each is named [0] or [1]. [0] is further subdivided into [00] and [01]. T maps [00] onto [0] and [01] onto [1]. If the subdivision of [10] is named [100] or [101] as in the figure, T maps [100] onto [00] and [101] onto [01]. If we recursively use this prescription, each point in $[0, 1]$ becomes correspondent to a particular 01 infinite sequence. This is nothing but the binary expansion of a number in $[0, 1]$ (however, $[111\cdots]$ is not identified with [0]).

How prediction becomes difficult: If we follow the way to construct small intervals in Fig. 2.15C, we see that, for example, the bundle of trajectories named as [0011010011] (there are ten digits) is mapped by T successively as $[011010011] \to [11010011] \to [1010011] \to [010011] \to \cdots \to [011] \to [11] \to [1]$ (digits are lost one by one from the left end). One more application of T makes the trajectories 'all over' $[0, 1]$. The lesson we have learned is that even if the initial condition is in the interval [0011010011] of width $1/2^{10} \sim 10^{-3}$, after ten consecutive applications of T we will totally lose the location information. Perhaps the reader might guess that we will do much better if we specify the initial point (the cylinder set) more accurately. If we wish to keep some knowledge of the initial position for 20 seconds, we need the accuracy of unrealistic $1/2^{20} \sim 10^{-6}$. In short, sooner or later we will fail to predict the behavior of the system.

Even if we fail to make any prediction, it does not imply the end of the world. Beyond the predictable time range what determines x_n? The system is deterministic, so it is determined by the far right portion of the 01 sequence obtained by coding of the initial condition, which we can never know beforehand. Isn't it virtually the same as an arbitrary 01 sequence? Then, after a while, chaotic behavior would be indistinguishable from the head-tail sequence obtained by tossing a coin. This is the intuitive motivation for the definition of chaos given in Section 2.3.

Chaos is a random deterministic behavior: After all, it is a natural idea that the essence of chaos is randomness with a tint of initial condition effects due to determinism. However, without carefully reflecting on the concept of randomness, we cannot express this intuition precisely. After a rather 'heavy' preparation (Sections 2.8-11), eventually we will arrive at the conclusion that chaos is a phenomenon that deterministic trajectories exhibit randomness (Section 2.12).

Quantitative correspondence of chaos and randomness: The reader may think the above argument provides only a qualitative characterization of chaos. However, there is a way to quantify randomness, which allows us to make a quantitative correspondence between chaos and randomness. This quantification is realized through quantifying the needed information to predict the future with a prescribed fixed accuracy. This is not hard in terms of the model being considered. How the information (quantified in Section 2.5) we know at the initial time becomes insufficient to describe the system behavior can be seen almost explicitly from Fig. 2.15C and the loss of digits from the 'cylinder sets' due to the application of T; the information is lost by 1 bit every time T is applied. Suppose we wish to predict the position of the point at time t in the future with the same accuracy we describe the system now ($t = 0$). The information we must prepare now increases by 1 bit, if we push the future time t further to $t + 1$. The increasing rate of the needed information (1 bit per unit time in the present example) is called the Kolmogorov-Sinai entropy (Section 2.7). On the other hand, to describe an arbitrary 01 sequence we of course need 1 bit per digit. That is, the needed information to describe a trajectory is 1 bit per unit time. This equality of the amounts of information needed to predict the future and to describe the trajectory is the assertion of Brudno's theorem (Section 2.12).

2.2 Dynamical systems

Before going into our main topics, let us summarize what the dynamical system is. 'Path space' and 'shift dynamical system' are important concepts.

The most complete description of the time series of events occurring in a system (i.e., the history of the system) is to make a table (chronological table) of all the events arranged in chronological order. However, for a given system many different histories are possible, depending on the initial condition, its environment, etc. Therefore, a complete description of the time evolution of a system may be given by the totality of the histories possible for the system (i.e., by the book of all the possible chronological tables allowed to the system). The book describes the time evolution of the system completely, so we may define a dynamical system by this book. However, usually a dynamical system is defined by the 'principle' (law or rule) to make this book.

Let Γ be a topological space and $\boldsymbol{T}$ an additive group.[33] Let $\varphi : \Gamma \times \boldsymbol{T} \to \Gamma$ be a map and for each $t \in \boldsymbol{T}$ $\varphi_t : \Gamma \to \Gamma$ is defined as $\varphi_t(x) = \varphi(x, t)$. If the family of maps satisfies the following conditions, we say φ (or φ_t) defines a *dynamical system* on Γ:
(a) $\varphi_t \circ \varphi_s = \varphi_{t+s}$ (i.e., $\varphi_t(\varphi_s(x)) = \varphi_{t+s}(x)$),
(b) $\varphi_0 = 1$, where 1 is the identity map on Γ.

In intuitive terms, Γ is the *phase space* where the states $x(t) = \varphi_t(x)$ of the system move around, and φ_t is the time evolution operator that evolves the system into the future by time t. If there is no passage of time, this operator does not alter the state (this is (b)). To apply this operator successively is equivalent to a single time evolution corresponding to the total passage of time (this is (a)).

The history of the discretized version of the earthquake model discussed in the preceding section was determined by a map ϕ from its phase space $\Gamma = [0, 2]$ into itself. This dynamical system does not explicitly depend on absolute time; the future is determined by the present position in the phase space and the time passage between the future and the present. Historically, the theory of dynamical systems[34] originated from the study of global properties of autonomous differential equations,[35] so it is natural to consider only such systems without any dependence on absolute time.[36]

$\{\varphi_t(x) : t \in \boldsymbol{T}\}$ is called the trajectory (or orbit) that passes through x. A history ω allowed to a dynamical system is a map $\omega : \boldsymbol{T} \to \Gamma$ for which there exists $x \in \Gamma$ such that $\omega(t) = \varphi_t(x)$, where x may be called the initial condition (Fig. 2.16). $\{(t, \omega(t)) : t \in \boldsymbol{T}\})$ is a particular chronological table. The totality of such tables is the book mentioned at the beginning of the section.

---

[33] A topological space is a set where the concept of neighborhood is defined. An additive group $\boldsymbol{T}$ is the same as a commutative or Abelian group. That is, if $s, t \in \boldsymbol{T}$, then $s+t$ is defined as an element of $\boldsymbol{T}$ and $s+t = t+s$ (commutative). That $\boldsymbol{T}$ is a group implies that for this $+$ operation the following three conditions hold: (i) If u is also an element of $\boldsymbol{T}$, $t + (s+u) = (t+s) + u$, (ii) there is a zero element 0 satisfying $t + 0 = t$ in $\boldsymbol{T}$, and (iii) for any t there is an inverse $-t$ in T such that $t + (-t) = 0$.
It is often the case that $\boldsymbol{T}$ is not a group but a monoid (i.e., condition (iii) is not required). In this case the resultant dynamical system cannot uniquely reverse its histories (as the one-sided Bernoulli process below).

[34] A standard (and not thick) introductory textbook of the theory of dynamical systems is, e.g., D. Ruelle, *Elements of Differentiable Dynamics and Bifurcation Theory* (Academic Press, 1989).

[35] Although it is not a history reference, M. W. Hirsch, "The dynamical systems approach to differential equations," Bull. Amer. Math. Soc. **11**, 1 (1984) is useful.

[36] If the dynamical rules may depend on absolute time, the rule may change arbitrarily as time passes, so we cannot expect very interesting mathematics. Therefore, we do not usually consider non-autonomous systems (except for, e.g., periodic modifications).

The totality $\mathcal{D}$ of histories allowed to a given dynamical system is called its *path space*. It may be better to call it the history space. In practice, however, not $\mathcal{D}$ itself, but a space containing it and easy to specify is often

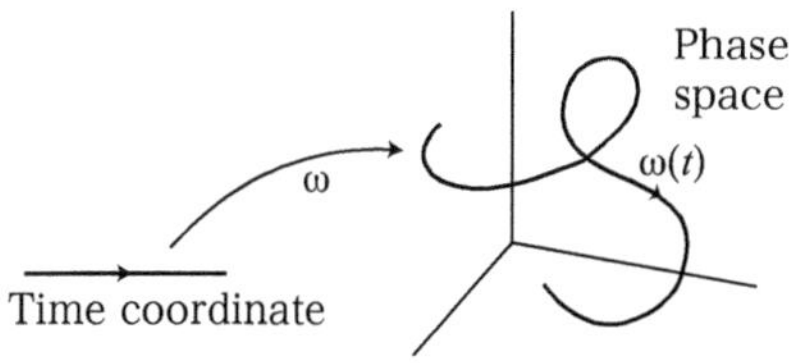

Fig. 2.16 Following a history is to trace an image of the time coordinate in the space of events (phase space) $\omega(\boldsymbol{T})$ according to the time order. The curve $\{\omega(t)\}$ parameterized with t is the image of the time coordinate, describing a possible history. It need not be a continuous curve.

called the path space of a given dynamical system.

Discussion 2.3. Can you distinguish deterministic and non-deterministic systems from the path space point of view of dynamical systems?[37] □

It is possible to regard the difference between the deterministic and non-deterministic dynamical systems is due to the difference of viewpoints. Take a discrete dynamical system, whose history $\omega \in \mathcal{D}$ may be written as $\cdots\omega(-2)\omega(-1)\omega(0)\omega(1)\omega(2)\cdots$. $\omega(0)$ is interpreted as the current state. Let us define the *shift operator* (or simply, *shift*) $\sigma : \mathcal{D} \to \mathcal{D}$ as

$$(\sigma\omega)(t) = \omega(t+1). \tag{2.6}$$

The shift is a vehicle to experience the history in chronological order. Suppose, for example, the phase space is $\Gamma = \{0, 1\}$ and its one possible history ω is $\cdots 00101001\,\dot{1}\,101001101\cdots$, where the dotted number is interpreted as the state we observe at present. Its time evolution is

$$\omega = \cdots 00101001\dot{1}101001101\cdots, \tag{2.7}$$

$$\sigma\omega = \cdots 01010011\dot{1}010011010\cdots, \tag{2.8}$$

$$\sigma^2\omega = \cdots 10100111\dot{0}100110101\cdots. \tag{2.9}$$

The currently observed state evolves step by step. The state $\omega(t+1)$ may not be determined by the states up to time t: $\cdots, \omega(t-1), \omega(t)$, so for the observer who is observing the current and the past states only the system behavior may not look deterministic.

Instead of interpreting the shift as a vehicle to experience a history chronologically, if we interpret ω as a whole to be one chronological table, the shift maps one table to another: $\sigma\omega = \omega_1$, where ω_1 gives the chronological table

[37] For a deterministic system, its behavior up to time t uniquely determines its future behavior. If we regard a record of casting a die successively as a history, it apparently lacks such a deterministic nature. As we will see just below, the conclusion depends on how we look at the system.

one time unit ahead of ω; indeed, for all time $t \in \boldsymbol{T}$ $\omega_1(t) = \omega(t+1)$. σ is understood as a map from $\mathcal{D}$ into itself; it defines a discrete dynamical system on the path space: $(\sigma, \mathcal{D})$.[38] (2.7)-(2.9) illustrated its time evolution. Here, we observe the history not around the present, but observe histories from God's point of view. Notice that ω completely determines ω_1. The dynamical system $(\sigma, \mathcal{D})$ is obviously deterministic.

More generally, we may define a shift operator σ acting on a set of certain both-side infinite sequences defined on a finite set Λ of discrete symbols just as above: $\sigma\omega(t) = \omega(t+1)$. Let Ω be a subset of $\Lambda^{\boldsymbol{Z}}$ (this is the totality of the both-side infinite sequences consisting of the symbols in Λ; $\boldsymbol{Z}$ is the standard notation for the totality of integers) invariant under σ. (σ, Ω) is generally called a *shift dynamical system*. In particular, if $\Omega = \Lambda^{\boldsymbol{Z}}$, (σ, Ω) is called the *full shift*. $\sigma\omega$ is determined uniquely by ω, so a shift dynamical system is deterministic.

So far we have considered shifts defined on both-side infinite sequences for which the time span is from $-\infty$ to $+\infty$. It is also common to consider a shift defined on one-sided infinite sequences for which time runs from 0 to $+\infty$. For a certain element ω of $\Lambda^{\boldsymbol{N}}$, i.e., a one-sided infinite sequence consisting of symbols taken from Λ, the shift σ is defined as $(t \geq 0)$

$$(\sigma\omega)(t) = \omega(t+1). \tag{2.10}$$

This looks exactly the same as (2.6), but the definition is only for $t \geq 0$, and the symbol shifted to the left eventually disappears after it reaches the leftmost position $t = 0$. For example, for $\omega = 001010011101001101\cdots$

$$\omega = 001010011101001101\cdots, \tag{2.11}$$

$$\sigma\omega = 010100111010011010\cdots, \tag{2.12}$$

$$\sigma^2\omega = 101001110100110101\cdots. \tag{2.13}$$

In the above, usually the leftmost symbol is interpreted as the state being observed at present. In contrast to the shift on both-side infinite sequences, the inverse σ^{-1} of the shift is not defined as a map for the one-sided case. Let Ω be a subset of $\Lambda^{\boldsymbol{N}}$ invariant under σ. (σ, Ω) is generally called a *one-sided shift dynamical system*. $\sigma\omega$ is determined uniquely by ω, so a one-sided shift dynamical system is deterministic (but not reversible).

Example 2.1. Coin-tossing. Tossing a coin many times, we can construct a sequence of heads and tails. If we denote 'head' by 1 and 'tail' by 0, we have a 01 sequence. In this case, the phase space is $\Gamma = \{0, 1\}$ and time is $\boldsymbol{T} = \boldsymbol{N}^+ \equiv \{1, 2, \cdots\}$. As its path space Ω we may choose the totality of one-sided 01 infinite sequences $\Omega = \{0, 1\}^{\boldsymbol{N}^+}$, because heads and tails can appear in any order. The resultant dynamical system (σ, Ω) is called the

[38] Notice that since σ maps a path to another path, $\mathcal{D}$ is regarded as the phase space of this dynamical system.

coin-tossing process. Let $\omega \in \Omega$. If $\omega(n)$ happens to be 0, this implies that the n-th outcome is a tail. Even if we know a history of this system up to time n: $\omega(1), \cdots, \omega(n)$, no one can predict anything beyond. Therefore, "the system is not deterministic." We know already we cannot simply declare this. More discussion follows below. □

As can be seen from the above discussion, we know that every discrete time dynamical system can be interpreted as a shift dynamical system defined on its history space (path space). Let us think about its significance.[39] Suppose a person has decided to initiate a coin-tossing sequence. Then, God selects and gives a certain one-sided 01 sequence (history) in $\{0,1\}^{\boldsymbol{N}^+}$, or equivalently, a particular map $\omega : T \to \Gamma$ to her. At time n she experiences $\omega(n)$. The shift we encountered above becomes a vehicle for her to trace her fate: $(\sigma\omega)(n) = \omega(n+1)$. However, from God's point of view, the coin tossing sequence she is 'producing' now is perfectly deterministic ($\omega \to \sigma\omega$). The 01 sequence given to her is predetermined up to the end of the time, but she really never knows what is next.

As we have already discussed, whether a dynamical system is deterministic or not depends on how it is interpreted/observed. As we will see soon, in the case of chaos to choose an initial condition corresponds to God's selecting a 01 sequence. We play God in chaos. Only we cannot precisely designate the initial condition infinitely. We can specify at most a finite 01 sequence, because we are not omnipotent. Consequently, the future of a chaotic system eludes our grasp sooner or later. In other words, we cannot play God perfectly, so the history looks non-deterministic.

The above correspondence of stochastic processes and deterministic processes contains a lot to learn, but one fundamental question lingers. How did God choose a particular ω that He gave to her? If God does not follow some other 'random number generator'[40] (say, a fortune), isn't God Himself a 'random number generator' (a capricious being)?[41] This is not the usual view (someone even said, "I am convinced that He does not play dice."), and is understood as God's Will (i.e., behind it is a rational decision we can never fathom).

However, it is a rudiment of science to recognize that not everything has a cause or reason: "Superstition is the belief in the causal nexus."[42] How-

---

[39] The following explanation follows P. Billingsley, *Ergodic Theory and Information* (Wiley, 1960), Section 1. In this book 'Tyche' is used instead of God.

[40] It has not yet been explained what the term 'random number' means, but here the reader has only to rely on her common-sense image of it.

[41] Feuerbach affirmed, "Religion denies, repudiates chance, making everything dependent on God, explaining everything by means of him; but this denial is only apparent; it merely gives chance the name of the divine sovereignty." (Chapter 19 of L. Feuerbach, *The Essence of Christianity* (translated by George Elliot (! of *Middlemarch*) (1854)). `http://www.marxists.org/reference/archive/feuerbach/works/essence/ec19.htm`.

[42] L. Wittgenstein, *Tractatus Logico-Philosophicus*, **5.1361**.

ever, we must clearly recognize that the science mind that pursues structures and laws of the Universe and the mind susceptible to superstition have deep commonality, because to regard (or to believe) correlation as causality is one of the cores of superstition. Keynes said about Newton: "He was the last of the magicians, the last of the Babylonians and Sumerians, the last great mind which looked out on the visible and intellectual world with the same eyes as those who began to build our intellectual inheritance rather less than 10,000 years ago."[43] His last published book was on the chronology of the Bible. The author is extremely sympathetic (or resonant) to such a mind. Only the presence or absence of a lining of skepticism distinguishes science from superstition.[44]

We have virtually assumed that the world is classical (i.e., not quantum-mechanical). For example, we have assumed that a history ω is fixed irrespective of its observation by us. In quantum mechanics the definite occurrence of an event in the classical sense is not certain until the event is 'registered' ('put on record') (even for past events). History is not like a sample that God has selected from the source population of histories (= path space). Even God cannot select ω beforehand in quantum mechanics.[45]

Discussion 2.4. Reading the interview of J. A. Wheeler (1911-2008) in *The Ghost in the Atom* (edited by P. C. W. Davis and J. R. Brown (Cambridge University Press, 1986); also see, e.g., T. Hellmuth, H. Walther, A. Zajoc, and W. Schleich, "Delayed-choice experiments in quantum interference," Phys. Rev. A **35**, 2532 (1987), if needed), study 'delayed choice experiment' and consider its significance. □

Therefore, isn't the idea and the concept of dynamical systems totally meaningless? It is, however, a very important empirical fact that we human beings did not notice that this world was quantum mechanical. That is, the classical world view is empirically 'almost correct.' For example, the theory of probability is constructed totally on the premise. Just as geocentrism (or the Ptolemaic theory), what has been believed for a long time has good empirical (and theoretical) reason to be 'true' with high precision.[46] Therefore, the task

---

[43] J. M. Keynes, "Newton, the Man," (1946). `http://www-groups.dcs.st-and.ac.uk/~history/Extras/Keynes_Newton.html`.

[44] However, we should not stop asking whether everything may be subjected to skepticism. "Don't be bound by dogmas" may be a dogma.

[45] Quantum mechanics contradicts the usual anthropotheistic monotheism more fundamentally than evolution. This is not a popular source of debate, probably because people feel that biology is easily understandable, but because they do not feel so for quantum mechanics.

[46] If we take these plain facts frankly into account, we realize that we must be cautious about the paradigmatic point of view that stresses discontinuities in science. (cf. T. Kuhn, *The Structure of Scientific Revolutions*, 3rd Ed. Univ. of Chicago Press, 1996). A mechanics that cannot explain why geocentrism often looks natural is not correct. If the general theory of relativity cannot produce Newtonian gravity in the weak gravity limit, then it is the relativity that must be rejected. In natural sciences,

to build a consistent world view may consist of two parts: one is to understand why the old classical view looks plausible, and the other is to make the classical view consistent with the whole world view with a minimal revision. The first part does not seem to have been understood completely. Probably, the fundamental issues of statistical mechanics are close to the problem above; we have not clearly understood statistical mechanics, either.[47] The second part of the task is to study the influence of quantum effects on classical pictures. At minute scales of the world quantum effects seem to dominate, and they mostly likely influence the classical nonlinear world as noise.

This book relies on this point of view and discusses salient nonlinear features in the classical world, the world directly accessible by us.

2.3 Characterizing chaos

Let us return to the main topic of this chapter: conceptual analysis. We wish to make a working definition of chaos through formulating our intuition in order to illustrate conceptual analysis. We consider only discrete time systems, but this does not lose any generality as we will see soon. The discrete time dynamical system (f, Γ) we consider is quite general; Γ is a set, and f is a map from Γ into itself (called an *endomorphism* of Γ). We do not require any structure for Γ and f can be any function. Usually, however, Γ is a topological space and f is a differentiable map, but the concept of chaos itself does not require such 'infrastructures.'

Intuitively speaking, as we have seen in Section 2.1, 'chaos' is a random behavior exhibited by deterministic dynamical systems. It is a consequence of expansion of the unknowable due to locally unstable but globally stable dynamics. Although we do not have any problem with the term 'deterministic,' we will not be able to close in on the essence of 'chaos' without clarifying the meaning of the word 'random.' If we could make the concept 'random' clear, it would be ideal, but this route is, as we will learn, not very easy. One strategy to go around this difficulty can be to build our approach around an example unanimously regarded as random. In this section we use this quick

---

it must be very rare that the observed facts change when the paradigm changes. It is because our cognition devices are not dependent on the paradigms of the day, but on our 4-billion-year legacy.

If everything that has been believed for a long time has a reason to be recognized as 'truth,' erroneous views/superstitions such as the existence of an anthropotheistic God must have 'rational reasons.' Such misconceptions must have probably biological reasons. An evolutionary biological explanation is attempted in R. Dawkins, *The God Delusion* (Houghton Mifflin, 2006), esp., Chapter 5. Perhaps, M. Shermer, *The Believing Brain* (Times Books Henry Holt and Co., New York, 2011) may be better.

[47] See, for example, S. Sugiura and A. Shimizu, "Thermal pure quantum states at finite temperature," Phys. Rev. Lett. **108**, 240401 (2012) and the papers quoted therein.

and easy approach. However, recognize that this strategy is based on a conceptually very important judgement.[48]

The coin-tossing process was introduced in the preceding section. If we produce 01 sequences by throwing a coin repeatedly, these sequences will almost surely be (intuitively) random (especially if the coin is fair). Occasionally, a regular sequence such as 010101010··· may be realized, but to encounter such a sequence is quite unlikely. Periodic histories of the coin-tossing process correspond to binary expansion sequences of rational numbers in $[0, 1]$, so they are a sheer minority (measure zero; see Appendix 1A **A1**). Therefore, it is intuitively plausible that 01 sequences produced by tossing a fair coin is almost surely random. Even if the coin is not fair, still we may ignore the possibility of encountering periodic histories, and perfect prediction of future is in any case impossible.[49]

We wish to respect our intuition that the coin-tossing process is 'random,' and build our theory on this foundation.[50] Thus, our starting point is the observation: anything 'naturally related' to the one-sided shift on 01 is more or less 'random.' Crudely summarized, if there is a 'natural correspondence' between the totality of all the histories of the coin-tossing process and a certain subset of trajectories of a deterministic discrete dynamical system, it is plausible to conclude that the dynamical system exhibits chaos.[51]

The phase space of a discrete dynamical system is not necessarily a discrete set as the case of one-sided shift dynamical systems. Therefore, we must make a certain correspondence between the phase space and a set of symbols (i.e., we need a coding of the phase space), if we wish to make a correspondence between a discrete time dynamical system and a one-sided shift $(\sigma, \{0, 1\}^{\boldsymbol{N}})$ (we can see a prototype of coding in Appendix 2.1A). Take an invariant subset A of the phase space Γ under f (i.e., $f^{-1}(A) = A$), and then divide it

[48] ⟨⟨**The true significance of examples**⟩⟩ Such approaches appear again and again in this book. Behind them is the author's firm conviction that we can know clearly much more than we can clearly speak of, or that our natural intelligence that is much older than our language has the capability to grasp intuitively the essence from examples directly. The true significance of examples lies in their direct appeal to our natural intelligence *without* intervention of languages.

[49] There is a way to construct unbiased random sequences from biased stochastic sequences. For example, the von Neumann extractor 'extracts' from sample sequences of $B(p, q)$ (shorter) sample sequences of $B(1/2, 1/2)$. Here, $B(p, q)$ denotes the Bernoulli process explained in Section 2.4.

[50] The argument given here is faithful to the basic idea of Y. Oono, "Period $\neq 2^n$ implies chaos," Prog. Theor. Phys., **59**, 1029 (1978). A similar idea can be found in D. S. Ornstein,"In what sense can a deterministic system be random?" Chaos, Solitons & Fractals **5**, 139 (1995) (the author does not mean to claim his priority at all; he only wishes to say the idea is quite natural).

[51] If a dynamical system is continuous, we may make a discrete system by periodic sampling of the continuous trajectories. If the resultant sequences look random, it is hard to imagine that the original continuous trajectories do not have any random elements. Therefore, we need not define chaos directly for continuous time dynamical systems.

into two sets 0 and 1. The system initially in A is either in set 0 or in set 1 at any given time, so we can map all the trajectories in A to 01 sequences. If the resultant shift dynamical system behaves 'strangely,' then the original system cannot be simple. In particular, if all the 01 sequences appear by this coding, then the time evolution of the system has a portion that may be mapped to the one-sided full shift on the 01 sequences (i.e., the coin-tossing process), so isn't it legitimate to say the system is chaotic? Perhaps, it may be more convenient to code the system behavior observed with a certain time interval. This idea leads to the following definition of chaos:

Definition 2.1. [Chaos] Let f be a map from the phase space Γ into itself (i.e., an endomorphism of Γ).[52] Choose $n \in \boldsymbol{N}^+$ and construct f^n that applies f n-times,[53] and restrict it to its invariant set $A \subset \Gamma$ ($f^{-n}(A) = A$). If it is isomorphic to the shift dynamical system $(\sigma, \{0,1\}^{\boldsymbol{N}})$, then we say the dynamical system (f, Γ) exhibits *chaos* (or the system is chaotic). That is, if φ is a one-to-one map and the following diagram becomes commutative,[54] we say the dynamical system (f, Γ) exhibits chaos.

$$\begin{array}{ccc} A & \xrightarrow{f^n} & A \\ \downarrow \varphi & & \downarrow \varphi \\ \{0,1\}^{\boldsymbol{N}} & \xrightarrow{\sigma} & \{0,1\}^{\boldsymbol{N}} \end{array}$$

□[55]

In other words, if an appropriate one-to-one coding scheme φ of points (i.e., states) in A (we can decode uniquely the sequences) allows transformation of the original dynamical system (restricted to A) into the full shift on two symbols, we say the dynamical system is chaotic. Roughly speaking, if the behavior of a dynamical system (restricted on an invariant set) is coded with symbols 0 and 1, and the result cannot be distinguished from the totality of the outcomes of the coin-tossing process, we wish to say the system is chaotic or is a chaotic dynamical system.

A continuous time dynamical system is chaotic, if a discrete dynamical system constructed from the original system in a 'natural' fashion is chaotic in the above sense. We will see such dynamical systems are really chaotic later (Theorem 2.3 in Section 2.7, Appendix 2.7A).

How good is the definition of chaos given above? In the context of this book, it is a mere working definition, so this question is very important. It is

---

[52] In this definition, maps are not understood measure-theoretically; they are pointwise transformations.

[53] $f^n(x) = (f \circ f \circ \cdots \circ f)(x) = f(f(f \cdots (f(x)) \cdots))$. n f's show up in each expression.

[54] This means in the present context that the results obtained by following various combinations of arrows never yield different results.

[55] Even if we use a more general shift already introduced in the preceding section in the definition, the definition does not become more general.

not easy to find criteria for a definition to be good, but consistency with intuition, close connection to fundamental concepts, equivalence to definitions based on very different points of view, etc., may be counted among them. Our definition relies on our intuition: as long as we assume that "(apparent) random behavior is a fundamentally important characteristic of chaos," consistency with intuition is built into the definition. Randomness must be a fundamental concept, so the definition has a close connection with a fundamental concept. This can be seen further from the theorem quoted below. It is also important to check the consistency and relations with other definitions of chaos to confirm the naturalness of the definition. The comparison is summarized in Appendix 2.3A.

The following theorem[56] is about the interval dynamical systems (dynamical systems whose phase space is an interval of real numbers), but it demonstrates the naturalness of our definition of chaos (technical terms are explained just below the theorem).

Theorem 2.1. *Let I be a finite interval and $F : I \to I$ be a continuous endomorphism. Then, the following (1)-(4) are equivalent.*
(1) F exhibits chaos.
(2) F has a periodic orbit whose period is not equal to the power of 2.
(3) There is a positive integer m such that F^m has a mixing invariant measure.
(4) F has an invariant measure whose Kolmogorov-Sinai entropy is positive.
□[57]

The needed concepts (mixing, invariant measure, Kolmogorov-Sinai entropy) will be defined later, but they are roughly as follows. 'Invariant measure' is a steady distribution. 'Mixing' implies that the system relaxes toward some steady state. 'Kolmogorov-Sinai entropy' is the required extra information to predict the next time-step state as accurately as the current state is described. Its positivity implies that (since more information is needed to determine the future state) the system behavior in the future becomes increasingly difficult to predict as the future is further away. A 'periodic orbit' with period n must be intuitively clear, but is a set of n distinct points $\{x_i\}_{i=1}^n$ such that for $i = 1, \cdots, n-1$ $f(x_i) = x_{i+1}$ and $f(x_n) = x_1$ hold.

Practically, the following proposition equivalent to (1)-(4) in the theorem is useful:

(5) There are two closed intervals J_1 and J_2 in I that share at most one point such that $f^p(J_1) \cap f^q(J_2) \supset J_1 \cup J_2$ holds for some positive integers p and q.

---

[56] M. Osikawa and Y. Oono, "Chaos in C^0-diffeomorphism of interval," Publ. RIMS **17**, 165 (1981); Y. Oono and M. Osikawa, "Chaos in nonlinear difference equations. I," Prog. Theor. Phys. **64**, 54 (1980) (There is no part II, since the author was 'expelled' from Japan in 1979). Also see: L. Block, "Mappings of the interval with finitely many periodic points have zero entropy," Proc. Amer. Math. Soc. **67**, 357 (1978); "Homoclinic points of mappings of the interval," *ibid.* **72**, 576 (1978).

[57] (1) ⇒ (2), (3), (4) is trivial. (4) ⇒ (1) is also almost trivial.

That Ito's earthquake model in Section 2.1 exhibits chaos is immediately seen from the 'folded paper' model (Fig. 2.14). We may draw a periodic orbit whose period is not a power of 2, but to check (5) may be the easiest.

Note 2.2. Formulas for chaotic sequences
Suppose $x_{n+1} = f(x_n)$ produces a sequence $\{x_n\}$ starting with an initial condition x_0. Even when f exhibits observable chaos (see Appendix 2.3A) for almost all initial conditions, the solutions may be explicitly written as $x_n = g(n, x_0)$ in terms of a fairly simple function g. A famous example is $f(x) = 4x(1-x)$, for which $g(n, x_0) = \sin^2[2^n \mathrm{Arcsin}\sqrt{x_0}]$. There are many other examples.[58] The reader might think that these sequences can be predicted for any far future time thanks to such analytic expressions, but try to evaluate this for three digits with $n = 100$.

The following, including Appendix 2.3A, is supplementary (so *the reader can go to the next section*). We saw that chaotic behavior of the earthquake model was closely connected to the sensitive dependence of its behavior on the initial condition. In Chapter 1, magnification of small effects was regarded as a fundamental significance of chaos. This property is emphasized by Ruelle, Guckenheimer[59] and others, and is mentioned first as the characteristic feature of chaos. Intuitively, this is easy to understand in terms of (one-sided) shifts, because, in contrast to the digits on the far right that describes minute differences in states, the digits near the left end correspond to global differences. Thus, movement of the digits to the left by shift corresponds to magnification of small structures. Two initial states whose codes are different only in digits on the far right are very close in the phase space, and time evolution magnifies the difference.

Sensitive dependence on initial conditions does not necessarily imply that the dynamical system is chaotic. This must be obvious to those who are familiar with roulette; the ball jumps around awhile, but eventually it settles down to a fixed point (the system is a multiply stable system). The set of the points to which a given dynamical system eventually settles down is called the *ω-limit set.*[60] In our example, it consists of discrete fixed points. The totality of points belonging to the trajectories eventually reaching a fixed point

---

[58] K. Umeno, "Method of constructing exactly solvable chaos," Phys. Rev. E **55**, 5280 (1997).

[59] J. Guckenheimer, "Sensitive dependence on initial conditions for one-dimensional maps," Commun. Math. Phys., **70**, 133 (1979).

[60] A precise definition is as follows. First, we define the ω-limit set $\omega(x)$ of a point x as follows:

$$\omega(x) = \{y : \text{there is a partial time sequence } n_i \to \infty \text{ such that } f^{n_i}(x) \to y\}.$$

Collect all such points and construct $L_+ \equiv \cup_{x \in \Gamma}\omega(x)$. A transitive invariant subset of L_+ is called the ω-limit set. Here, a 'transitive set' is a set that has a dense orbit; a dense orbit of a set A is an orbit that can pass through any neighborhood of any point in A. Therefore, roughly speaking, an ω-limit set is a set A satisfying $f(A) = A$ with an orbit densely filling A. Fixed points and limit cycles are examples of the ω-limit set.

is called its *basin of attraction*.[61] A fixed point that has a basin of attraction is called an *attractor*.[62] If several attractors coexist, the ultimate fate of the system is determined by which basin its initial condition lies in. Even if the long-time (eventual) behavior of the system is not chaotic, if the boundaries between basins are extremely complicated, then sensitive dependence on initial conditions can exist.[63] Obviously, roulettes, dice, and coins must be (at least approximately) such systems.

Appendix 2.3A Various definitions of chaos

This note does not explain technical terms, but the concept of 'observability' is important, so its definition is given at the beginning. In short, we wish to call events observable if we can observe them with a positive probability when we sample them (or sample required initial conditions for them) uniformly according to the Lebesgue measure (see Appendix 2.4A) of the phase space. It may be understood as one characterization of observability by numerical experiments. Events (sets) are assumed to be observable, if they can be reached from a Lebesgue-measure-positive initial condition set. We say an event (set) B is reachable from an initial condition, if the trajectory starting with the initial condition goes into B after a finite time (it may go out later) or if it is eventually absorbed into B.

Definition 2.2 (*Observability*). We say a set B is observable with respect to a given dynamical system (f, Γ), if the totality of the points on the trajectories that can reach B has a positive Lebesgue measure. In other words, B is observable, if B has a basin of attraction with a positive Lebesgue measure or the set $\{x : \exists n \geq 0, f^n(x) \in B, x \in \Gamma\}$ has a positive Lebesgue measure. □

The chaos defined in this section is sometimes called *formal chaos*, because there is no guarantee of its observability. For chaos of one-dimensional map systems, the observability of chaos and the existence of an absolutely continuous invariant measure seem to be equivalent.

The equivalence of (1) and (2) in Theorem 2.1 is the well-known theorem: *Period* $\neq 2^n$ *implies chaos*. This is apparently very similar to the famous Li-Yorke's theorem: *Period three implies chaos*[64] and unfortunately gives an impression that the former is a mere extension of the latter. However, the similarity is superficial; the definitions of chaos are distinct in these two theorems.

The Li-Yorke chaos is defined as follows. Let X be a compact metric space. A map $f : X \to X$ exhibits the *Li-Yorke chaos*, if there is an uncountable set of non-periodic orbits $R \subset I$ called the *scrambled set* defined by the following three conditions:
(A) $\forall x \in R, \forall y \in R, x \neq y \Rightarrow \limsup_{n\to\infty} |f^n x - f^n y| > 0$,
(B) $\forall x \in R, \forall y \in R, x \neq y \Rightarrow \liminf_{n\to\infty} |f^n x - f^n y| = 0$,

---

61 A precise definition is as follows. The basin (of attraction) of an ω-limit set A is the join of all the open sets U satisfying the following conditions:
(i) U satisfies $f(U) \subset U$.
(ii) $\cap_{n\geq 0} f^n(U) = A$.

62 More generally, any ω-limit set with a basin of attraction is called an attractor.

63 See, for example, H. E. Nusse and J. A. Yorke, "Basin of attraction," Science **271**, 1376 (1996).

64 T.-Y. Li and J. A. Yorke, "Period three implies chaos," Am. Math. Month. **82**, 985 (1975).

(C) For any periodic orbit y $\forall x \in R \Rightarrow \limsup_{n\to\infty} |f^n x - f^n y| > 0$.

For a continuous map f on (i.e., a C^0-endomorphism of) an interval, if the chaos in the sense given in this book occurs, then the Li-Yorke chaos also occurs (nothing can be said generally, if the map is not continuous).[65] However, the converse is not true. The Li-Yorke chaos lacks the equivalence to the positivity of the Kolmogorov-Sinai entropy. Therefore, Theorem 2.1 does not follow from the Li-Yorke theorem.

The most serious defect of the Li-Yorke chaos is that it does not pay any attention to randomness explicitly. Non-periodic motions are paid due attention, but they can be quite different from chaotic behaviors. To emphasize the existence of non-periodic orbits to characterize chaos misplaces the emphasis. It is likely that genuine chaos happens to be observed when there is the Li-Yorke chaos merely due to some topological constraints common to many dynamical systems. Another problem is that the scrambled set R is usually not an observable set. For example, if a piecewise smooth continuous map exhibits observable chaos, then the inner measure of its scrambled set is always zero.[66] That is, the chaotic behavior we encounter with computer experiments are not chaos characterized by the Li-Yorke theorem. S.-H. Li proposed a revised version of scrambled sets called ω-scrambled sets to make the Li-Yorke chaos equivalent to ours.[67]

The most popular definition of chaos at present may be that of Devaney.[68] If we take into account S.-H. Li's result,[69] this definition may be stated as follows. Let (f, X) be a discrete time dynamical system, and D ($\subset X$) be a closed invariant set (in this definition, $f^{-1}(D) \supset D$). If the following two conditions hold, the dynamical system exhibits chaos.

(D1) $f|_D$ (the restriction of f to D) is topologically transitive on D (i.e, $f|_D$ is surjective on D and has an orbit dense in D),

(D2) The totality of the periodic orbit of f is dense in D.

D is called a chaos set. The set A in Theorem 2.1 is a chaos set. S.-H. Li showed the equivalence of this definition and our definition, if the phase space is a compact metric space.

We have seen that our definition in this section is equivalent to the definitions given later. We will gradually realize that the essence is the positivity

---

[65] To make this assertion hold even without continuity, the definition of chaos must be narrowed slightly with more conditions. An example is in M. Osikawa and Y. Oono, Publ. RIMS **17**, 165 (1981) cited earlier.

[66] Y. Baba, I. Kubo and Y. Takahashi, "Li-Yorke's scrambled sets have measure 0" Nonlinear Analysis **26**, 1611 (1996). If a map is only continuous, then there are examples of scrambled sets that are measurable and with positive measure: see for example, I. Kan, "A chaotic function possessing a scrambled set with positive Lebesgue measure," Proc. Amer. Math. Soc. **92**, 45 (1984) or J. Smital, "A chaotic function with a scrambled set of positive Lebesgue measure," Proc. Amer. Math. Soc. **92**, 50 (1984).

[67] S.-H. Li, "ω-chaos and topological entropy," Trans. Amer. Math. Soc. **339**, 243 (1993).

[68] R. Devaney, *An introduction to chaotic dynamical systems* (Benjamin/Cummings, 1986).

[69] S.-H. Li, "Dynamical properties of the shift maps on the inverse limit space," Ergodic Theor. Dynam. Syst. **12**, 95 (1992). See also J. Banks, J. Brooks, G. Cairns, G. Davis and P. Stacey, "On Devaney's definition of chaos," Am. Math. Month. **99**, 332 (1992). In the latter X need not be bounded.

of the Kolmogorov-Sinai entropy. As has been quoted, already in 1968[70] Alekseev defined quasirandom dynamical systems in terms of a Markov chain with a positive Kolmogorov-Sinai entropy. It is an example showing that Russian dynamical systems study was far ahead of its Western counterpart (actually, the use of entropy to classify dynamical systems was the Russian starting point of 'modern' dynamical systems study). The term 'chaos' may have been good for popularization of the concept, but the term 'quasirandom' summarizes the essence.

2.4 How to quantify 'history'

In the preceding section, a (working) definition of chaos is given. That is, we have considered how to formulate mathematically our intuition that chaos is an apparently random motion of a deterministic dynamical system. As seen in Appendix 2.3A, since it has turned out to be equivalent to various definitions proposed later, it is not a bad definition. Indeed, although it is said to be 'a (working) definition,' it is virtually the most widely accepted definition. Still, if we wish to feel the definition to be 'good,' the more consistency with our feeling and intuition, the better. To this end we should look for other characteristics of chaos that are intuitively appealing. If an apparently disparate characterization may be deduced from the definition discussed above (or better, can be proved equivalent), then it is a sign of the goodness of our definition. Senses and intuitions have been polished for us to survive in this world, so, in a very serious sense, the real world must have been impressed onto (imprinted on) our senses and intuitions. Therefore, the spirit of respecting our senses and intuitions is identical to the spirit of respecting experiments, and is one of the important bases of science.[71]

In the preceding section in order to incorporate our intuition that chaos is closely related to random motion into the characterization of chaos, we started with a typical chaotic process, and utilized it to define chaos. There still remains a big question: what is meant by 'being random'? What is 'being random' intuitively? 'Being random' means that something is scrambled, arranged in an unbiased or impartial fashion. Therefore, in this section, we prepare a device to quantitatively check that the 'occurrence of various histories is not biased.' The concept of a cylinder set we will encounter is important. Appendix 2.4A is an outline of measure and probability. This appendix may

70 [1968: Prague Spring, The Tet Offensive, Down to the Countryside Movement, Nuclear nonproliferation treaty, Assassinations of Rev. M. L. King and Senator R. F. Kennedy.]

71 As far as classical physics goes, it seems we can do fundamental physics only with the aid of introspection and meditation (without any empirical facts; experiments are to convince 'the feeble-minded'). This is to rely on the results of phylogenetic learning that have been built into our body. Needless to say, our power of meditation is not sufficiently strong, so experiments and observations are always extremely helpful.

be read independently as another illustration of conceptual analysis.

In the case of the coin-tossing process, that the coin is fair implies that all the possible histories, i.e., all the 01 sequences can occur with equal likelihood (probability).[72] The path space Ω is the totality of all the 01 sequences. To measure the 'ease' of the occurrence of a certain history = sequence, we must introduce a kind of weight. It is not easy to handle infinitely long histories at once, so let us consider a finitary method to check the fairness of a coin. For example, prepare numerous 01 sequences of length 10. That is, we try many length 10 coin-tossing processes. If the coin is fair, then any 01 sequence of length 10 is expected to appear with probability 2^{-10}. In short, we may conclude that the coin is not biased, if the empirical probabilities of any finite length trials do not significantly deviate from fair expectations. Therefore, if we wish to model a particular coin, we could specify all the probabilities of bundles of 01 sequences characterized by finite subsequences at particular positions (for example, a bundle of histories that are characterized by 01010 starting at the 52nd trial; this is an example of a cylinder set as we will see soon).

Let us formalize the above explanation in a more abstract fashion. Let us start looking at an arbitrary history.[73] We wish to specify the probability of a history to be in an appropriate subset $A \subset \Omega$, where Ω is the path space of the system under consideration. A standard method for specifying a probability measure on the path space Ω is to assign the probabilities of 'cylinder sets' consistently. Here, the term, 'measure' may be understood as something like the volume. A probability measure is a measure whose total volume is normalized to be unity. A more respectable explanation is in Appendix 2.4A.

A *cylinder set* (in the context of shift dynamical systems) is a subset of the path space defined as follows: Let m, k be positive integers and set

$$[\omega_m, \cdots, \omega_{m+k-1}] \equiv \{\omega' : \omega'_i = \omega_i \text{ for } i = m, \cdots, m+k-1,\ \omega' \in \Omega\}, \quad (2.14)$$

where we have used an abbreviation: $\omega'_i \equiv \omega'(i)$. In short, the cylinder set $[\omega_m, \cdots, \omega_{m+k-1}]$ is a bunch of histories whose states from time m up to $m+k-1$ are $\omega_m, \cdots, \omega_{m+k-1}$, respectively. In the above definition, the specified states are in [ ]. For example, one-sided infinite sequences $110101\cdots$ and $010110\cdots$ share 101 beginning at the second digit, so they both belong to the cylinder set $c \equiv [\omega(2) = 1, \omega(3) = 0, \omega(4) = 1]$. This cylinder set may be partitioned into two subsets $c_0 \equiv [\omega(2) = 1, \omega(3) = 0, \omega(4) = 1, \omega(5) = 0]$ and $c_1 \equiv [\omega(2) = 1, \omega(3) = 0, \omega(4) = 1, \omega(5) = 1]$. Therefore, the probability

72 This conclusion is harmonious with our feeling. We have been selected to feel that such a coin is fair. A related discussion will be found at the end of Appendix 2.4A.

73 Let us proceed with common sense, but the word 'arbitrary' is, in actuality, not simple. Usually, a probability measure is assumed, but it is a really deep question how to explicitly formulate the word 'arbitrary' without such an assumption.

that a system history is found in c_0 plus that in c_1 must be equal to the probability that a history is found in the cylinder set c. In other words, the probabilities assigned to c_0 and c_1 must be consistent with that to c. This is the meaning of 'consistency' mentioned above. If we can specify a consistent assignment of probabilities to cylinder sets (this requires a countably many conditions), we may specify a probability on this infinite dimensional space.[74]

For example, the simplest consistent probability measure on the path space $\Omega = \{0,1\}^{\mathbf{Z}}$ may be specified by setting the probability to have 1 to be p ($0 \le p \le 1$) independent of whatever has happened before or will happen after it. Then, the probability of the cylinder set c is $p^2(1-p)$, that for c_0 is $p^2(1-p)^2$ and that for c_1 is $p^3(1-p)$. Obviously, the assignment is consistent: $p^2(1-p)^2 + p^3(1-p) = p^2(1-p)$. The resultant probability measure on the path space is called a *Bernoulli measure.*

Example 2.2. Bernoulli process
Let the phase space Γ have N elementary events $a_1, \cdots, a_N$, and the path space Ω is the totality of both-side infinite sequences of the form $\omega = \{\omega_n\}_{n=-\infty}^{+\infty}$ ($\omega_n \in \{a_1, \cdots, a_N\}$). The probability to find the elementary event a_i is specified as p_i irrespective of its absolute time of occurrence and the events at other instants, where $\sum_{i=1}^{N} p_i = 1$, $p_i > 0$. The shift dynamical system on the path space with the probability measure defined as above is called the *Bernoulli process* $B(p_1, \cdots, p_N)$. The coin-tossing process with a fair coin that is going on from an infinite past is $B(1/2, 1/2)$. If a coin-tossing process is described by $B(p, q)$ with $p \neq q$, we say that the coin is not fair. Throwing a fair die is $B(1/6, 1/6, 1/6, 1/6, 1/6, 1/6)$. If the discrete time n takes only non-negative integers, the above processes are called the *one-sided Bernoulli processes.* □

Appendix 2.4A What is measure, what is probability?

The concept of measure does not appear in elementary calculus, but it is a fundamental and important concept. It is not very difficult to understand, since it is important. Besides, the introduction of the Lebesgue measure by Lebesgue is a good example of conceptual analysis, so let us look at its elementary part. A good introductory book for this topic is the already quoted one by Kolmogorov and Fomin. It is desirable that those who wish to study fundamental aspects of statistical mechanics and dynamical systems have proper understanding of the subject.

What is the volume?

For simplicity, let us confine ourselves to 2-space. Thus, the question is: what is the area? Extension to higher dimensions should not be hard. If the

---

[74] This is guaranteed by *Kolmogorov's extension theorem.* See, for example, R. Durrett, *Probability, Theory and Examples* (Wadsworth & Brooks, 1991), Appendix 7. This textbook has an excellent mathematical taste and is not very difficult. Recommended. However, it is not a book for those who have never studied probability theory at all.

shape of a figure is complicated, whether it has an area could be a problem,[75] so let us begin with an apparently trivial case.

"The area of the rectangle $[0, a] \times [0, b]$ is ab."

Is this really so? If so, why is this true? Isn't it strange that we can ask such a question before defining 'area'? Then, if we wish to be logically conscientious, we must accept the following definition:

Definition. The area of a figure congruent to the rectangle $\langle 0, a \rangle$ (along the x-axis) $\times$ $\langle 0, b \rangle$ (along the y-axis) is *defined* as ab.[76] Here, '$\langle$' implies '[' or '(', '$\rangle$' is ']' or ')', that is, we do not care whether the boundary is included or not. Notice that the area of an rectangle does not depend on whether its boundary is included or not. This is already incorporated in the definition.

The area of a fundamental set

A figure made as the direct sum (that is, join without overlap except at edges and vertices) of a finite number of rectangles (whose edges are parallel to the coordinate axes and whose boundaries may or may not be included) is called a fundamental set (Fig. 2.17). It should be obvious that the join and the intersection (common set) of two fundamental sets are both fundamental sets. The area of a fundamental set is defined as the total sum of the areas of the constituent rectangles.

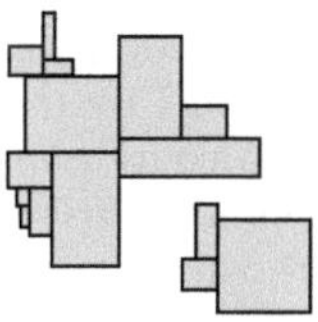

Fig. 2.17 Fundamental set: it is a figure made of a finite number of rectangles whose edges are parallel to a certain Cartesian coordinate axes and the rectangles do not overlap except at edges and vertices. Its area is the total sum of the areas of the constituent rectangles.

How to define the area of more complicated figures; a strategy

For a more complicated figure, a good strategy must be to approximate it by a sequence of fundamental sets allowing increasingly smaller rectangles. Therefore, following Archimedes, we approximate the figure from inside and from outside (that is, the figure is approximated by a sequence of fundamental sets enclosed by the figure and by a sequence of fundamental sets enclosing the figure). If the areas of the inside and the outside approximate sequences agree in the limit, it is rational to define the area of the figure by the limit.

Let us start from outside.

Outer measure

---

[75] (Under the usual axioms of mathematics = the ZFC axiomatic system) we encounter figures without areas.

[76] ⟪**Adaptor between different categories**⟫ Area connects the world of figures and that of numbers. Therefore, we need explicitly an '*adaptor*' connecting these two disparate categories that must be free from ambiguous interpretation. The definition of the area of an rectangle given here is a typical example of the adaptor. Notice that it is operationally unambiguous. Recall how the units of various physical quantities are defined. A unit connects something extra-numerical to numbers, so it is a kind of adaptor, requiring an unambiguous specification. Thus, an explicit specification of an operational procedure is the core of the definition of a unit. The reason why physics appreciates operational definitions is that there is no room for interpretation.

Let A be a given bounded set (that is, a set that may be enclosed in a sufficiently large disk). Using a finite number of (or countably many) rectangles P_k $(k = 1, 2, \cdots)$, we cover A, where the boundaries of the rectangles may or may not be included, appropriately. If $P_i \cap P_j = \emptyset$ $(i \neq j)$ and $\cup P_k \supset A$, $P = \{P_k\}$ is called a finite (or countable) cover of A by rectangles (Fig. 2.18O). Let the area of the rectangle P_k be $m(P_k)$. We define the *outer measure* $m^*(A)$ of A as

$$m^*(A) \equiv \inf \sum_k m(P_k). \tag{2.15}$$

Here, inf[77] is taken over all the possible finite or countable covers by rectangles.

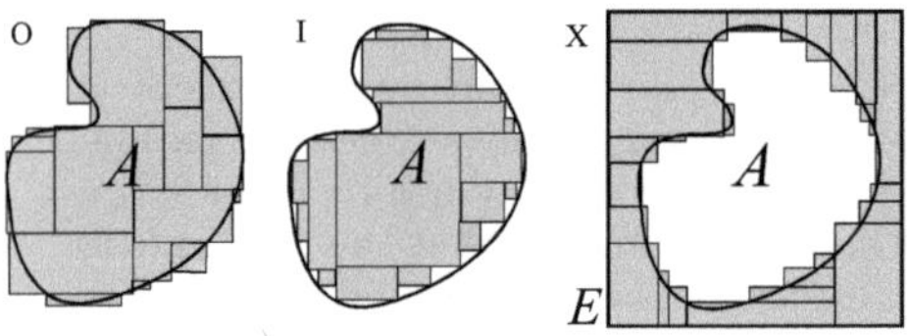

Fig. 2.18 Let A be the set enclosed by a closed curve. O denotes a finite cover by rectangles. If there is an area of A, it is smaller than the sum of the areas of these rectangles. The outer measure is defined by approximating the area from outside. In contrast, the inner measure is computed by the approximation shown in I by the rectangles included in the figure A. In the text, by using a large rectangle E containing A, $E \setminus A$ is made and its outer measure is computed with the aid of finite covers; the situation is illustrated in X. The relation between I and X is just the relation between negative and positive films. If the approximation O from outside and the approximation I from inside agree in the limit of refinement, we may say that A has an area. In this case, we say A is measurable, and the agreed area is called the area of A.

Inner measure

For simplicity, let us assume that A is a bounded set. Take a sufficiently large rectangle E that can enclose A. Of course, we know the area of E is $m(E)$. The *inner measure* of A is defined as[78]

$$m_*(A) = m(E) - m^*(E \setminus A). \tag{2.16}$$

It is easy to see that this is equivalent to the approximation from inside (Fig. 2.18I). Clearly, for any bounded set A $m^*(A) \geq m_*(A)$ holds.

Area of figure, Lebesgue measure

Let A be a bounded set. If $m^*(A) = m_*(A)$, A is said to be a *measurable set* (in the present case, a set for which its area is definable) and $\mu(A) = m^*(A)$ is called its area (two-dimensional *Lebesgue measure*).

At last the area is defined. The properties of a fundamental set we have used are the following two:

---

[77] The infimum of a set of numbers is the largest number among all the numbers that are not larger than any number in the set. For example, the infimum of positive numbers is 0. As is illustrated by this example, the infimum of a set need not be an element of the set. When the infimum is included in the set, it is called the minimum of the set. The above example tells us that the minimum need not exist for a given set.

[78] $A \setminus B$ in the following formula denotes the set of points in A but not in B, that is, $A \cap B^c$.

(i) It is written as a (countable) direct sum of the sets whose areas are defined.
(ii) The family of fundamental sets is closed under $\cap$, $\cup$ and $\setminus$ (we say that the family of the fundamental sets makes a set ring.[79])
An important property of the area is its additivity: If P_i are mutually non-overlapping rectangles, $\mu(\cup P_i) = \sum \mu(P_i)$. Furthermore, the *σ-additivity* for countably many summands also holds.[80]

Notice that such a summary as that the area is a translationally symmetric σ-additive set-theoretical function which is normalized to give unity for a unit square does not work, because this does not tell us on what family of sets this set-theoretical function is defined.[81] The above abstract summary does not state the operational detail about how to measure the areas of various shapes, so no means to judge is explicitly given what figures are measurable. Lebesgue's definition of the area outlined above explicitly designates how to obtain the area of a given figure. □

Discussion 2.5. Read S. Wagon, *The Banach-Tarski Paradox* (Cambridge University Press, 1993). Is there any possibility for the Banach-Tarski theorem to be meaningful in understanding natural phenomena?[82] □

General measure (abstract Lebesgue measure)

The essence of characterization of the area is that there is a family of sets closed under certain 'combination rules' and that there is a σ-additive set-theoretical function on it. Therefore, we start with a σ-additive family $\mathcal{M}$ consisting of subsets of a set X: A family of sets satisfying the following conditions is called a *σ-additive family*:
(s1) $X, \emptyset \in \mathcal{M}$,
(s2) If $A \in \mathcal{M}$, then $X \setminus A \in \mathcal{M}$,
(s3) If $A_n \in \mathcal{M}$ $(n = 1, 2, \cdots)$, then $\cup_{n=1}^{\infty} A_n \in \mathcal{M}$.
$(X, \mathcal{M})$ is called a *measurable space*. A non-negative and σ-additive set-theoretical function m defined on a measurable set that assigns zero to an empty set is called a *measure*, and $(X, \mathcal{M}, m)$ is called a *measure space*. Starting with this measure m, we can define the outer measure on a general set $A \subset X$, mimicking the procedure already discussed above. The inner measure can also be constructed. When these two agree, we can define a set-theoretical function μ as $\mu(A) = m^*(A)$, and we say A is μ-measurable. Thus, we can define μ that corresponds to the Lebesgue measure explained above in the context of the area. μ is called the *Lebesgue extension* of m (this is called an abstract Lebesgue measure, but often this is also called a Lebesgue measure). This construction of μ is called the completion of m. In summary, if $(X, \mathcal{M}, m)$ is a measure space, we define a new family of subsets of X based on $\mathcal{M}$ as

$$\overline{\mathcal{M}} = \{A \subset X : \exists B_1, B_2 \in \mathcal{M} \text{ where } B_1 \subset A \subset B_2, m(B_2 \setminus B_1) = 0\}. \quad (2.17)$$

---

[79] More precisely, that a family $\mathcal{S}$ of sets makes a ring implies the following two:
(i) $\mathcal{S}$ includes $\emptyset$,
(ii) if $A, B \in \mathcal{S}$, then both $A \cap B$ and $A \cup B$ are included in $\mathcal{S}$.

[80] Indeed, if $A = \cup_{n=1}^{\infty} A_n$ and A_n are mutually exclusive (i.e., for $n \neq m$ $A_n \cap A_m = \emptyset$), for an arbitrary positive integer N $A \supset \cup_{n=1}^{N} A_n$, so $\mu(A) \geq \sum_{n=1}^{N} \mu(A_n)$. Taking the limit $N \to \infty$, we obtain $\mu(A) \geq \sum_{n=1}^{\infty} \mu(A_n)$. On the other hand, for the external measure $m^*(A) \leq \sum_{n=1}^{\infty} m^*(A_n)$, so $\mu(A) \leq \sum_{n=1}^{\infty} \mu(A_n)$.

[81] If we assume that every set has an area, under the usual axiomatic system of mathematics, we are in trouble. See Discussion 2.5.

[82] Professor Lee Rubel (1927-1995) told the author in 1988 that any serious theorem will be needed by serious physics (according to history). S. Banach (1892-1945), A. Tarski (1901-1983).

If μ is defined as $\mu(A) \equiv m(B_2)$ for $A \in \overline{\mathcal{M}}$, $(X, \overline{\mathcal{M}}, \mu)$ is a measure space, and is called the completion of $(X, \mathcal{M}, m)$.[83]

The final answer to the question, "What is the area?" is: the area is the completion of the Borel (1871-1956) measure, where the *Borel measure* is the σ-additive translation-symmetric measure that gives unity for a unit square and is defined on the Borel family of sets which is the smallest σ-additive family of sets including all the rectangles.

Generally speaking, a measure is something like a weighted volume. However, there is no guarantee that every set has a measure. It is instructive that a quite important part of the characterization of a concept is allocated to an 'operationally' explicit description (e.g., how to measure, how to compute). Recall that Riemann's definition of the integral was based on this operational spirit, so it can immediately be used to compute integrals numerically.

Jordan measure

Before Lebesgue, Jordan (1838-1922) defined his measure. His idea was to tessellate small elementary figures (e.g., squares of edge length ε) in a given set A as much as possible to estimate the area (from below). This estimate gives $a_\varepsilon(A)$ = the upper bound of the number of the elementary figures in A $\times$ the area of the elementary figure. Then, take the limit $\varepsilon \to 0$ to define the inner measure $\underline{a}(A)$ of A. The outer measure of A may be analogously defined. Then, if $\overline{a}(A) = \underline{a}(A)$, we say A is (Jordan) measurable, and the agreed value is the area of A. Since Lebesgue's method of covering is more flexible than that by Jordan, we have the following inequalities:

$$\underline{a}(A) \le \underline{m}(A) \le \overline{m}(A) \le \overline{a}(A). \tag{2.18}$$

Therefore, Jordan measurability implies Lebesgue measurability. However, the converse is not generally true. Since Jordan does not allow the use of all the sizes at once, we cannot say, for example, that the outer measure of any countable set is zero within Jordan's framework. σ-additivity cannot be asserted, either.

It is very interesting to note that the argument based on a certain unit is, even if the unit size is taken infinitesimal eventually, definitely weaker than the argument that allows the use of all the sizes at once. Since, eventually, we allow indefinitely small units, we might expect that the conclusions must be the same, but it is not. Is there any implication of this observation for the physical world or for physics?

Incidentally, we must clearly pay attention to the fact that humankind recovered the refined mathematical and logical level of Archimedes (287-212 BCE) of more than 2000 years ago only around or slightly before the time of Jordan. We must not forget the warning that culture can easily retrogress (medievalization occurs all too easily) with a sense of impending crisis.

What is probability?

Kolmogorov (1903-1987) defined probability as a measure whose total mass

[83] The completion is unique. In a complete measure space, if A is measure zero ($\mu(A) = 0$), then its subsets are all measure zero. Generally, a measure with this property is called a complete measure. Completion of $(X, \mathcal{M}, m)$ may be understood as the extension of the definition of measure m on the σ-additive family generated by all the sets in $\mathcal{M}$ + all the measure zero set with respect to m.

is normalized to unity.[84] Since a long time ago 'What is probability?' has been a difficult problem.[85] An interpretation of the probability of an event is that it is a measure of our confidence in the occurrence of the event (*subjective probability*). It is something like a weight of the event, if we express the event as a set of elementary events compatible with the event. That is, if an event is interpreted as a set, its probability should be handled just as a measure of the set. Therefore, without going further into the problem of interpretation of probability, to specify only how to handle it clearly is the approach adopted by Kolmogorov. This approach may not squarely answer the question: "What is the probability?". For example, there is no obvious relation to relative frequencies. An important lesson is that we can construct a theory of probability that is sufficiently rich and practical without answering any 'philosophically (apparently) deep' questions.

Avoiding the discussion of the meaning of probability and constructing only the algorithm for it may be admissible as mathematics, but the approach is incomplete if we wish to apply it to study Nature, reality. If we wish to apply the concept of probability to reality, we need its interpretation. Even Kolmogorov's definition is not aloof from the interpretation of probability, although he apparently avoids it. This approach contains the crucial idea that for the subjective probability (the extent of confidence) to be rational it must be interpreted as a measure.

For the cases of casting dice and tossing coins the numbers we call probability are based on our experience about frequencies and are consistent with the law of large numbers. It is not hard to accept intuitively that the empirical probabilities thus obtained obey the same logic as measures do. Such probabilities are understood as objective (and can be empirically confirmed with the aid of the law of large numbers). There is, however, a deep-rooted opinion that subjective probability is distinct from empirical probability (based on frequency). Such an opinion is a typical example of the humanistic fallacy that our logic and language are unrelated to our empirical world.

Suppose there are two mutually exclusive events 1 and 2 with objective probabilities (relative frequencies) p_1 and p_2 ($> p_1$), respectively. If the subjective probability p' of a gambler for these events becomes $p'_1 > p'_2$, then his gain on the average must be smaller than the gambler with $p = p'$ (i.e., whose subjective assessment is consistent with the objective reality).[86] Thus, the agreement of subjective probability and empirical probability based on relative frequency is forced upon us (i.e., the subjects who choose), when we are subjected to natural selection. The probability based on relative frequency satisfies measure-theoretical axioms. Therefore, the subjective probability molded by natural selection follows, as long as it is useful for our survival, measure-theoretical axioms. Consequently, the assertion that subjective probability = extent of confidence behaves as volume or weight looks very natural. Or, we should say that our nervous system/emotion has been made to evolve so that

---

[84] A. N. Kolmogorov, *Foundations of the Theory of Probability* (2nd edition) (Chelsea, 1957); `http://www.mathematik.com/Kolmogorov/index.html`. Recommended. The third edition in Russian contains a reprint of this article with a nice outline of the history of probability theory by A. N. Siryaev.

[85] A summary can be found in D. Gillis, *Philosophical theories of probability* (Routledge, 2000), but the argument given here is not described in this book.

[86] The reader might say what matters is not the subjective probability, but the probability of the person to undertake appropriate behaviors, because actual behavior is important, not the belief. However, it must be disadvantageous that the correspondence between thoughts and actions is not simple.

this looks natural.[87] The essence of probability is the amount of confidence backed by relative frequency, so even apparently subjective probabilities can be effective in empirical sciences.

There have been many philosophers who oppose the frequentist theory of probability because, e.g., the probability of a unique event is hard to think of. For example, Carnap thought that probability has a logical meaning independent of empirical facts, and tried to found probability on "the degree of confirmation" that is based on the logical relations among events. However, such attempts are typical humanistic fallacies totally forgetting about the fact that logical capacity has been formed under natural selection. When one tries to think about probability, even an apparently unique event is not considered as an event that occurs only once. It is embedded into the totality of experienced events (and logically inferred conclusions from them) during the evolution process that generated our nervous system. In other words, the probability of an event is gauged against what has been embodied by the phylogenetic learning.[88]

There have been numerous attempts to relate probability to randomness. The feeling of randomness comes from the experience when we choose items (events) in which there is no way to reduce damage by any suitable bias. Consequently, if the world is 'uniform,' it is equivalent to equal probabilities for the events. However, as we will see later, it is hard to define randomness precisely, so it is not easy to found probability on randomness.[89]

2.5 How to quantify information

Let us continue preparing for checking how 'chaos' defined in Section 2.3 is intuitively appealing. Through quantification of 'history' we can characterize the extent of 'unpredictability' of the history (trajectory). We would be happy if we could formulate the statement: if the extent of unpredictability is positive, the dynamics should be chaotic. Those who already know the

[87] R. T. Cox, "Probability, frequency and reasonable expectation," Am. J. Phys. **14**, 1 (1946) is a highly interesting paper that deduces additivity of rational expectations axiomatically. N. Chapter, J. B. Tenenbaum and A. Yuille, "Probabilistic models of cognition: Conceptual foundations," Trends Cognitive Sci. **10**, 287 (2006) contains many related topics.

[88] Interestingly enough, we seem to have an ability to take statistics unconsciously from our experiences. Not only we but various animals have the capacity for Bayesian learning under natural conditions as summarized in the following paper: T. J. Valone, "Are animals capable of Bayesian updating? An empirical review," Oikos **112**, 252 (2006). There is also an unconscious process of slow learning through trial and errors: P. J. Bayley, J. C. Frascino and L. R. Squire, "Robust habit learning in the absence of awareness and independent of the medial temporal lobe," Nature **436**, 550 (2005).

[89] G. Shafer and V. Vovk, *Probability and Finance: its only a game!* (Wiley-Interscience, 2001) is a notable attempt to develop probability theory not depending on the Kolmogorov axiomatic system. For example, it is possible to prove that if a gambler plays against Nature, and if there is no way to accumulate his wealth unboundedly, then the strategy of Nature (or the output of Nature) must satisfy the strong law of large numbers.

Kolmogorov-Sinai entropy *can go to and browse through Section 2.7 and then go to Section 2.8.*

Something similar to shuffling cards occurs where the bundle of trajectories spread in the great earthquake model in Section 2.1 as illustrated in Fig. 2.9. If we could trace individual cards, shuffling cards makes no essential change. Shuffling cards is meaningful because we cannot follow each card; we must use, e.g., correlations among cards to get some knowledge about a particular card. Our spatial resolving power cannot be infinitely powerful, so even if the system is deterministic, especially if correlations are destroyed, we cannot know which trajectory is the actual trajectory. Therefore, if the trajectories are spread and then mixed up, we can no more predict what would happen next as accurately as we can know the current state of the system. That is, the information we currently have is not enough for prediction of the next state with the same accuracy with which we know the current state. Thus, the future becomes increasingly unpredictable. If the extent of this 'increase of unpredictability' is larger, the system behavior should look more chaotic. For example, in the earthquake model if $b > 1$ is larger, the system behaves more randomly.

Notice clearly, as stressed repeatedly, that the future becoming unpredictable and the determinacy of a dynamical system are perfectly compatible. Unpredictability is not because determinacy fails, but because we cannot know trajectories accurately. Then, there might be a reader who concludes that whether a system is chaotic or not depends on how we look at the system. However, it is an objective fact (a fact everybody must admit) that our observation capability is limited,[90] so whether a system exhibits chaos or not is a property of the system independent of who observes it. In other words, whether we feel it chaotic or not is a property of the system.

The discussion up to this point suggests the idea that the extent of 'chaos' can be quantified by the amount of the deficiency of '*information*' to specify the future of the system. If at any time point the information to specify the state at the next time step is not enough, we may feel this as a loss of information due to dynamics (especially if we consider steady dynamics). Thus, the idea mentioned above suggests that information loss rate may be used to quantify the extent of chaos. This point of view is easy to understand for one-sided shift dynamical systems. As seen in (2.11)-(2.13) the symbols are lost at the left end after moving to the left from the right side.

To formulate the idea mentioned above precisely, we must know how to quantify our feeling of 'having information' or 'not having information.' To this end, we pursue the quantification of information, suspending the question of what the information is; if we could quantify something, we could feel

[90] This is a mere empirical fact. If the world is really discrete and we can observe everything without any error, what we have been discussing here is meaningless. However, the world we live in is not like that.

at least that we know about it to some extent. Energy is a typical example. We know how to quantify it, but it is difficult to give a general answer when we are asked what energy is. A method to quantify information was given by Shannon and is now the foundation of information theory.[91] *Those who are familiar with the Shannon formula can go to the next section from here.*

Let us consider how to quantify the 'information' carried by a message. Here, again it is our task to formulate explicitly or mathematically what we recognize intuitively as plausible. Suppose we can use n kinds of symbols. When we use the symbols evenly and without any particular relations among them (i.e., for a sufficiently long message all the symbols appear with equal relative frequency without correlation), let us denote the 'information content' per symbol of a message as $I(n)$; it is natural to expect that this quantity depends only on n, because every symbol is used evenly without correlation. If we could use more kinds of symbols, the message with the same length must have more information, so $I(n)$ must be a monotone increasing function of n.

Next, let us assume that we can use two sets of symbols with n kinds and m kinds of symbols, respectively. We assume that each symbol set is used evenly without correlation and also there is no correlation between these two sets of symbols. Let us use a sequence of symbol pairs (a_i, b_j) $(i = 1, \cdots, n,$ $j = 1, \cdots, m)$ to send a message. This is equivalent to using a symbol set consisting of $m \times n$ kinds of symbols, so the expected information content per symbol pair must be $I(mn)$. The same symbol sequence may be sent by sending only a_i first and then only b_j later. Therefore, we conclude[92]

$$I(mn) = I(m) + I(n). \tag{2.19}$$

This and the monotonicity of I tells us that for some positive number c (see Note 2.3)

$$I(n) = c \log n. \tag{2.20}$$

The choice of c is equivalent to choosing the unit of information, but we could also interpret it as the choice of the base of the logarithm. If the base of the logarithm is a positive integer k, that is, if $I(n) = \log_k n$, the unit quantity of information is the information contained in the correct answer to a k-choice problem. If $k = 2$, the information is said to be measured in bits.

---

[91] T. M. Cover and J. A. Thomas, *Elements of Information Theory* (Wiley, 1991) is an excellent textbook. About Shannon himself, see S. W. Golomb et al., "Claude Elwood Shannon (1916-2002)," Notices AMS **49**, 8 (2002). Shannon's thesis was about genetics. See J. F. Crow, "Shannon's brief foray into genetics," Genetics **159**, 915 (2001).

[92] Precisely speaking, we must also send a rule specifying how to combine symbols and how to interpret composite symbols, etc., but the length of the needed sequence (message) to send such a rule is finite and constant unrelated to the length of the main body of the message itself. Therefore, for a sufficiently long message, the information content per symbol is asymptotically independent of this finite (overhead) portion.

Note 2.3. Cauchy's equation
Writing $n = e^N$, $m = e^M$ and $I(n) = f(N)$, (2.19) reads

$$f(N+M) = f(N) + f(M). \tag{2.21}$$

If this holds for any real numbers N and M, this is called *Cauchy's equation.* At rational points, it is easy to see $f(x) = cx$, where $c = f(1)$. To say more, we need extra conditions. If f is monotonic as in our case, we can say this is true for any real numbers.[93]

The above conclusion for the case with all the symbols being used evenly may be rewritten as $I(n) = -\log_2(1/n)$. Therefore, if a symbol appears with relative frequency p (we may say it appears with probability p), it is plausible to quantify the information carried by this symbol by $-\log_2 p$.[94] Then, if each symbol a_i appears with probability p_i independently, the expectation value of the *information* carried by a message per letter should be

$$H(p_1, \cdots, p_n) = -\sum_{i=1}^{n} p_i \log p_i, \tag{2.22}$$

where $x \log x$ is defined to be zero at $x = 0$. This is the famous *Shannon's formula.* For a history generated by a Bernoulli process $B(p_1, \cdots, p_n)$, the expectation value of information that a certain event occurs at an instant is given by (2.22). $I(n)$ is the result for $p_1 = \cdots = p_n = 1/n$.[95]

Have we understood what information is by the above quantification

[93] `http://enc.tfode.com/Cauchy's_functional_equation` has a succinct summary.

[94] ⟪**Surprisal and information**⟫ $-\log p$ is sometimes called *surprisal*, i.e., the extent of surprise we feel when an event with probability p actually happens. If a symbol with large surprisal actually appears, since this is very rare, we will feel a large amount of surprise. However, do not forget that the 'extent of surprise' is measured relative to the situation where all the events occur evenly (that is, this 'even case' is regarded as the most commonly expected situation in the world); it is not an absolute measure of surprise. In the first place surprise is always relative to what we have already known.

Surprisal may be axiomatically characterized. It may be easier to understand information as the expectation value of surprisal. Let $f(p)$ be the extent of surprise we feel when we know that the event with probability p actually happens. It should have the following properties:
(i) $f(p) > 0$.
(ii) $f(p)$ is a monotone decreasing function of p.
(iii) When independent events happen, the extent of surprise is the sum of that of each event: $f(p_1 p_2) = f(p_1) + f(p_2)$.
These imply $f(p) \propto -\log p$.

[95] For actual symbol sequences, e.g., actual English sentences, alphabetic symbols A, $\cdots$, Z are arranged with strong correlations, so the information per symbol must be far smaller than $\log_2 27 \simeq 4.7$ (27 instead of 26, because we take the space into account). The actual estimate is about 1.3 bits. A good summary may be found in T. M. Cover and R. C. King, "A converging gambling estimate of the entropy of English," IEEE Trans Inf. Theor. **IT-24**, 413 (1978).

scheme? Hardly. However, we should see Shannon's genius in asking only how to quantify it instead of asking what it is.

2.6 Measure-theoretical dynamical systems

To quantify the insufficiency of information to predict the dynamics of a dynamical system, or to quantify the information loss rate due to dynamics, we need probabilities of events, so we must introduce probability to the phase space Γ. *Those who know measure-theoretical dynamical systems can go to the next section.*

Prepare numerous clones of the dynamical system under consideration, and plot their states in the same phase space. Then we would see a cloud of points that describe individual clones in the phase space (Fig. 2.19).

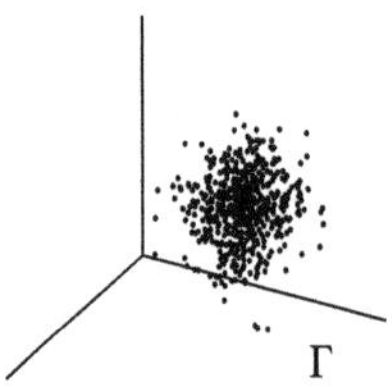

Fig. 2.19 Distribution or ensemble. The space is the phase space Γ. Each point represents an instantaneous state of each system. An invariant measure corresponds to a steady state for which the cloud as a whole becomes time-independent, although individual points may keep wandering around.

The distribution described by the cloud, if the total mass is normalized to be unity, may be understood as a probability measure on the phase space. Following the time evolution of the system, the cloud may change its shape. The set of clones is called an *ensemble*, so a cloud is the representation of an ensemble in the phase space. An ensemble cannot be determined unless its statistical properties are stated (we must specify how the points are distributed). For the time being, let us understand that individual clones are appropriately scattered over the phase space.

Discussion 2.6. When we consider an ensemble in general, what determines its 'initial distribution'? □

After a sufficiently long time, often the cloud representing an ensemble ceases to change its shape. Then, we say that the ensemble has reached its steady state. Individual points corresponding to the members of the ensemble may still keep wandering around in the cloud, but the cloud as a whole is balanced and the distribution on the phase space becomes invariant. The (probability) distribution corresponding to this invariant cloud is called an *invariant measure* of the dynamical system. More precisely, measure μ on the phase space Γ is an invariant measure of a dynamical system (T, Γ) defined by a map $T : \Gamma \to \Gamma$, if

$$\mu = \mu \circ T^{-1} \tag{2.23}$$

holds. That is, for an arbitrary (μ-measurable[96]) subset $A \subset \Gamma$

$$\mu(A) = \mu(T^{-1}A) \tag{2.24}$$

holds, where $T^{-1}A$ is the totality of points that comes to A after a unit time step (Fig. 2.20).

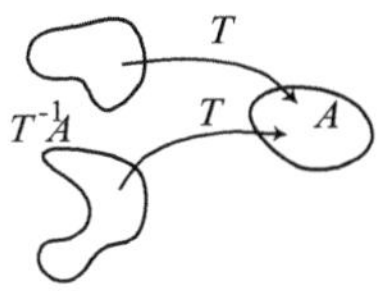

Fig. 2.20 The preimage $T^{-1}A$ of A by the time evolution operator T need not be connected (in this figure it consists of two connected components). Since $T(T^{-1}A) = A$, $T^{-1}A$ is the totality of the points that come to A after one time step.

For an arbitrary measurable set A, if we know $T^{-1}A$ at present, we know everything we can discuss probabilistically that will occur after one time step. The individual points in the cloud describing the ensemble correspond to individual systems, so they are neither created nor annihilated by the dynamics. Therefore, (if the systems are reversible[97]) the points in $T^{-1}A$ and that in A must be one-to-one correspondent. (2.24) expresses the condition that the evaluation of the weights of A and $T^{-1}A$ by μ is consistent with this conservation law. Generally speaking, an invariant measure for a given dynamical system is not unique. For example, $Tx = \{2x\}$ discussed in Appendix 2.1A has actually uncountably many distinct invariant measures. Or, if a dynamical system has several unstable periodic orbits, we can easily imagine an invariant measure that has the weight only on one of them. In this case, all the other members of the ensemble that are not on this special orbit are totally ignored.

Let T be a map from Γ into itself, and μ be a T-invariant measure on Γ. The triplet (T, μ, Γ) is called a *measure-theoretical dynamical system*. This may be interpreted as a mathematical expression of a steady state allowed to the dynamical system (T, Γ); given a dynamical system (T, Γ), for each

[96] For a measure see Appendix 2.4A. If we take an arbitrary set, we may not be able to define its probability (i.e., not measurable). Hence, we always need the word 'measurable.' Physicists do not have any non-measurable sets in their mind, so such a word is never written. This seems to cause nothing serious in physics, probably because Lebesgue measurability is always there as a prerequisite for physical observability (according to Y. Takahashi).

[97] 'Reversible' means that the past is uniquely reconstructed from the future and the present information. If the system is not reversible, for example, if the map T merges two points into one, we must count the number of points with a suitable multiplicity.

invariant measure μ, a distinct measure-theoretical dynamical system is constructed.

If a measure μ satisfies $\mu(A) = 0$ for any measure zero set A (i.e., any set without volume[98]), μ is called an *absolutely continuous measure.* For a dynamical system with an absolutely continuous invariant measure, chaos may often be observed by numerical experiments. If a measure is absolutely continuous, its probability density g may be defined as $d\mu = gd\lambda$, where λ is the Lebesgue measure.[99]

For any invariant set $A \subset \Gamma$ (that is, any set satisfying $A = T^{-1}A$[100]), if $\mu(A) = 0$ or 1, the measure-theoretical dynamical system is said to be *ergodic.* Roughly speaking, if a dynamical system is ergodic, the trajectory starting from any initial point[101] can eventually go into any μ-measure positive set. In fact, if there is a positive measure set C to which any trajectory starting from a certain positive measure set D cannot reach, then there must be an invariant set B with $0 < \mu(B) < 1$ such that $B \supset C$ and $B^c \supset D$. Conversely, if there is an invariant set B such that $0 < \mu(B) < 1$. Then, we can choose a measure positive set C with $B \cap C = \emptyset$ such that any trajectory starting from C never visits B (if possible, $\mu(\cup_{k=1}^{\infty} T^{-k}B \cap C) > 0$, but $T^{-1}B = B$ and $B \cap C = \emptyset$, a contradiction, where T^{-k} implies to apply T^{-1} k-times).

Let Γ be a unit circle, and T be a rigid rotation around the center by an angle that is irrational multiple of 2π. Since the uniform distribution μ_U on the circle is rotation-invariant, we can make a measure-theoretical dynamical system (T, μ_U, Γ), which is ergodic. This not so interesting example illustrates that ergodicity is totally insufficient for modeling irreversible phenomena such as relaxation exhibited by usual many-body systems.

Note 2.4. Ergodicity is irrelevant to equilibrium statistical mechanics
When ergodicity is discussed, most people recall Boltzmann (1844-1906) and statistical mechanics. "The equality of the time average and the ensemble average" is one of the consequences of ergodicity conceived by Boltzmann, but ergodicity is irrelevant to statistical mechanics. The reason is simple: time average has nothing to do with thermodynamics.

How long does it take to observe a thermodynamic observable, say, magnetization? 1 msec must be more than enough. Macroscopic quantities we observe are usually interpreted as the average of observables over microscopic states. Boltzmann interpreted the required distribution (probability) to compute the average as the ratio of sojourn times in various microstates to the total observation time. Since sufficiently many microstates must be sampled evenly over the phase space to compute the expectation values, ergodicity appeared cru-

---

98 the ordinary Lebesgue measure; it is the length in 1D, area in 2D, etc. See Appendix 1A **A1**.

99 That is, g is the *Radon-Nikodym derivative* $d\mu/d\lambda$. See, e.g., Kolmogorov and Fomin cited already.

100 This condition must be, precisely speaking, the equality ignoring the μ-measure zero difference.

101 Precisely speaking, starting from μ-almost all initial points. 'μ-almost all' implies that all except for μ-measure zero sets.

cial, but for a macroscopic object the required time for this average is (more than) astronomical (the order of Poincaré recurrence time), so the time average cannot be used to compute thermodynamic quantities. The secret of statistical mechanics lies in the fact that almost all the microstates of a macroscopic system in equilibrium give the identical thermodynamic observable values for the macroscopic system.

There are people who think that for observables such as magnetization the relaxation time is very short, so the law of large numbers holds for even a very short time observation, and allows us to measure them accurately with a short time observation as a time average. This point of view forgets about the fact that during a short time the phase point stays close to the starting point in the phase space. Recall that thermodynamic observations give the same result even if the observation is made a year later. A short-time observation after a year would sample another very small portion of the phase space around a new starting point. Without some special assumption about the phase space, the agreement of these two observations cannot be guaranteed. The simplest interpretation of the reproducibility of observations at different times is the uniformity of the phase space. Under this condition the assumption about relaxation is superfluous. Thus, instantaneous observation is enough to obtain thermodynamic quantities, and there is no room for ergodic theory. The founders of statistical mechanics, especially Ehrenfest (1880-1933) and other successors totally misunderstood statistical mechanics, and that is why the ergodic problem was regarded as a fundamental issue of statistical mechanics.

The logical contradiction pointed out by Zermelo between the recurrence and Boltzmann's 'derivation of irreversibility' by mechanics was dismissed by Boltzmann, who pointed out the extreme length of the recurrence time, but, ironically, his own logic destroys his basic idea to found statistical mechanics on ergodicity (as was actually recognized by Boltzmann himself later).

For any sets[102] A and B (both $\subset \Gamma$), if

$$\lim_{n\to\infty} \mu(T^{-n}A \cap B) = \mu(A)\mu(B) \tag{2.25}$$

holds, the measure-theoretical dynamical system is said to be *mixing*. $T^{-n}A$ is the set that agrees with A after n time steps. $T^{-n}A \cap B$ is the totality of the points that is at present in B and will be in A after n time steps. Intuitively, the cloud of points starting from B spreads over the phase space evenly if n is sufficiently large, so the fraction of the cloud that overlaps with A is proportional to the 'statistical weight' of A. It is obvious that initial conditions cannot be used to predict the future. Physically, irreversible processes such as relaxation phenomena occur. In particular, the time-correlation function eventually decays to zero.

---

102 Precisely speaking, any μ-measurable sets. From now on, such statements will not be added.

2.7 How to quantify chaos

The preceding two sections have paved the road to quantification of randomness in chaos. That is, we can now study how much insufficient the 'information' we have at present is to specify 'the future' in order to quantify the 'degree of chaos' of deterministic dynamical systems.

When a dynamical system is given, we wish to describe its state at each instant with a constant precision. Suppose the data with information H is required to describe the current state with a prescribed precision. If we could predict the state with the same precision after one unit time with the aid of this initial data, all the states in any future can be predicted with the same precision in terms of the information H contained in the initial data. However, this is not usually the case. As we have seen in the explanation of mixing in the preceding section, for chaotic systems the cloud of the ensemble is scrambled and the precise locations of the individual points in the phase space are lost. If we wish to locate the state of the system after one time unit as precisely as we can locate it now, we have to increase the precision of the description of the present state. Consequently, we need extra information ΔH, which we could call the extra information required for equi-precision (equi-precise description). Our basic idea is that this extra information ΔH for an equi-precise description of the next time step may be quantified by the logarithm of the 'local expansion rate' of the ensemble cloud.

In this section, first we compute the information deficit as intuitively as possible for a 1D map. This turns out to be a calculation of the Kolmogorov-Sinai entropy. If this simple case tells the reader the basic idea sufficiently, *she can go from there to the next section.* However, because the Kolmogorov-Sinai entropy is the key concept to understanding chaos, its detailed explanation is given in Appendix 2.7A. This will help the reader to understand the concept intuitively. Relevant theorems are also summarized at the end of this appendix. To understand the culmination of our conceptual analysis of chaos, 'partition' and 'generator' must be understood, so at least scanning the first two pages of Appendix 2.7A is desirable.

Consider a discrete measure-theoretical dynamical system (F, μ, I) on the interval I with a piecewise continuous and differentiable map[103] F and an absolutely continuous invariant measure μ. The average amount $h_\mu(F)$ of the extra information required for the equi-precise description of the states is given by the following formula:

[103] 'piecewise continuous' implies that a map consists of a several continuous pieces. T in Appendix 2.1A is an example. 'Piecewise continuously differentiable' implies that each piece is differentiable inside, and, at its ends, one-sided derivatives from inside are well defined. A unimodal piecewise linear map discussed in Section 2.1 is an example.

$$h_\mu(F) = \int_I \mu(\mathrm{d}x) \log |F'(x)|. \tag{2.26}$$

Henceforth, we will often write $\mu(\mathrm{d}x) = \mathrm{d}\mu(x)$.[104]

If the slope of the graph of the function F is large, the bundle of trajectories is expanded widely, so more information is required for the equi-precise prediction. Thus, (2.26) is plausible.

Let us demonstrate (2.26). If the reader thinks it highly plausible, she need not read the proof and *can go to the next section.* For simplicity, let us assume that F is unimodal (its graph has only one peak and no valley; Fig. 2.21). The invariance condition (2.24) for the measure μ reads, in this case (see Fig. 2.21),

$$\mu(\mathrm{d}x_1) + \mu(\mathrm{d}x_2) = \mu(\mathrm{d}x). \tag{2.27}$$

Here, $\mathrm{d}x$ denotes a small interval and its preimage (the set mapped to $\mathrm{d}x$ by F) is written as $\mathrm{d}x_1 \cup \mathrm{d}x_2$ ($F^{-1}(\mathrm{d}x) = \mathrm{d}x_1 \cup \mathrm{d}x_2$). For example, if F is applied once to $\mathrm{d}x_1$, it is stretched and mapped onto $\mathrm{d}x$, so without any doubt the precision to specify the location is reduced. That is, with the equal probability $\mu(\mathrm{d}x_1)/\mu(\mathrm{d}x) = \mu(\mathrm{d}x_1)/\mu(F(\mathrm{d}x_1))$ they are spread over $\mathrm{d}x$. Consequently, the information was lost by $-\log(\mu(\mathrm{d}x_1)/\mu(\mathrm{d}x))$. Therefore, the expectation value of the information deficit per unit time must be obtained by averaging this over x_1 with the aid of the (probability) measure μ. We must also do an analogous calculation for $\mathrm{d}x_2$. Consequently, the rate of the information deficit is written as

$$h_\mu(F) = -\int_I \mu(\mathrm{d}x) \log \frac{\mu(\mathrm{d}x)}{\mu(F(\mathrm{d}x))}. \tag{2.28}$$

This loss must be compensated for with the extra information ($\varDelta H$ in the above) to realize an equi-precise description of the system.

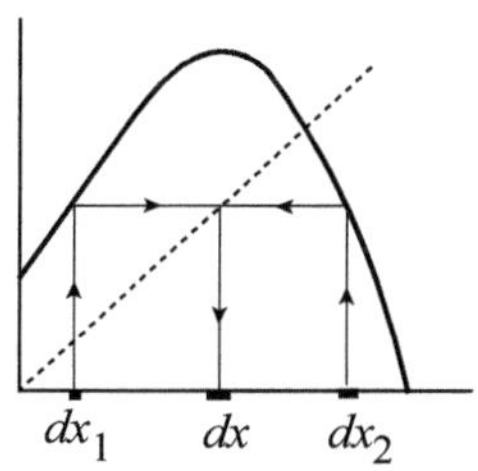

Fig. 2.21 An explanation of (2.27). The points in a small interval $\mathrm{d}x$ were in $\mathrm{d}x_1$ or in $\mathrm{d}x_2$ one time step ago. These intervals are stretched and superposed onto $\mathrm{d}x$.

[104] Here, a formal calculation is explained as given in Y. Oono, "Kolmogorov-Sinai entropy as disorder parameter for chaos," Prog. Theor. Phys. **60**, 1944 (1978). To obtain this final form (which is correct for absolutely invariant measures) H. Fujisaka's (1949-2007) help was crucial. The formula had been obtained by Rohlin, so let us call it *Rohlin's formula.* The generalized form that is also correct for non-absolutely continuous invariant measures is (2.28). As noted at the beginning of this chapter with a quotation from Sinai's article, from the Russian point of view, the paper stated only the obvious.

If μ is absolutely continuous, we can introduce the invariant density g as $g(x)\lambda(\mathrm{d}x) = \mu(\mathrm{d}x)$, where λ is the Lebesgue measure. Hence,

$$\frac{\mu(\mathrm{d}x_i)}{\mu(\mathrm{d}x)} = \frac{\mu(\mathrm{d}x_i)}{\lambda(\mathrm{d}x_i)}\frac{\lambda(\mathrm{d}x)}{\mu(\mathrm{d}x)}\frac{\lambda(\mathrm{d}x_i)}{\lambda(\mathrm{d}x)} = \frac{g(x_i)}{g(x)|F'(x_i)|}. \tag{2.29}$$

Noting that $F(x_i) = x$ with the aid of (2.29), we may rewrite (2.28) as

$$h_\mu(F) = \int_I \lambda(\mathrm{d}x) g(x) \log \frac{g(F(x))|F'(x)|}{g(x)}. \tag{2.30}$$

Here, the integral with respect to the Lebesgue measure is explicitly written as $\int \lambda(\mathrm{d}x)$, but it is simply $\int \mathrm{d}x$ with the usual notation. With the aid of the elementary property of the δ-function,[105] regarding y as the independent variable, we have

$$\delta(x - F(y)) = \frac{1}{|F'(x_1)|}\delta(y - x_1) + \frac{1}{|F'(x_2)|}\delta(y - x_2), \tag{2.31}$$

where $F(x_1) = F(x_2) = x$. With the help of this formula (2.27) can be rewritten as

$$g(x) = \int_I \mathrm{d}y\, g(y)\delta(x - F(y)), \tag{2.32}$$

which is called the *Perron-Frobenius equation*. Using this, we see

$$\int_I \mathrm{d}y\, g(y) \log g(F(y)) = \int_I \mathrm{d}y \int_I \mathrm{d}z\, g(y)\delta(z - F(y)) \log g(z) = \int_I \mathrm{d}z\, g(z) \log g(z), \tag{2.33}$$

so indeed (2.28) is nothing but (2.26).

For an ergodic Markov chain[106] the extra information required for equi-precision may be computed by the same idea as explained above. If its transition matrix is given by $\Pi \equiv \{p_{i\to j}\}$, then

---

[105] Let a be an isolated (real) zero point (i.e., $f(a) = 0$), and f be differentiable around it. In a sufficiently small neighborhood of a with the aid of the variable change: $y = f(x)$ we get for sufficiently small positive ε

$$\int_{a-\varepsilon}^{a+\varepsilon} \delta(f(x))\varphi(x)\mathrm{d}x = \frac{1}{|f'(a)|}\varphi(a).$$

Here, φ is a sufficiently smooth test function. Therefore, if f has a several isolated real zeros a_i, we can add each contribution to get

$$\int_{-\infty}^{\infty} \delta(f(x))\varphi(x)\mathrm{d}x = \sum_i \frac{1}{|f'(a_i)|}\varphi(a_i).$$

[106] See, for example, Durrett, *ibid.*; Z. Brzeźniak and T. Zastawniak, *Basic Stochastic Processes, a course through exercises* (Springer, 1998) is a very kind measure-theoretical introduction to stochastic processes.

$$h(\Pi) = -\sum_{i,j} p_i p_{i\to j} \log p_{i\to j}, \tag{2.34}$$

where p_i is the invariant measure (stationary state) (i.e., $\sum_i p_i p_{i\to j} = p_j$). In particular, for a Bernoulli process $B(p_1, \cdots, p_n)$, we have

$$h(B(p_1, \cdots, p_n)) = -\sum_{i=1}^{n} p_i \log p_i. \tag{2.35}$$

Applying the same idea to computing the increasing rate (2.28) to other dynamical systems (e.g., the Axiom A systems[107]) is easy. For the examples discussed or mentioned so far, the quantity h agrees with the Kolmogorov-Sinai entropy. Its general definition is given in Appendix 2.7A.

Note 2.5. Condition for Rochlin's formula
Rigorously speaking, the agreement between the Kolmogorov-Sinai entropy and the 'information deficiency rate' computed in terms of the 'expansion rate' above is, for the general case, a conjecture. As noted already (2.26) was proved by Rohlin long ago.[108] Ledrappier[109] proved the following theorem for C^2-maps:[110]

Theorem 2.2. *Let f be a C^2-endomorphism on an interval. A necessary and sufficient condition for an ergodic invariant measure to be absolutely continuous is that the Kolmogorov-Sinai entropy is given by Rohlin's formula (2.26).* □

If a system is chaotic, the bundle of histories close to a given history thins quickly due to expansion, so it is often the case that the external noise makes a chaotic system 'more chaotic.' However, with our crude description of dynamical systems so far given, we cannot claim anything general as to the noise response of a chaotic system because the 'effectiveness' of noise strongly depends on the system details such as which trajectories actually come close in the phase space. We cannot conclude that a system with a larger Kolmogorov-Sinai entropy is more sensitive to noise. For example, due to noise transition into a particular set $A \in \Gamma$ can become disproportionately frequent. If this element is embedded by the intrinsic dynamics into a small set, the spread of the 'history bundle' could become smaller due to noise. This is indeed the essence of the noise-induced order discovered by Tsuda.[111]

Discussion 2.7. Research *pseudo-orbit tracing property.* This is related to the observability of chaos. □

---

[107] For Axiom A systems we can obtain the so-called Pesin's equality (see Appendix 2.7B).

[108] V. A. Rohlin, "Lectures on the entropy theory of transformations with an invariant measure," Russ. Math. Surveys **22**(5), 1 (1967).

[109] F. Ledrappier, "Some properties of absolutely continuous invariant measures on an interval," Ergodic Theor. Dynam. Syst. **1**, 77 (1981).

[110] Actually, for C^1-maps with some conditions.

[111] K. Matsumoto and I. Tsuda, "Noise-induced order," J. Stat. Mech. **31**, 87 (1983).

Appendix 2.7A Kolmogorov-Sinai entropy, an introduction

The identity of the information deficiency rate = the extra information required for the equi-precise description and the Kolmogorov-Sinai entropy is generally expected, so the definition of the *Kolmogorov-Sinai entropy* for a measure-theoretical dynamical system (T, μ, Γ) is given. This appendix gives an elementary introduction to the concept and some basic theorems facilitating its intuitive understanding. Some preparation is needed. The reader may browse through the following, but the elementary part of the theory is explained in more detail than in the standard textbooks.

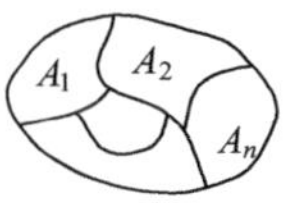

Fig. 2.22 Partition $\mathcal{A} = \{A_1, \cdots, A_n\}$. A finite partition $\mathcal{A}$ of a set Γ is a family of finitely many subsets of Γ such that its members A_i do not have any overlap with each other and the total sum perfectly agrees with Γ.

A (finite) *partition* $\mathcal{A}$ of a set Γ (Fig. 2.22) is a family of subsets $\{A_1, \cdots, A_n\}$ of Γ satisfying the conditions:
(1) For any i and j $(\neq i)$ $A_i \cap A_j = \emptyset$, and
(2) $\cup_{i=1}^n A_i = \Gamma$.
$\{A_i\}$ may be interpreted as the totality of the mutually exclusive observation results.[112] Our observation is always under finite precision. Therefore, if the phase space is continuous, we can never specify a particular point in it by observation. Therefore, it is reasonable to introduce such discrete (coarse-grained) observables.[113] The composition operation $\vee$ of two partitions of Γ, $\mathcal{A} \equiv \{A_1, \cdots, A_n\}$ and $\mathcal{B} \equiv \{B_1, \cdots, B_m\}$, is defined as follows (Fig. 2.23):

$$\mathcal{A} \vee \mathcal{B} \equiv \{A_1 \cap B_1, A_1 \cap B_2, \cdots, A_n \cap B_m\}. \tag{2.36}$$

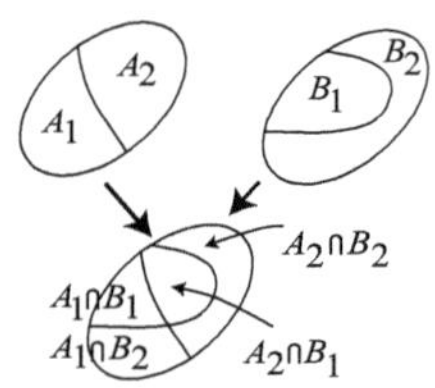

Fig. 2.23 Composition of partitions: $\{A_1, A_2\} \vee \{B_1, B_2\} = \{A_1 \cap B_1, A_1 \cap B_2, A_2 \cap B_1, A_2 \cap B_2\}$

On the right-hand side appear all the nonempty intersections of A_i and B_j

[112] Here, each set A_i may be a collection of discrete pieces or with holes (i.e., it need not be (singly) connected).

[113] We could not tell which A_i the phase point is actually in, so mustn't A_i be fuzzy? Here, we adopt an interpretation that the relation between the values of a macro-observable that can be observed with a finite precision and the actual microstates of the dynamical system is given by the system itself independent of our capability of observing each microstate. That is, an element of a partition $\mathcal{A}$ consists of a definite set of microstates as an intrinsic property of the system, although we cannot determine this with our finite precision observations.

(note that $\mathcal{A} \vee \mathcal{B} = \mathcal{B} \vee \mathcal{A}$). The elements of $\mathcal{A} \vee \mathcal{B}$ may be interpreted as the mutually exclusive outcomes obtained by performing two macroscopic observations corresponding to $\mathcal{A}$ and $\mathcal{B}$ simultaneously.

Let us return to the general measure-theoretical dynamical system (T, μ, Γ). The average information we can obtain about the system by a single observation of macroscopic observable $\mathcal{A}$ may be written with the aid of Shannon's formula as[114]

$$H(\mathcal{A}) = - \sum_{A \in \mathcal{A}} \mu(A) \log \mu(A). \tag{2.37}$$

This information implies that unless we have, on the average, this amount of information about the system, we cannot infer in which A_i the observation result is in.

When a system evolves according to the dynamical law T, how much information do we need to infer its coarse-grained state (one of $\mathcal{A}$) at the next time (time $t+1$) under the condition that we know the coarse-grained state at present (time t)? To infer both the states at time t and time $t+1$ without any prior knowledge, we need information of $H(\mathcal{A} \vee T^{-1}\mathcal{A})$, where $T^{-1}\mathcal{A} = \{T^{-1}A_1, \cdots, T^{-1}A_n\}$. We can understand this as follows. An element of $T^{-1}\mathcal{A}$ coincides with some element of $\mathcal{A}$ after one time step. That is, describing the state of the system after one time step in terms of the observable $\mathcal{A}$ is equivalent to knowing in which element of $T^{-1}\mathcal{A}$ the current state is. Therefore, to specify both the states at $t+1$ and at t is to specify a single element in $\mathcal{A} \vee T^{-1}\mathcal{A}$. Consequently, on the average without $H(\mathcal{A} \vee T^{-1}\mathcal{A})$ of information, we cannot infer in which A_i at time t and in which A_j at $t+1$ the system is in. If we already know the macrostate at time t, to predict the macrostate at $t+1$ we need $H(\mathcal{A} \vee T^{-1}\mathcal{A}) - H(\mathcal{A})$ extra information on the average. This is the amount of extra information ΔH required for equi-precise prediction of the one time step future.

We wish to know what happens in a steady state. We need not stick to a particular time t and $t+1$; we should consider the average deficiency for a long time:

$$\frac{1}{n}[H(\mathcal{A} \vee T^{-1}\mathcal{A} \vee \cdots \vee T^{-n}\mathcal{A}) - H(\mathcal{A})] \to h_\mu(T, \mathcal{A}), \tag{2.38}$$

where the existence of this limit in $n \to \infty$ is guaranteed by the (intuitively plausible) subadditivity of H, $H(\mathcal{A} \vee \mathcal{B}) \le H(\mathcal{A}) + H(\mathcal{B})$ (see Notes 2.6 and 2.7) and $H(T^{-n}\mathcal{A}) = H(\mathcal{A})$ ($n = 1, 2, \cdots$; this can be seen from the invariance of the measure (2.24)).

Note 2.6. Subadditivity of entropy: $H(\mathcal{A} \vee \mathcal{B}) \le H(\mathcal{A}) + H(\mathcal{B})$.
The inequality must be intuitively natural, because in order to describe two observables $\mathcal{A}$ and $\mathcal{B}$ it is better to use information about the relation between these two as well than to use information separately from each of the observables (if the inequality does not hold, we must conclude that the definition of H is not good).

$$H(\mathcal{A} \vee \mathcal{B}) = - \sum_{i,j} \mu(A_i \cap B_j) \log \mu(A_i \cap B_j), \tag{2.39}$$

$$= - \sum_{i,j} \mu(A_i \cap B_j) \left\{ \log \left(\frac{\mu(A_i \cap B_j)}{\mu(A_i)\mu(B_j)} \right) + \log \mu(A_i)\mu(B_j) \right\}, \tag{2.40}$$

[114] The partition must be a measurable partition; $\mu(A_i)$ must be meaningful. Such a statement will not be written explicitly in the following.

$$= -\sum_{i,j} \mu(A_i \cap B_j) \log \left(\frac{\mu(A_i \cap B_j)}{\mu(A_i)\mu(B_j)} \right) + H(\mathcal{A}) + H(\mathcal{B}). \tag{2.41}$$

If the first term on the right-hand side of the last line (2.41) is nonpositive, the proof of $H(\mathcal{A} \vee \mathcal{B}) \leq H(\mathcal{A}) + H(\mathcal{B})$ is over. This is the following important inequality: Let p and q be probabilities ($p_i \geq 0$ and $\sum_i p_i = 1$, etc., hold). We have

$$\sum_i p_i \log \frac{p_i}{q_i} \geq 0. \tag{2.42}$$

To show this we use the inequality $x \log x \geq x - 1$ for $x \geq 0$.[115] Introduce $x = p_i/q_i$ into this and sum over i after multiplying q_i:

$$\sum_i q_i \left(\frac{p_i}{q_i} \log \frac{p_i}{q_i} \right) \geq \sum_i p_i - \sum_i q_i = 0. \tag{2.43}$$

Note 2.7. Fekete's lemma
If $\{f(n)\}$ is subadditive (i.e., $f(n+m) \leq f(m) + f(n)$ holds for any positive integers, n, m), $\lim_{n\to\infty} f(n)/n = \inf_m f(m)/m$.[116] □

Obviously, $\liminf f(n)/n \geq \inf f(m)/m$. Writing $n = s + km$ ($m > 0$, $s \geq 0$ are integers), we get

$$\frac{f(n)}{n} = \frac{f(s+km)}{s+km} \leq \frac{f(s) + kf(m)}{s+km} \to \frac{f(m)}{m}. \tag{2.44}$$

Therefore, $\limsup f(n)/n \leq \inf f(m)/m$. Hence, the infimum and supremum limits agree, and the limit exists.

We need not stick to a particular observable corresponding to $\mathcal{A}$. We should look for 'the best' observable' to observe the system. Here, 'the best observable' should imply the observable that allows us to observe the system maximally in detail (with the least effort). Such an observable must be sensitive to the time evolution of the system, so the increasing rate of the information deficiency for this observable must be the largest. With this idea the Kolmogorov-Sinai entropy (or measure-theoretical entropy) is defined as follows:

$$h_\mu(T) \equiv \sup_{\mathcal{A}} h_\mu(T, \mathcal{A}), \tag{2.45}$$

where the supremum is taken over all the finite partitions of Γ (roughly speaking, we try all the finite-resolving power observations). If all the future data of a certain macro-observable determines the future history (trajectory) uniquely, we do not need any more detailed observations. Take an arbitrary history ω. Let us write the element of $\mathcal{A} \vee T^{-1}\mathcal{A} \vee \cdots \vee T^{-n+1}\mathcal{A}$ containing ω as $A^n(\omega)$ (which

---

[115] The minimum value of $f(x) = x \log x - x + 1$ for $x \geq 0$ is zero as can be seen from its graph.

[116] **⟪Infimum limit (lim inf), supremum limit (lim sup)⟫** $\liminf_{n\to\infty} x_n = \lim_{n\to\infty} \inf\{x_n, x_{n+1}, \cdots\}$. That is, we make the lower bound y_n of the sequence beyond x_n and then take its limit $n \to \infty$. Since $\{y_n\}$ is monotone increasing, the limit is well defined (may not be bounded); similarly, $\limsup_{n\to\infty} x_n = \lim_{n\to\infty} \sup\{x_n, x_{n+1}, \cdots\}$.

is an example of a cylinder set).[117] Since $\omega \in A^n(\omega)$ for any $n = 0, 1, 2, \cdots$, $\omega \in \cap_{n=0}^{\infty} A^n(\omega)$. If this common set does not contain any history other than ω, in other words, if $\cap_{n=0}^{\infty} A^n(\omega) = \{\omega\}$, any further detailed observation is superfluous. If such a relation holds for μ-almost all ω (i.e., except for μ-measure zero set), the partition $\mathcal{A}$ is called a *generator*. If $\mathcal{A}$ is a generator, as expected,[118]

$$h_\mu(T) = h_\mu(T, \mathcal{A}). \tag{2.46}$$

For example, for $Tx = \{2x\}$ in Appendix 2.1A $\{[0, 1/2], (1/2, 1]\}$ is a generator.

If a measurable partition $\mathcal{A}$ separates all points (precisely speaking, for μ-almost all points), in other words, for any two points $x \neq y \in \Gamma$, there is a positive integer n and an element $A \in \mathcal{A}$ such that $T^n(x) \in A$ and $T^n(y) \notin A$, $\mathcal{A}$ is a generator. Actually, if $\mathcal{A}$ were not a generator, there are two histories $T^n(x)$ and $T^n(y)$ that are always contained in a single element of $\mathcal{A} \vee T^{-1}\mathcal{A} \vee \cdots \vee T^{-n+1}\mathcal{A}$ for any $n = 0, 1, 2, \cdots$. This contradicts the requirement that each point is separated.

By now the meaning of the statement in Theorem 2.1 in Section 2.3, that chaos $\Leftrightarrow$ there is an invariant measure with a positive Kolmogorov-Sinai entropy, should be clear. A chaotic dynamical system is a dynamical system for which the extra information required for equi-precise description increases linearly in time.[119] Consequently, chaotic motion looks random. Thus, we have further confirmed that our definition of chaos is consistent with our intuition.

For chaos, if we wish to predict its state after N time steps with sufficient precision, we need a tremendous amount of information at present (we need $\sim e^{Nh}$ times as much precision as required to describe the present state; see just below), so there is no wonder that sooner or later the behavior of the system becomes unpredictable.

Ornstein and Weiss proved the following theorem:[120]

Theorem 2.3. *Not completely predictable systems (= systems with a positive Kolmogorov-Sinai entropy) have Bernoulli flows (continuous dynamical systems whose periodically sampled sequences become Bernoulli processes) as their factor dynamical systems.* □

Here, "to have A as a factor dynamical system" implies that the original dynamical system behaves as dynamical system A if it is reduced to a certain space (more precisely, there is a homomorphism from the system onto A). They simply equate the chaotic system and the system with a positive Kolmogorov-Sinai entropy in the quoted review article.

From here to the end of this appendix, important theorems about the

---

117 An element of $\mathcal{A} \vee T^{-1}\mathcal{A} \vee \cdots \vee T^{-n+1}\mathcal{A}$ has the form: $A_i \cap T^{-1}A_j \cap \cdots \cap T^{-n+1}A_k$, which is the totality $\{x : x \in A_i, Tx \in A_j, \cdots, T^{n-1}x \in A_k\}$. Therefore, around here in the text, $x \in \Gamma$ and a history $\omega = \{x, Tx, \cdots\}$ are identified.

118 P. Walters, *An Introduction to Ergodic Theory* (Springer, 1982) is an excellent textbook of the Kolmogorov-Sinai entropy (but students outside mathematics may not be able to read it with ease). P. Billingsley, *Ergodic Theory and Information* (Wiley, 1960) is also excellent.

119 In contrast to the above explanation, there were people who wished to introduce the Kolmogorov-Sinai entropy as the rate of generation of information by the dynamical system. However, generally speaking, the generating rate of information just considered is larger than the Kolmogorov-Sinai entropy. This fact is captured by Ruelle's inequality (2.57), which we will encounter later.

120 D. S. Ornstein and B. Weiss, "Statistical Properties of Chaotic Systems," Bull. Amer. Math. Soc. **24**, 11 (1991), Theorem 1.4.3.

Kolmogorov-Sinai entropy, especially theorems that help us to understand it intuitively, will be summarized without proof, although their intuitive meanings will be explained. Also Lyapunov indices, etc., are introduced in the following Appendix 2.7B, but *the reader can go to the next section without paying much attention to details* (but we will use the Shannon-McMillan-Breiman theorem 2.5 in Section 2.12).

Theorem 2.4 (Krieger[121]). *Let (T, μ, Γ) be an ergodic dynamical system. If its Kolmogorov-Sinai entropy satisfies $h_\mu(T) < \log k$ for some integer $k > 1$, there is a generator with k elements.* □

The theorem roughly tells us that if the Kolmogorov-Sinai entropy of the system is $\log k$, then we can encode its dynamics using k symbols without losing any information.

Theorem 2.5 (Shannon-McMillan-Breiman). *Let (T, μ, Γ) be an ergodic dynamical system and $\mathcal{A}$ a finite partition of Γ. Let $A^n(x)$ be an element of $\mathcal{A} \vee T^{-1}\mathcal{A} \vee \cdots \vee T^{-n+1}\mathcal{A}$ containing $x \in \Gamma$. For μ-almost all x*

$$\lim_{n\to\infty} \left[-\frac{1}{n} \log \mu(A^n(x)) \right] = h_\mu(T, \mathcal{A}). \tag{2.47}$$

□[122]

In the above formula $A^n(x)$ is, as before, the bundle of histories (i.e., cylinder set) that cannot be distinguished from the history with the initial condition x for n time steps with a coarse-grained observation corresponding to the partition $\mathcal{A}$. $\mu(A^n(x))$ is the volume of the initial conditions for the history satisfying the condition (we could interpret it as the volume of the cylinder set). If the observation time span n is increased, the condition becomes increasingly stringent, so the volume decreases exponentially. (2.47) measures how fast the volume of the cylinder set decreases. The more chaotic the system is, the harder trajectories to stay close to a particular one starting from x, so this value must become larger. It may be more intuitive to rewrite (2.47) as

$$\mu(A^n(x)) \sim \mathrm{e}^{-nh_\mu(T,\mathcal{A})}. \tag{2.48}$$

The Brin-Katok theorem explicitly counts the number of histories in the ε-neighborhood of the trajectory starting from x.[123] Let the totality of the initial conditions that stay in the ε-neighborhood of the trajectory starting from $x \in \Gamma$ for N time steps be (the theorem holds for continuous dynamical systems as well)

$$B_N(x, \varepsilon) = \{y \in M : d(T^n x, T^n y) \le \varepsilon, 0 \le n \le N\}. \tag{2.49}$$

Theorem 2.6 (Brin-Katok). *Let (T, μ, Γ) be an ergodic dynamical system. For μ-almost all $x \in \Gamma$,*

$$h(\mu) = -\lim_{\varepsilon\to 0}\lim_{N\to\infty} \frac{1}{N} \log \mu(B_N(x, \varepsilon)). \tag{2.50}$$

□

[121] W. Krieger, "On entropy and generators of measure-preserving transformations," Trans. Amer. Math. Soc. **149**, 453 (1970); corrections *ibid.* **168**, 519 (1972).

[122] For a proof of the Shannon-McMillan-Breiman theorem, see, for example, W. Parry, *Entropy and Generators in Ergodic Theory* (Benjamin, New York 1969).

[123] M. Brin and A. Katok, "On local entropy" Lecture Notes Math. **1007**, 30 (1983).

Chaos may be predictable for a certain time span thanks to its deterministic nature, in contradistinction to noise, but not for a long time. We can estimate during how many steps n we can predict dynamics with the aid of the Kolmogorov-Sinai entropy h. According to the Brin-Katok theorem $n \sim (\log \delta x)/h$, where δx is our resolving power of the initial condition. However, there are chaotic dynamical systems with very small h, for which accurate prediction of considerable future is possible.[124]

A special case of the Shannon-McMillan-Breiman theorem is the *asymptotic equipartition* (AEP) theorem of information theory:[125] Let $\{X_i\}$ be a sequence of independently and identically distributed stochastic variables. Let us write the probability for (consecutive) n samples, $X_1, X_2, \cdots, X_n$, as $p(X_1, \cdots, X_n)$. Then,

$$-\frac{1}{n} \log p(X_1, \cdots, X_n) \to H(X), \tag{2.51}$$

where $H(X)$ is the entropy of the individual stochastic variables. Notice that this is nothing but the weak law of large numbers (footnote 24 in Section 1.2) for the logarithm of probability. $X_1, X_2, \cdots, X_n, \cdots$ may be interpreted as a history, so the Shannon-McMillan-Breiman theorem is the extension of the asymptotic equipartition theorem to correlated histories.

Note 2.8. Kolmogorov-Sinai entropy as isomorphism invariant
The Kolmogorov-Sinai entropy was originally proposed as an invariant under *isomorphism* of measure-theoretical dynamical systems. Suppose there are two measure-theoretical dynamical systems (T, μ, Γ) and (T', μ', Γ'). Let $\phi : \Gamma \to \Gamma'$ be a one-to-one (except for measure zero sets) correspondent and measure-preserving map: $\mu = \mu' \circ \phi$ (that is, the measure of any measurable set measured by the measure μ on Γ and measured by the measure μ' on Γ' after mapping from Γ by ϕ are identical).[126] If the following diagram:

$$\begin{array}{ccc} \Gamma & \xrightarrow{T} & \Gamma \\ \downarrow \phi & & \downarrow \phi \\ \Gamma' & \xrightarrow{T'} & \Gamma' \end{array}$$

is commutative (except on measure zero sets of Γ and Γ'), that is, if $T = \phi^{-1} \circ T' \circ \phi$, these two measure-theoretical dynamical systems are said to be isomorphic.

It is almost obvious that the Kolmogorov-Sinai entropies of isomorphic measure-theoretical dynamical systems are identical, because for a partition $\mathcal{A}$ of Γ $H(\mathcal{A}) = H(\phi\mathcal{A})$ and $\phi \circ T = T' \circ \phi$. We say that the Kolmogorov-Sinai entropy is an *isomorphism invariant*. If two dynami-

[124] For asteroid 522 Helga the Lyapunov characteristic time (the time needed to magnify the initial error by e; see Appendix 2.7B) is 6,900 years, so accurate prediction is possible. See A. Milani and A. M. Nobili, "An example of stable chaos in the Solar System," Nature, **357**, 569 (1992). J. J. Lissauer, "Chaotic motion in the Solar System," Rev. Mod. Phys. **71**, 835 (1999) is a review.

[125] It is quite important in information theory. See T. M. Cover and J. A. Thomas, *Elements of Information Theory* (Wiley, 1991) p50-

[126] Precisely, we must assume not only that $\phi : \Gamma \to \Gamma'$ is one to one as a map between measurable spaces, but also that measurable subsets of Γ are mapped on those of Γ' by ϕ, and vice versa by ϕ^{-1}.

cal systems have different Kolmogorov-Sinai entropies, then they cannot be isomorphic. For example, as can be seen from (2.35) $B(1/2, 1/2)$ and $B(1/3, 2/3)$ cannot be isomorphic.

An isomorphism invariant that takes the same value if and only if dynamical systems are isomorphic is called a *complete isomorphism invariant*. If we find such an invariant, the classification of dynamical systems according to isomorphism is reduced to the computation of the invariant. Is the Kolmogorov-Sinai entropy such an invariant? When Meshalkin[127] demonstrated the isomorphism between $B(1/4, 1/4, 1/4, 1/4)$ and $B(1/2, 1/8, 1/8, 1/8, 1/8)$ (both have the Kolmogorov-Sinai entropy $2\log 2$), the affirmative answer was expected (also there was a crucial contribution of Sinai: the weak isomorphism theorem for Bernoulli processes). In 1970[128] Ornstein proved that for Bernoulli processes the Kolmogorov-Sinai entropy is a complete invariant. Furthermore, completeness was proved for any finite mixing Markov chain with finite Kolmogorov-Sinai entropy.[129,130]

Appendix 2.7B Lyapunov index and Oseledec's theorem

As seen in examples above and from the Brin-Katok theorem, exponential separation of nearby trajectories may be regarded as a characteristic feature of chaos. Consider two trajectories starting from x and a nearby point $x + \varepsilon v$, where ε is a small positive number and v a directional vector. They tend to separate exponentially as time increases. At time t let us write the separation distance between these two trajectories as $\exp(t\lambda(x, v))$. The exponent $\lambda(x, v)$ is called the *Lyapunov exponent* (or Lyapunov characteristic exponent; A. M. Lyapunov 1857-1918) for the vector v at $x \in \Gamma$. Its precise definition is

$$\lambda(x, v) = \limsup_{n \to \infty} \frac{1}{n} \log \|D\phi^n(x)v\|, \tag{2.52}$$

where D is differentiation with respect to x[131] and $\| \ \|$ is the norm in the tangent vector space of Γ. The map T defining the dynamical system is written as ϕ to avoid confusion in this appendix. At each x $\lambda(x, v)$ as a function of v takes q ($\leq$ the dimension of Γ) distinct values:

$$\lambda^{(1)}(x) > \cdots > \lambda^{(q)}(x). \tag{2.53}$$

Its existence is guaranteed by

---

127 L. D. Meshalkin, "A case of Bernoulli scheme isomorphism," Dokl. Acad. Sci. USSR, **128**(1) 41 (1959).

128 [1970: Bertrand Russell died (b. 1872), Aswan High Dam completed, Allende became the President of Chile.]

129 I. P. Cornfeld, S. V. Fomin, and Ya. G. Sinai, *Ergodic Theory*, (Springer, 1982); D. S. Ornstein, *Ergodic Theory, Randomness, and Dynamical Systems* (Yale University Press, 1974).

130 However, the cases with zero entropy are different; there are numerous non-isomorphic dynamical systems with zero entropy.

131 This gives a Jacobi matrix in general.

Theorem 2.7 (Oseledec). [132] *At each $x \in \Gamma$ the tangent vector space $T_x\Gamma$ ($\simeq \boldsymbol{R}^n$) of Γ may be decomposed into a direct sum of the form*

$$T_x\Gamma = \bigoplus_{i=1}^{q(x)} H_i(x) \tag{2.54}$$

and for $v \in H_j(x)$

$$\limsup_{n\to\infty} \frac{1}{n} \log \|D\phi^n(x)v\| = \lambda_j(x). \tag{2.55}$$

If the dynamical system is ergodic, this does not depend on x. □

The directions that give positive exponents are the directions sensitive to perturbation. The sum of all the positive Lyapunov exponents

$$\lambda_+(x) = \sum_{i:\lambda_i(x)>0} \lambda_i(x). \tag{2.56}$$

satisfies *Ruelle's inequality:*[133]

$$h_\mu(\phi) \le \int \lambda_+(x)d\mu. \tag{2.57}$$

Therefore, if the dynamical system is ergodic, its Kolmogorov-Sinai entropy is bounded by the sum of positive Lyapunov exponents. The equality holds (*Pesin's equality*)[134] for many examples of well-behaved observable chaos.

2.8 Preparation for characterizing randomness

In the preceding section, we attempted to quantify our feeling of 'being random' through quantifying the extent of unpredictability, and found that the Kolmogorov-Sinai entropy is a good (i.e., consistent with our intuition) measure of unpredictability, a key feature of randomness. However, what is 'randomness'? How can we characterize it? From this section we start to illustrate another deep conceptual analysis. However, do not forget that this is a part of the analysis of concept 'chaos,' or more precisely, a part of illustration of conceptual analysis using 'chaos' as a vehicle.

This section is a preparatory section bridging the story up to this point and the rest of this chapter; we will see that the essence of randomness is

---

[132] For a proof, see, for example, A. Katok and B. Hasselblat, *Introduction to the Modern Theory of Dynamical Systems* (Cambridge University Press, 1996) p665. The original theorem is about the general cocycle of a dynamical system and is called Oseledec's multiplicative ergodic theorem, but here it is quoted only in the form directly relevant to our topic.

[133] D. Ruelle, "An inequality for the entropy of differentiable maps," Bol. Soc. Brasil. Mat. **9**, 83 (1978).

[134] Ya. B. Pesin, "Characteristic Lyapunov exponents and smooth ergodic theory," Russ. Math. Surveys **32**(4), 55 (1977).

related to 'esoteric topics' such as the theory of computation.[135] Within this chapter, our final destination is Brudno's theorem. This theorem teaches us that the Kolmogorov-Sinai entropy has a precise algorithmic implication. In the end, we will see that the characterization of randomness of chaos by the positivity of the Kolmogorov-Sinai entropy and that by algorithmic consideration perfectly agree. That is, our definition of chaos is further demonstrated to be natural.

The reader might not think this is the desirable end. An algorithmic characterization of randomness is admissible, but are computers and computation sufficiently fundamental concepts? There are various opinions. On the one hand, we could admit that just like continuity or smoothness, randomness is not a concept as basic as it looks. There are people who may even claim that the universe is a computer. For them 'computation' and 'algorithm' are more basic concepts than 'randomness.' On the other hand, there are people who wish to regard 'randomness' as a (more) basic concept, and feel that to characterize 'randomness' with artificial concepts such as computation fails to recognize the key issue, because randomness is ubiquitous. "Existence of definite procedures' might sound more natural than 'computation,' but it is hard to imagine that natural phenomena are influenced by the presence or absence of 'algorithms.' These people never think Nature is a big computer. We will come back to this discussion in Section 2.13.

A standard computational theoretical characterization of randomness will be discussed in the next and subsequent sections. This section is a general introduction to the topic.

What is a random number table? Intuitively, it is a table on which numbers are arranged without any regularity. A random number table is a table tabulating a number sequence that is recognized to have no regularity, passing all the statistical tests (cf. footnote 139). In other words, it is a table of number sequences such that the null hypothesis that they are random has not been rejected by any of the statistical tests available to humankind.[136] The testing methods progress daily, so the random number tables should also evolve daily.[137,138]

---

[135] In the US, the relation between chaos and algorithmic randomness was enthusiastically advocated by J. Ford (1927-1995) in the 80s, but, needless to say, in Russia the relation was there from the very beginning in the 60s (or even in the 50s in Kolmogorov). In Japan the relation was clearly recognized by the end of 70s. For example, when the author started to write papers on chaos, the relation was pointed out by his father (Yosiro Oono [of n-port circuit synthesis]). Also Brudno's theorem, which is the main topic of Section 2.12, was well-known among Japanese researchers thanks to Yoichiro Takahashi. The theorem was not known even in the late 80s in the US among physicists.

[136] For example, see A. L. Rukhin, "Testing randomness: a suite of statistical procedures," Theory Probab. Appl., **45**, 111 (2001).

[137] The author's math mentor Professor H. Watanabe once said, "Random number is like God. Its existence might be admissible, but, if you are shown 'this is It,' it is quite doubtful."

[138] The best random number is supposedly the natural random number that can be made by counting the number of, e.g., β-decay. Thus, we might say that without quantum phenomena there are no truly random numbers, but even radioactive decay can be influenced by extranuclear environments, so we cannot simply say they are

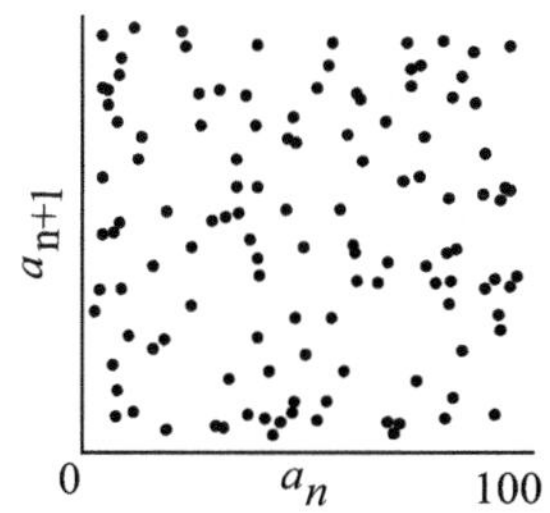

3	0	2	1	0	2	0	2	2	1
3	0	2	1	0	1	2	2	3	0
2	1	1	1	3	1	1	1	0	0
1	1	0	3	0	0	1	2	0	3
1	1	1	2	1	1	0	0	1	2
2	1	1	2	1	0	3	1	2	1
3	0	1	1	0	0	2	1	3	2
1	2	0	1	2	1	0	0	1	1
2	0	0	1	1	1	0	1	1	0
2	1	1	2	3	2	3	1	1	1

Fig. 2.24 The famous Kitagawa table is not a good-quality random number table. The numbers in the first 5 rows in the table are interpreted as the sequence of two-digit numbers as $a_1, a_2, \cdots$, and a_{n+1} is plotted against a_n in the left figure. The square is divided into 100 square boxes of 10 × 10, and the numbers found in each square are counted as shown in the right figure. If the sequence is random, the points must be distributed uniformly in the figure, so in the right figure, we expect 79 points inside the square and 45 on the periphery. Actually, there are 65 inside and 59 outside, failing the uniformity test with the P value less than 0.5%. In short, $\{a_n\}$ has a tendency to change in an oscillatory fashion.

As an example, take the well-known Kitagawa random number table.[139]

---

really random.

Incidentally, it is said that, when a person is asked to write down as random a number sequence as possible, the randomness of the produced number sequence and the person's intelligence are positively correlated.

[139] See T. Kitagawa, *Statistical Inference* (Suisoku-Tokei Gaku) (Iwanami, 1958).
⟪**How to make a random number table**⟫ The Kitagawa table was constructed as follows [in O. Miyataka and T. Nakayama, *Monte Carlo Method* (Nikkan-Kogyo, 1960) `http://www.sci.kagoshima-u.ac.jp/~ebsa/miyatake01/`]. First, on each page of the random number table of R. A. Fisher and F. Yates, *Statistical Tables for Biological, Agricultural and Medical Research* (Oliver and Boyd, 1938), two consecutive numbers were paired, and the resultant 25 × 50 pairs were scrambled. Then, the columns on different pages of the resultant scrambled table were exchanged. Next, the following statistical tests were performed on each page, and the best 4 pages were kept:
(1) Frequency test: using the χ^2-test, to check whether all the digits appear equally frequently,
(2) 'Pair' test: the table is considered as the table of two consecutive number pairs, and the frequencies of the pairs were tested just as in (1).
(3) Poker test: the frequencies of the various patterns of the consecutive, e.g., five-digit blocks (say, *abcde, abacd, aabac*, etc.) were tested.
(4) Gap test: Reading the table column-wise, the spacing between consecutive identical digits was tested.

The Fisher-Yates table was constructed as follows: from the 20-digit logarithm table (of A. J. Thompson, *Logarithmetica Britannica: Logarithms to 20 Decimal Places 10,000-100,000* (This work of Dr. Thompson's is an attempt to commemorate in a worthy manner the first great table of common logarithms, which was computed by Henry Briggs and published in London in 1624. It brings together the series of nine separate parts, issued between 1924 and 1952 from University College, London, in Karl Pearson's Tracts for Computers series; reprinted from Cambridge University

Is this random? The numbers in the first 5 rows are interpreted as the sequence of two-digit numbers as $a_1, a_2, \cdots$, and a_{n+1} is plotted against a_n in the left figure (we have already seen a similar plot in Section 2.1). If the Kitagawa table is really random, then the points on the 100×100 square must distribute uniformly. Figure 2.24 left is the figure created by this method applied to the 125 two-digit numbers in the first 5 rows of the table (there are 124 points). Now, the square is divided into 100 small square boxes and the number of points is counted in the peripheral 36 boxes and the inside 64 boxes. 59 points are on the periphery and 65 inside. This implies that the random number has a periodic component. If there is no such regularity, we can expect 45 points on the periphery and 79 inside. If we perform the χ^2-test, the null hypothesis that the distribution is uniform is rejected with the 0.5%-level.

Statistical tests check the non-existence of particular patterns in the number sequence. The Kitagawa table failed to be a random number table, because it has a significant pattern. Therefore, it seems to be a good idea to declare that the sequence without any feature (lawlessness) is a random number sequence. However, within our usual logical system that admits the exclusion of a middle (i.e., there are only two possibilities A or non-A), 'that there is no feature' becomes a respectable feature; we fall into an impasse that the random number sequence is a sequence with a characteristic that it has no characteristics.

von Mises (1883-1953) wanted to systematize probability theory based on randomness, but it was difficult because formalizing lawlessness was difficult. However, if we could positively characterize the feature that there is no feature, in other words, if we can define 'being without features' by an explicitly specifiable property, then 'there is no characteristic feature' is no more the negation of 'there are characteristic features.' This is the same spirit as we discussed in Chapter 1: if 'nonlinearity' is characterized by 'scale interference,' then it is not a mere negation of linearity. However, the characteristics such as 'lawlessness' or 'featurelessness" are ambiguous, allowing various interpretations.

If we can find a feature (regularity) in something (say, a number sequence), we could save the phone charge by exploiting the regularity when we wish to send it to the second person. For example, although it is a trivial example, if we wish to send $10101010\cdots10101010$ that has one million 10's, it is far better to send the message, "repeat 10 1,000,000 times," than to send the raw sequence itself. The number of digits required to describe a number N itself is asymptotically proportional to $\log N$, so such a regular sequence may be sent with the cost proportional to the logarithm of the original message length.[140] This is, of course, far more money-saving than the raw message.

Let us consider another example:

0273900749729736354964533288869844061196496162773449518273695588220757
3551766515898551909866653935494810688732068599075407923424023009259007
0173196036225475647894064754834664776041146323390565134330684495397907

Press, 2008), 15,000 digits were selected randomly, and then they were randomly arranged. However, the digit 6 appeared slightly more frequently than others, so 50 of them were randomly selected and replaced with other digits randomly.

[140] A student suggested that to send the number $\log N$, we could take its logarithm to compress it further and save money. Is it a good idea? If not, what is wrong?

$$\begin{aligned}&0903023460461470961696886885014083470405460742958699138296682468185710\\&3188790652870366508324319744047718556789348230894310682870272280973624\\&8093996270607472645539925399442808113736943388729406307926159599546262\\&4629707062594845569034711972996409089418059534393251236235508134949004\\&3642785271\end{aligned} \quad (2.58)$$

This sequence may look random, but it is the 500 digits starting from the 10,501st digit of π. If we wish to send this sequence, we can send a message, "the <u>500</u> digits starting from the 10,501st digit of π," and it is already shorter than the original message.[141] In this case as well, the message we must send is asymptotically proportional to the length of the part specifying the length of the sequence (the underlined part).

If we can compress a message, it is obviously non-random. Therefore, can we characterize the randomness of a message by the fact that it cannot be compressed (made shorter for communication) however we may try? To compress a given sequence of symbols we must use its regularity and meaning, so whether we can recognize them or not is the key issue of the randomness of the sequence.[142] However, by whom should the regularities be recognized?

The basic idea of the *algorithmic randomness* due to Solomonov, Kolmogorov and Chaitin is that this recognition should be done by the most powerful computer. Then, basic questions arise such as 'What is the most powerful computer?" and, in the first place, "What is computation?"

A computer is a machine to perform computation. For this statement to make sense, we must know what 'computation' is. Computation is to process a number into another number. The process of computation is not haphazard, but is understood to obey certain rules strictly. Therefore, we may say intuitively that computation is to transform one finite number sequence into another, using finitely many definite procedures. We will go into another deep conceptual analysis example.

---

[141] However, the recipient of the message must perform a considerable procedure to obtain the actual message. Compressed information is often costly to expand. An interesting topic related to this is D. Bailey, P. Borwein and S. Plouffe, "On the rapid computation of various polylogarithmic constants," Math. Computat. **66**, 903 (1997) (`http://www.cecm.sfu.ca/ pborwein/`) and V. Adamchik and S. Wagon, "π: a 2000-year search changes direction," Mathematica in Education and Research, **5**(1), 11 (1996), D. H. Bailey, J. M. Borwein, P. B. Borwein, and S. Plouffe, "The quest for Pi," Math. Intelligencer **19**(1), 50 (1997).

[142] However, we do not discuss the vague concept called 'meaning.'

2.9 What is computation?

This section outlines the proposal by Church (1903-1995) about 'computation.' If we use the expression everyone understands by now, his proposal is essentially as follows. A function whose program can be accepted by a digital computer is a 'partial recursive function,' and if the computation specified by the program is guaranteed to be completed within a finite time, the function is a 'recursive function' = computable function. Computation is a process to obtain the value of a computable function. If the reader feels this explanation is good enough, *she can jump from here to the next section.*

We consider only computation of a number with finitely many digits without any roundoff errors. Consequently, we can understand computation as a kind of map from a non-negative integer to another such number. A map that maps non-negative integers $\boldsymbol{N}$ into itself is called a *number-theoretic function.* We consider only the computation of number-theoretic functions. Church's idea for characterization (definition) of computation is as follows. First, take a few functions that everyone intuitively accepts to be obviously computable as the starting point. The totality of computable functions is constructed from these starting functions with a finite number of applications of the procedures that everyone agrees to be executable. To obtain values of the computable functions is called computation. Therefore, let us consider the characterization of 'computable functions.'[143] If the reader thinks this outline of the strategy is enough, *she can jump from here to the last 2 or 3 paragraphs of this section.*

In the following, in contrast to the ordinary analysis, when we speak of a function, it need not be a map but can be a partial function. That is, $f(x_1, \cdots, x_n)$, where $x_1, \cdots, x_n$ are non-negative integers (i.e., $x_i \in \boldsymbol{N}$), need not be meaningful (need not be defined) for all the n-tuples $\{x_1, \cdots, x_n\}$ of non-negative integers (that is, the domain is not specified beforehand). For those tuples for which f is not defined, f is not evaluated. If f is defined on the totality of $\{x_1, \cdots, x_n\}$, it is called a *total function.*

As the functions to start with, which everyone must agree to be computable, the following three functions S, P and C are adopted:
(A) $S(x) = x + 1$,
(B) $P_i^n(x_1, \cdots, x_n) = x_i$,
(C) $C_m^n(x_1, \cdots, x_n) = m$.
S is a function to give the successor of x in $\boldsymbol{N}$. P_i^n is a 'projection operator' to read the i-th variable out of n variables. C_m^n is a constant function assigning a constant m to all the n-tuples $\{x_1, \cdots, x_n\}$.

As unambiguous 'procedures' (basic operations) that everyone should

---

143 A classic introduction to the topic is M. Davis, *Computability and Unsolvability* (Dover, 1982). Newer textbooks include D. S. Bridges, *Computability, a mathematical sketchbook* (Springer, Graduate Texts in Mathematics 146, 1994), for example.

agree to be applicable to any function let us accept the following I–III:
I *Composition*: From functions $g_1, \cdots, g_m$ and h we can make another function

$$f(x_1, \cdots, x_n) = h(g_1(x_1, \cdots, x_n), \cdots, g_m(x_1, \cdots, x_n)), \tag{2.59}$$

where h is an m-variable function and g_i $(i = 1, \cdots, m)$ are n-variable functions.
II *(Primitive) recursion*: Starting with $f(x_1, \cdots, x_n, 0) = g(x_1, \cdots, x_n)$, we can construct $f(x_1, \cdots, x_n, m)$ recursively as follows:

$$f(x_1, \cdots, x_n, m) = h(x_1, \cdots, x_n, m-1, f(x_1, \cdots, x_n, m-1)), \tag{2.60}$$

where g and h are, respectively, n- and $(n+2)$-variable functions.
III *Minimalization* (or minimization) or *unbounded search*: Let $f(x_1, \cdots, x_n)$ be a total function. For each $\{x_1, \cdots, x_{n-1}\}$ we can determine the smallest x_n satisfying $f(x_1, \cdots, x_n) = 0$.[144]

A function that can be constructed from the basic functions (A)–(C) with a finite number of applications of the basic procedures I–III is called a *partial recursive function.* There is no problem with I being a computable procedure. II is the same. Perhaps it may be tedious, but applying finitely many steps patiently step by step can complete the procedure.[145]

However, Procedure III (minimalization) is tricky. Since f is a total function, for any $\{x_1, \cdots, x_{n-1}\}$ we can certainly evaluate $f(x_1, \cdots, x_{n-1}, m)$ for any m with a finitely many steps. Therefore, fixing $\{x_1, \cdots, x_{n-1}\}$, and putting m starting with 0 in the ascending order into f one by one, we can check whether $f(x_1, \cdots, x_{n-1}, m)$ is zero or not. If f becomes zero for the first time with $m = q$, then we define $h(x_1, \cdots, x_{n-1}) = q$. However, the existence of such a non-negative integer q is not known beforehand (in other words, we do not know beforehand whether $h(x_1, \cdots, x_{n-1})$ is a total function or not). Therefore, we cannot know beforehand whether the minimalization process even ends or not. Indeed, there is a way to check whether a given m is an answer or not however large it may be, but no one knows the upper bound of m such that if there is no answer up to the value there is really no answer.

In summary, a partial recursive function is a number-theoretical function which construction procedure can be described unambiguously (i.e., its algorithm is given). However, whether it can be actually computed (constructed) cannot be known beforehand (due to minimalization). Informally (but actually in a not very inaccurate way), a function whose procedure to compute can be programmed on the usual digital computer is a partial recursive function. There is no guarantee that the program actually completes the computation and produces its value for all the inputs. Suppose we begin evaluating a par-

---

144 Here, it is crucial that f is a total function. Each step of the algorithm must end within a finite number of steps, so f must be total.

145 Functions that can be constructed only with the aid of these two procedures are called *primitive recursive functions.*

tial recursive function for variable x. If we have not obtained the value after some computation, this may imply that the function is not defined for this x (because the minimalization step does not have a solution) or it is defined but we must be much more patient. We hesitate to declare such a function computable.

The functions we can really compute must be such that not only its each computational step is explicitly and unambiguously specified, but also the whole computation is guaranteed to be completed with a finite number of steps. Such functions are called *recursive functions*. That is, total partial recursive functions are called recursive functions. A recursive function is a function with an algorithm that is guaranteed to be completed with a finite number of steps for any (admissible) inputs. Continuing the above informal expression, we can say that a recursive function is a function which can be programmed on the usual digital computer, and the program produces a number (with sufficient but finite computational time and memory) for any (admissible) input.

Church proposed that recursive functions are *computable* (*Church's thesis*). That is, Church proposed to identify computable functions with recursive functions.[146] Impeccably unambiguous computation is possible only when the computational procedures are given purely syntactically (that is, given just as symbol sequences that do not require any interpretation). This is a sort of ultimate reductionism.

The crucial points of this proposal are that the algorithm is explicitly given and that the whole process is completed with a finite number of steps. When Church's thesis was proposed, it was not immediately and generally accepted that being a recursive function is a convincing characterization of any computable function. The reason was that there was no clear feel for constructive procedures that may be explicitly written down; are there not non-recursive (that is, not I-III above) completely new types of algorithms with which a different class of functions may become computable? Isn't the above proposal under the restriction of the era (i.e., the level of mathematics of the day)? Furthermore, since such an intuitively appealing concept as continuity requires, to be defined clearly, the axioms of the topological space, it is possible that apparently intuitively obvious basic procedures and basic functions may not be logically simple.[147]

---

[146] The definition here is consistent with M. Davis, *Computability and Unsolvability* (Dover, 1982). However, names and definitions are different in different books. In M. Li and P. Vitànyi, *An Introduction to Kolmogorov Complexity and Its Applications* (Springer, 1993) and J. E. Hopcroft and J. D. Ullman, *Introduction to Automata Theory, Languages and Computation* (Addison Wesley, 1979) Church's thesis is the proposal that partial recursive functions are computable. Bridges calls them computable partial functions. Davis calls partial recursive functions partially computable functions.

[147] See R. Gandy, "The Confluence of Ideas in 1936," in *The Universal Turing Machine, a half-century survey* (Oxford University Press, 1988). Neither Gödel nor Post

Subsequently, various definitions of computability were proposed, but interestingly all the definitions were equivalent to Church's thesis. That is, it gained a certain signature of naturalness. Usually, this is the description of the relevant history, but actually, one of the reasons that Church confidently proposed his thesis was that various definitions were equivalent.[148] Still, as mentioned above, the oppositions could not be quenched. However, the characterization of computability in terms of the Turing machine explained in the next section (roughly speaking, a digital computer with indefinitely large memory capacity) silenced all the oppositions to Church's proposal. Church himself wrote that Turing computability did not require any preparation to relate constructibility and the existence of effective procedures in the ordinary sense. In short, for basic concepts clear consistency with our intuition is crucial, so the best characterization of basic concepts is often supplied by explicitly visualizable machinery.

2.10 Turing machine

We have already mentioned digital computers to explain Church's idea intuitively. A digital computer formalized as a logical tool is the Turing machine (1936). Historically, this is earlier than any digital computers. It requires only a sheet of paper and a pencil, and can do everything digital computers can do, so some say that it is an ultimate personal computer. We may regard the Turing machine as an ultimate abstraction of digital computers. Computability by the Turing machine and computability in Church's sense are equivalent. This assertion should not go beyond our intuition, so the reader who regards the assertion quite natural *may proceed to the next section.* However, she must know that a programmable Turing machine is called a *universal Turing machine.*

A *Turing machine* (TM) consists of an infinite tape, a read/write head, and a black box with a finite number of internal states (Fig. 2.25). The tape is divided into cells. The read/write head can scan only one cell on the tape at a time.[149]

---

accepted the proposal. [1936: remilitarization of the Rhineland, Spanish Civil War started, X'ian Incident; J. M. Keynes, *The general theory of employment, interest, and money*].

[148] The situation is explained in detail in W. Sieg, "Step by recursive step: Church's analysis of effective calculability," Bull. Symbolic Logic **3**, 154 (1997).

[149] There are many different versions, e.g., with many tapes or a tape that is only infinite in one direction, but the reader has only to be able to have a general notion of a TM. See J. H. Hopcroft and J. D. Ullman, *Introduction to Automata Theory, Languages, and Computation* (Addison Wesley, 1979), for example.

C. Petzold, *The annotated Turing* (Wiley, 2008) is an amazing book annotating Turing's original paper.

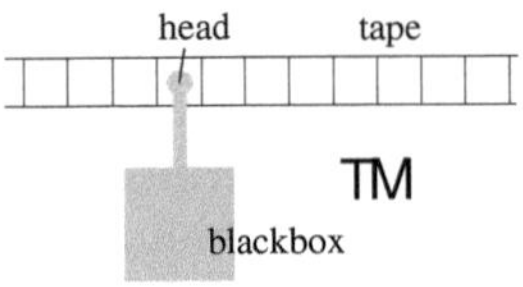

Fig. 2.25 A Turing machine consists of an infinite tape that is divided into cells, a read/write head that can scan a single cell at a time, and a black box with a finite number of the internal states.

Definition 2.3. A *Turing program* is a finite set of four tuples $(q, S, *, q')$, where
(1) q and q' are the internal states of the black box,
(2) S is the symbol in the cell being scanned now (it may be 1 or blank B, or 0, 1, B, etc., depending on authors, but in any case there are only finitely many of them),
(3) $*$ denotes R, L or S': R (respectively, L) implies that the head moves to the right (respectively, to the left) in the next time step (or it may be better to say the tape is moved in the opposite direction); S' implies that the head does not move but rewrites the symbol in the cell being scanned as $S \to S'$.
The implication of $(q, S, *, q')$ is as follows:
If the internal state of the black box is q and if S is written in the tape cell being scanned by the head, then the head performs $*$, and the internal state of the black box is changed to q'. □

A *Turing machine* may be identified with its Turing program.[150] That is, the machine is understood as a single-task machine. The input non-negative integer x is written on the tape according to a certain rule as, e.g., a 01 sequence (there are many different schemes, but here no details are important). The Turing machine (its black box) is initially in the *initial state* q_I, and the head is located at the leftmost non-blank cell on the tape (the portion where numbers are written is bounded). The machine has a special final state called the *halting state* q_H. The sequence written on the tape when the machine state reaches q_H is interpreted as the computational result y of the Turing machine. Thus, the Turing machine defines a function $x \to y$. If the Turing machine halts for any input, it defines a total function. However, a Turing machine may not define a total function, because it may not halt (q_H may never be reached) for particular inputs.[151] A function defined by a Tur-

[150] To identify the program and the machine is not as simple as we think. The program lives in the world of symbols, which is distinct from the real world, so how to specify the correspondence between the symbols and the actual movements of the parts of the actual machine is nontrivial. We have already mentioned the concept called 'adaptors.' This will be discussed in Chapter 4.

[151] Will the Turing machine ever halt for input x? This is the famous *halting problem*. There is no algorithm to answer this question. This is a typical *decision problem* that cannot be decided. See a footnote in Appendix 2.14A.

ing machine that halts for any input is called a *Turing computable function.* The fundamental theorem is:

Theorem 2.8 (Turing). *Turing computable functions are computable functions in Church's sense and vice versa.* □

This theorem is proved through demonstrating that the three basic functions S, P and C and the fundamental operations I-III can be Turing-programmable. For example, Davis' textbook explains detailed construction of the programs. Reading the programs to check their functions (to check that indeed they work) is a matter of patience; to write such a program is something like writing a program in a machine language (less efficient, actually), so for the ordinary scientists checking the demonstration is probably useless. Therefore, no further discussion will be given on the equivalence demonstration. Now that digital computers are everywhere, the theorem should be almost obvious.

Turing chose the restrictions built in the basic mechanism/function of the Turing machine through taking account of the limits of our sensory organs and intelligence. Our sensory organs can distinguish only finitely many distinct signals, and the number of distinguishable states of our brain is also finite. The number of kinds and quantities of tasks our brain and effectors can perform is also finite. These conclusions may be reasonable, because even if the signals and states are continuous, they are meaningful only after being 'quantized' in the actual noisy world, and because our brain is a finite object. In short, Turing conceived our brain as a finite automaton[152] = an automaton that relies on finitely many symbols, finitely many rules and operations, and finitely many internal states. Thus, he required an 'artificial brain' as follows:[153]
(i) Only finitely many kinds of symbols can appear on each cell (on the tape).
(ii) It can survey only finitely many cells at once.
(iii) At each instance, it can rewrite only one cell.
(iv) Only a finite range of the whole tape can be used (scanned).
(v) There are only finitely many states of the black box, and there are only finitely many instructions that can be performed.
The only idealization, from Turing's point of view, is that there is no memory capacity limit. However, since computers and our brains can indefinitely

---

[152] Do not confuse this with the so-called finite state automaton (or finite state machine) in computer science.

[153] Whether we may assume this or not is a crucial question. See Section 5.3. Incidentally, Thomas Huxley, Darwin's bulldog, gave in the Belfast meeting of British Association for the Advancement of Science in 1874 a talk entitled "On the hypothesis that animals are automata, and its history," (The Huxley File: `http://aleph0.clarku.edu/huxley/guide8.html`): "we are conscious automata," "brutes share consciousness with humans," and "all states of consciousness in us, as in them, are immediately caused by molecular changes of brain substance."

increase their external memory capacity, it is a benign idealization.[154]

The equivalence of Church's computability and Turing's computability convinced many people of the naturalness of Church's thesis. However, this does not mean the acceptance of Turing's basic idea, which seems to have motivated the formulation of Turing machines, that we are finite automata. Gödel (1906-1978) and Post (1897-1954) always believed that our mathematical intelligence was not mechanical. Especially, Gödel argued that our capability to manipulate abstract concepts was not restricted by the finiteness that Turing respected literally; the restrictions apply only when we manipulate (potentially) concrete objects such as symbol sequences; we had to take into account non-finitary creative thoughts to understand our mathematical capabilities.[155]

A single Turing machine computes a single partial recursive function. That is, it is a single-task machine. Notice that a Turing machine can be encoded as a number sequence (i.e., we can make an integer that can be deciphered to give a Turing program[156]— the Gödel number of the Turing program). Now, it is possible to make a machine (= compiler) that spits the four tuples of the Turing program upon reading the corresponding Gödel number. Then, if we subsequently feed the ordinary input to this master Turing machine, it can perform any computation that can be done by any Turing machine (it is a by-now common programmable digital computer). A formal proof of this statement may not be simple, but we are not at all surprised thanks to our daily experience. This machine is called a *universal Turing machine.* We may understand it as an ideal digital computer without any memory capacity limit. Turing demonstrated that everything a machine with constraints (i)-(v) above can perform can be done by a universal Turing machine. Thus, we may regard a universal Turing machine as the most powerful computer. If we accept Turing's analysis of our intelligence, any intelligent task we can do can be done by a universal Turing machine (with a suitable input).

A universal Turing machine is not a parallel computer. However, whether a machine is parallel or not is not fundamental. It is still a finite automaton.

---

[154] It is known that the computational capability of the neural network is equivalent to that of the universal Turing machine: W. S. McCulloch and W. A. Pitts, "A logical calculus of the ideas imminent in nervous activity," Bull. Math. Biophys. **5**, 115 (1943) [W. S. McCulloch, *Embodiments of Mind* (MIT Press, 1988)].

[155] ⟪**Natural intelligence and finiteness constraints**⟫ Perhaps, Gödel may have wished to say that our natural intelligence is not restricted by finiteness constraints. Our natural intelligence is 'embodied.' It is open to the external physical world through our body. It may be more sensible to idealize our brain as a nonfinitary system. Thus, after all, Gödel's intuition may well be correct. Indeed, abstract concepts are much more concretely supported materialistically (or, we could say, by molecular biology) than concrete concepts. That is, abstract concepts are directly connected to our body and are phylognetically ancient. Abstract concepts are more embodied than concrete concepts such as 01 sequences.

[156] q, L, etc., appearing in the Turing program is encoded into numbers.

Therefore, the universal Turing machine being not parallel is an unimportant restriction. In the theory of computation we totally ignore computational speed; what matters is whether a required computation can be performed within a finite time.[157]

With this fairly long preparation, we are now ready to formulate the idea unambiguously: let us call a number sequence that cannot be compressed by the most powerful computer a random number.

2.11 Characterizing randomness

Since there are many ways to formulate Turing machines, universal Turing machines cannot be unique. We wish to have the most powerful computer, so perhaps we have to look for the most powerful universal Turing machine. Actually, all the universal Turing machines are equally powerful, so we may choose any of them for our purpose:

Theorem 2.9 (Solomonov-Kolmogorov). *Let M and M' be two universal Turing machines, and $\ell_M(x)$ (resp., $\ell_{M'}(x)$) be the length of the shortest program for M (resp., M') to produce output x (measured in, say, bits). Then, we have*

$$\ell_M(x) \preceq \ell_{M'}(x), \ \ell_{M'}(x) \preceq \ell_M(x), \tag{2.61}$$

where $A(x) \preceq B(x)$ implies that there is a positive constant c independent of x such that $A(x) \le B(x) + c$; c may depend on A and B. □

The key to prove this theorem is that M can emulate M' and vice versa. For example, the program for M to emulate M' must be finite, however long it may be. Therefore, if we disregard the length of the program for this overhead (that is, if we write this length as c in the definition of $\preceq$), the length of the needed program does not change whether x is computed directly on M' or computed on the emulated M' on M. The meaning of $\preceq$ is just the inequality disregarding an additive constant (corresponding to the overhead), so the theorem should hold. Thus, from now on, when we wish to write the shortest program length $\ell_M(x)$, we will not explicitly specify the used universal Turing machine M and write simply $\ell(x)$.

Now, we can unambiguously discuss the asymptotic randomness of a long number sequence with the aid of an arbitrary universal Turing machine. Let $\omega[n]$ be the first n digits of a binary sequence ω. How does $\ell(\omega[n])$ behave generally? For example, for the uninteresting $1111\cdots$, asymptotically the information required specifying the number of digits n dominates the program, so $\ell(11\cdots[n])$ behaves as $\log n$. For a number sequence ω with an obvious

[157] Quantum computers may drastically change the required time, but even they cannot compute Turing noncomputable functions. Thus, when we ask the fundamental question what computers can ever do, quantum computers need not be considered.

regularity the shortest program to specify $\omega[n]$ is independent of n except for the portion needed to specify n itself. There are only countably many such regular sequences. On the other hand, in the $n \to \infty$ limit if $\ell(\omega[n])/\log n$ can be indefinitely large, the pattern in the sequence cannot be specified asymptotically by a finite length program, so it should not be simple. However, such sequences include many subtle sequences.[158] If no regularity is discernible at all in ω, to specify $\omega[n]$ requires specifying almost all the n digits, so we expect $\ell(\omega[n]) \sim n$.[159] In other words, a typical random sequence must be such ω that for infinitely many n $\ell(\omega[n]) \sim n$. Let us call such sequences (typically) *algorithmically random sequences*. For them, randomness can be quantified as follows:

Definition 2.4.

$$K(\omega) \equiv \limsup_{n\to\infty} \ell(\omega[n])/n \tag{2.62}$$

is called the *randomness* of a binary sequence $\omega \in \{0,1\}^{\boldsymbol{N}}$. Here, $\omega[n]$ is the first n digits of ω and $\ell(\omega[n])$ is the length of the shortest program (written in 01) to produce $\omega[n]$.[160]

Thanks to Kolmogorov and Solomonov the randomness does not depend on the choice of the universal Turing machine M (so it is not written already).

Kolmogorov used the word 'complexity' to describe the above quantity. In this book this word should be reserved for genuine complex systems, so we use the word 'randomness' instead.[161]

Notice that no program can be written to compute $K(\omega)$ (there is no algorithm for it). This should be easily inferred from the appearance of the expression as 'the shortest program' in the definition. For an arbitrarily chosen number sequence, except for almost trivial cases, it is hard to show that it is not a random sequence, because we must write down a short program to produce it. Worse still, it is harder to claim that the given sequence is random, because we must show that however hard one tries one cannot write a short program. Therefore, for a given number sequence, generally we cannot compute its randomness K (except for the case with $K = 0$, it is very hard, if not impossible.)

However, we can say something meaningful about a set of number se-

---

158 This situation is exactly the same as the entropy zero dynamical systems.

159 $\sim n$ implies $= O[n]$.

160 The reason we must use lim sup is, for example, we cannot ignore the appearance of meaningful sequences occasionally. That is why it was said $\ell(\omega[n]) \sim n$ for infinitely many n instead of all n.

161 M. Li and P. Vitányi, *An Introduction to Kolmogorov Complexity and Its Applications* (Springer, 1993) is the standard textbook of the Kolmogorov complexity. The reader will realize that there are many kinds of definitions and concepts, but here, to be simple, the most basic definition is used. The explanation in this section roughly follows A. K. Zvonkin and L. A. Levine, "The complexity of finite objects and the development of the concepts of information and randomness by means of the theory of algorithms," Russ. Math. Surveys **25**(6), 83 (1970).

quences. For example, for almost all (with respect to the Lebesgue measure) numbers in $[0, 1]$ their binary expansion sequences are algorithmically random. All the algebraic numbers[162] are not random. In the next section we will discuss average randomness.[163]

2.12 Understanding the essence of chaos

This section is the 'grand finale' of the conceptual analysis of chaos.

In the preceding section, in order to answer the question, "what is randomness?" algorithmic randomness was defined. We cannot judge whether a particular sequence is random or not in general. Whether this is due to the defect in the definition or due to the nature of randomness itself is debatable, but, in any case, obviously nonrandom sequences are not random, and it is intuitively certain that the ensemble of number sequences whose average algorithmic randomness is positive contains many highly random sequences. Therefore, to study a chaotic system as an ensemble of histories (trajectories) algorithmic randomness must be a good tool.

Let us take a measure-theoretical dynamical system (T, μ, Γ), where T is an endomorphism of a compact phase space Γ with an invariant measure μ (Section 2.6). We assume μ is ergodic (Section 2.6). We have seen in Section 2.7 that its Kolmogorov-Sinai entropy $h_\mu(T)$ is a good quantifier of chaos, or of the extent of difficulty in predicting its future, because $h_\mu(T)$ measures the needed extra information for equi-precise description of the states. The extent of the prediction difficulty must be related to the complicated nature of the dynamics (or trajectories). Then, it is a natural guess that the algorithmic randomness $K(\omega)$ of the symbol sequence ω obtained by encoding the trajectories of a dynamical system using a generator and its Kolmogorov-Sinai entropy $h_\mu(T)$ must have an intimate relationship. In fact, Brudno's theorem asserts that they are identical. This is the ultimate justification of our characterization of chaos given in Section 2.3. If the reader is contented with the statement, *she may go to the next section.* The rest of this section explains why the statement is natural.

Discussion 2.8. We have extensively used symbolic dynamics (or shift dynamics) isomorphic or homomorphic to the original dynamical system to analyze it. Some people bitterly criticize this strategy, saying that this ap-

---

[162] numbers that can be zeros of integer coefficient polynomials.

[163] There is an attempt to make a computable measure of randomness. One approach is to use much less powerful computers than the universal Turing machine. However, such an approach appears to be fundamentally off the mark to characterize the concept of randomness, because 'randomness' may well be a transcendental concept ('almighty noise'). There can be a point of view that within our mathematics it is natural that we cannot tell whether a given sequence is random or not in general.

proach does not respect how 'random sequences' are actually produced. We could produce the same '01 sequence' from a black box containing a person with a coin. Therefore, if one observes only the coded results, one can never infer the content of the black box (even if it is driven by a tent map). Hence, characterizing the dynamical system in terms of the coded result is impossible. *A fortiori* characterizing chaos with the randomness of the trajectories is flawed.

How do you respond? □[164]

Let $\mathcal{B} = \{B_1, \cdots, B_k\}$ be a generator (see Appendix 2.7A) for (T, μ, Γ). We can code a trajectory starting from x at time 0 as ω in terms of k symbols with the rule $\omega_n = a$ if the trajectory goes through B_a at time n ($T^n x \in B_a$). Define the randomness of the trajectory starting from x as

$$K(x, T) \equiv \limsup_{n \to \infty} \frac{1}{n} \ell(\omega[n]), \tag{2.63}$$

where $\omega[n]$ is, as before, the first n symbols of ω, and $\ell(z)$ is the code length of the minimal program for z in terms of k symbols as defined in the preceding section (there, k was 2).

In terms of the notations and quantities introduced just above, Brudno's theorem adapted to the current context reads:

Theorem 2.10. *For μ-almost all $x \in \Gamma$*

$$K(x, T) = h_\mu(T)/\log k.$$

Here, h_μ is the Kolmogorov-Sinai entropy of (T, μ, Γ) defined (as usual) in terms of the natural logarithm, but it is divided by $\log k$, so the right-hand side gives the entropy defined in terms of the base k logarithm. □

What is claimed is, qualitatively, the equality between the amount of the extra information required for prediction of the state at time t in the future and the amount of information to describe the trajectory for the time span t. It is quite a natural assertion. To predict a state after t we need an initial condition that can supply information that requires a k-symbol sequence of length $t \times h_\mu(T)/\log k$. On the other hand, this information should be enough to specify a trajectory for t, so the trajectory for the time span t must require this much of code length to describe. Thus, Brudno's theorem is quite natural. In the following an exposition is given of the reason why the theorem holds, but *the reader who thinks it is intuitively already plausible can jump to the next section from here.*

---

[164] Since this is a discussion topic, no comment should be added, but note that we do not simply treat dynamical systems as black boxes. The correspondence between a shift and a dynamical system must be at least homomorphic. There cannot be any deterministic dynamical system homomorphic to the package of a person + a coin.

The information needed to describe a trajectory may be considered in terms of the symbol sequence after coding. Therefore, the core of Theorem 2.10 is the following fact about the shift dynamical system (isomorphic to the dynamical system under consideration):

Theorem 2.11. *Let (σ, Ω) be a certain shift dynamical system with k symbols, and μ its ergodic invariant measure. Then, for μ-almost all $\omega \in \Omega$*

$$K(\omega) = h_\mu(\sigma)/\log k,$$

where $h_\mu(\sigma)$ is the Kolmogorov-Sinai entropy of the measure-theoretical dynamical system (σ, μ, Ω), and $K(\omega)$ is the randomness defined in (2.62) (as in Theorem 2.10, it is defined for k-symbol sequences instead of binary sequences; that is why $\log k$ *appears).* □

To show Theorem 2.11, we first show that the set of ω satisfying

$$K(\omega) < h_\mu(\sigma)/\log k \tag{2.64}$$

is μ-measure zero.[165] The number of $\omega[n]$ satisfying $K(\omega) \sim \ell(\omega[n])/n \le s$ is no more than k^{ns} (in our context ω is the k-symbol sequence). On the other hand, according to the Shannon-McMillan-Breiman theorem (Theorem 2.5 in Appendix 2.7A), the measure of the cylinder set specified by $\omega[n]$ is estimated as $e^{-nh_\mu(\sigma)}$. Therefore, the measure of all ω satisfying (2.64) is bounded from above by $e^{n(s\log k - h_\mu(\sigma))}$. This exponent is negative ($s\log k < h_\mu(\sigma)$; do not forget that h_μ is defined with the natural logarithm), so the upper bound converges to zero in the large n limit. The possibility of (2.64) is almost surely ignored.

Next, we wish to show that μ-almost surely (= for μ-almost all ω)

$$K(\omega) \le h_\mu(\sigma)/\log k. \tag{2.65}$$

If this is demonstrated, then, since we just showed that the cases with $K(\omega) < h_\mu(\sigma)/\log k$ may be ignored almost surely, only the equality remains. Since we have only to estimate the upper limit of $K(\omega)$, let us estimate the upper limit of $\ell(\omega[n])$. $\omega[n]$ is decomposed as follows in terms of q m-symbol-sequences ω_i^m ($i = 1, \cdots, M$, where M is the total number of distinct m-symbol-sequences; $n = mq + r$, i.e., $q = [n/m]$ and r is the residue):

$$\omega[n] = \omega_0^r \omega_{i_1}^m \omega_{i_2}^m \cdots \omega_{i_q}^m. \tag{2.66}$$

Here, ω_i^m is the ith kind of m-symbol-sequence, which is assumed to appear s_i times. With this representation, $\omega[n]$ can be uniquely specified by r, m, $s_1, \cdots, s_M$, ω_0^r and the arrangement of q m-symbol sequences $\omega_{i_1}^m \omega_{i_2}^m \cdots \omega_{i_q}^m$. Therefore, the needed information to specify $\omega[n]$ is given by (or, the length of

[165] The measurability of the set must be demonstrated first, but such a routine is omitted.

the shortest required program with k symbols, or the information measured with the base k logarithm is given by)

$$\ell(\omega[n]) \le \ell(r) + R + \ell(m) + \ell(q) + \sum_{j=1}^{M} \ell(s_j) + h(\omega_{i_1}^m \omega_{i_2}^m \cdots \omega_{i_q}^m), \qquad (2.67)$$

where $h(\omega_{i_1}^m \omega_{i_2}^m \cdots \omega_{i_q}^m)$ is the information needed to specify the arrangement of q m-symbol sequences $\omega_{i_1}^m \omega_{i_2}^m \cdots \omega_{i_q}^m$ and R is the information required to specify ω_0^r, which is bounded by a constant independent of n. That is, except for the last term, all the terms are $o[n]$ and unrelated to the randomness. Hence,

$$K(\omega) \le \limsup_{n\to\infty} h(\omega_{i_1}^m \omega_{i_2}^m \cdots \omega_{i_q}^m)/n. \qquad (2.68)$$

$h(\omega_{i_1}^m \omega_{i_2}^m \cdots \omega_{i_q}^m)$ is bounded by the information (in terms of base k logarithm) carried by the possible sequences under the assumption that all such sequences appear with equal probability. Therefore, it cannot exceed the logarithm (base k) of the number of sequences that can appear as $\omega[n]$. That is, $K(\omega) \le \lim_{n\to\infty}[\log_k N(n)]/n$. $N(n)$ is equal to the number of non-empty cylinder sets of length n. According to the Shannon-McMillan-Breiman theorem, for the cylinder sets contributing to entropy $\mu(\omega[n])/e^{-nh_\mu(\sigma)}$ must not vanish in the $n \to \infty$ limit. Therefore, the number of cylinder sets we must count must be, since the sampling probability of each cylinder set is $e^{-nh_\mu(\sigma)}$, the order of its inverse: $N(n) \sim e^{nh_\mu(\sigma)}$. Thus, we can understand (2.65). As can be seen from the explanation here, Brudno's theorem is based on very crude estimates, so it is a natural theorem. Such a theorem should have been discovered by theoretical physicists without any help of mathematicians.

In this chapter we have tried to characterize "a deterministic dynamical system behaving unpredictably," that is, a chaotic dynamical system, as unambiguously as possible as an example of conceptual analysis. The final result is that it is a deterministic dynamical system whose trajectories are algorithmically random. Crudely put, chaos is deterministic randomness.

Is this a satisfactory outcome? That a trajectory is random is, with the quantification in terms of the Kolmogorov-Sinai entropy, invariant under the isomorphism of the dynamical systems. Isomorphism (Note 2.7) is a crude correspondence ignoring even the topology of the phase space, so the characterization we have pursued has nothing at all to do with how the correlation function decays or what the shape of the attractor is. Even the observability of chaos by computer experiments is not invariant under isomorphism. Chaos as random behavior of deterministic dynamical systems is a common phenomenon and far more basic than the exponential decay of the correlation function, or the invariant sets being fractal.

It is interesting that a deterministic dynamical system does not allow us to predict its future despite its determinacy, and under superficial understanding of determinacy and predictability this could cause enough surprise. The significance of chaos is in its capability to illustrate with a simple system that the world on the scale we can directly observe is far from closed within itself and can be strongly influenced by events occurring on disparate scales that cannot be observed directly. Borel pointed out long ago that the motion of molecules in gas becomes unpredictable even in the very near future due to the motion of a small mass on a far star (through the change of gravitational field). It is very interesting that chaos is already possible in a simple system (chaos is actually found in the theoretically possible simplest system as we have seen in Section 2.1), but this was already recognized by Poincaré (1854-1912) with the three body problem (especially, the restricted three body problem).[166]

Since we started this chapter with a simple realizable example, the question of whether there is actually chaos may sound strange. However, it is difficult to tell whether the actual apparently chaotic phenomenon is really chaos or not due to the existing noise.[167] For example, it is easy to make an example that apparently exhibits observable chaos, even though there is no observable chaos without noise. It is difficult to answer affirmatively the question whether there is really chaos without external noise in a system for which the existence of chaos is experimentally confirmed (this is in principle impossible[168]). Therefore, whether the concept of chaos is meaningful in natural science or not depends on whether it is useful as an ideal concept to understand the real world just as points and lines in elementary geometry. The relevance of chaos to the instabilities in some engineering systems or instabilities in numerical computation shows that it is a useful ideal concept.

For actual systems, what is important is its response to small perturbations. For chaos, its practically important aspect is almost exhausted by the exponential separation of nearby trajectories. Whether the system is deterministic or not is unimportant. What is practically important is that the phase space is bounded and the trajectory itinerates irregularly various 'key' points in the phase space that are crucial to the system behavior. However, in order to model a system that easily exhibits such trajectories with small

---

[166] J. Moser, *Stable and random motions in dynamical systems* (Princeton University Press, 1973) is an excellent introductory textbook.

[167] Here, 'noise' need not mean the effect of the unknown scale, but any unwanted external disturbance as usual.

[168] For example, for a one-dimensional map, indefinitely small modification of the map can change it to have a stable fixed point. Such a system behaves just as before the modification, if a small noise is added. M. Cencini, M. Falcioni, E. Olbrich, H. Kantz, and A. Vulpiani, "Chaos or noise: difficulties of a distinction," Phys. Rev. E **62**, 427 (2000) recommend a more practical attitude toward chaos.

external perturbations, use of chaotic systems is at least metaphorically effective.[169]

2.13 Is the characterization of randomness satisfactory?

The starting point of our analysis of chaos was that the coin-tossing process almost surely generates 01 sequences that is intuitively random. To make the concept 'random' mathematical, we identified it with the lack of any computable regularity. This identification does not seem to contradict our intuition. However, is it really true that all the regularities are all those detectable by computers? There can be the following fundamental question: why can we say that if computers cannot detect any pattern, Nature herself cannot, either? Turing's analysis in Section 2.10 need not be the analysis of the limitations of the capability of Nature. In this characterization don't we admit that the Turing machine or computation itself is a more fundamental concept than randomness? Is this consistent with our intuition?

Randomness is always fundamental in sampling in experiments. Equilibrium statistical mechanics is based on the principle that almost any randomly chosen mechanical elementary state reproduces equilibrium thermodynamics. If we consider randomness as a fundamental concept, as long as there are no other concepts that we can accept as more fundamental, we cannot define it. In the end, randomness would be formalized only as a primitive concept of an axiomatic system for randomness, just as points and lines in Euclidean geometry. As far as the author is aware, the most serious approach in this direction is due to van Lambalgen.[170] As we have seen, randomness is algorithmically characterized by incompressibility. We needed the theory of computation to define incompressibility unambiguously. The axiomatic approach may be roughly interpreted as an attempt to axiomatize the concept corresponding to 'incompressibility.'[171]

As we have seen in Chapter 1, randomness is really significant only in nonlinear systems. We have also seen that the standard axiomatic system of sets is a legitimate heir of Fourier analysis, so to speak. A wild and heretical

[169] See for a chaotic itinerancy view of the brain by I. Tsuda, "Hypotheses on the functional roles of chaotic transitory dynamics," Chaos **19**, 015113 (2009).

[170] M. van Lambalgen, "The axiomatization of randomness," J. Symbolic Logic **55**, 1143 (1990).

[171] This axiomatic system is not compatible with the standard axiomatic system of sets (ZFC). The author does not understand how serious this incompatibility is. See M. van Lambalgen, "Independence, randomness and the Axiom of Choice," J. Symbolic Logic **57**, 1274 (1992). See also "Logic: from foundations to applications, European logic colloquium" edited by W. Hodges, M. Hyland, and J. Truss (Clarendon Press, 1996) Chapter 12 "Independence structures in set theory."

guess is that even on the foundation of mathematics is a shadow of linear systems, so it might not be very suitable to study nonlinear systems and the concept of randomness.

2.14 How is 'complexity' understood?

Kolmogorov named his algorithmic randomness as 'complexity,' but this and our intuitively grasped concept of 'complexity' do not overlap well. We do not wish to call something simply random (irregular) 'complex' (See the beginning of Chapter 5).

No one describes an ordered state as a complex state, nor a totally random system. Naturally, there appeared people who thought a complex state was something between these two extreme states. As we have seen, chaos is closely related to randomness, so it is not complex. On the other hand, periodic motions are not complex, either. Therefore, doesn't the motion close to the transition point from periodic to chaotic motions deserve to be called complex? 'Edge of chaos' is the keyword of this idea.[172] Crutchfield has made efforts to quantify the complexity around the edge in terms of the complexity of the grammar of the symbol sequences generated by the dynamics.[173] Indeed, the generated grammars are most complex around the edge. If complex systems have a similar intermediate character between chaos and periodic behaviors, the system must have subtle space-time correlations. Of course, many attempts have appeared to distinguish complex systems from random systems with the aid of correlations.[174]

These approaches certainly capture some aspects expected for complex systems such as subtle order, flexible response to external stimuli, etc. However, there is a serious problem. As was virtually stated already, the problem is whether we can judge such a number sequence as (2.58) (in this case the answer was given) as a part of a random number sequence or as a special sequence with (deep) meaning. It is hard to expect that such blind analysis methods as studying correlation functions can tell us that a sequence is not random but has meaning and is constructed in an intricate manner.

A moment's reflection would tell us that whether a signal sequence is com-

---

[172] C. G. Langton, "Computation at the edge of chaos: phase transition and emergent computation". Physica D **42**, 12 (1990) first used this term (according to T. Ikegami).

[173] J. Crutchfield, "The calculi of emergence: computation, dynamics and induction," Physica D **75**, 11 (1994)

[174] Summaries may be found in P. Grassberger, "Problems in quantifying self-generated complexity," Helv. Phys. Acta **62**, 489 (1989) and R. Wackerbauer, A, Witt, H. Atmanspacher, J. Kurths, and H. Scheingraber, "A comparative classification of complexity measures," Chaos, Solitons & Fractals **4**, 133 (1994). As explained in Chapter 5 'self-generated complexity' is an oxymoron.

pressible or not has nothing to do with the complexity of the sequence. For example, take a literary masterpiece[175] and prepare a 01 sequence constructed by compressing the work (compression is not needed, though, for the present purpose). Then, replace the 2^nth digit of the binary expansion of π with the nth digit of the literary work. The result is an extremely compressible sequence, but its complexity must be identical to that of the masterpiece.[176] Therefore, all the considerations so far given about 'complexity' must have overlooked something quite essential. The category of 'complexity' must have been fundamentally misunderstood.

Where did we start our analysis of chaos? We started it with a typical example, and then its conceptual analysis was based on the coin-tossing process that everyone should accept as a typical random process. To analyze an apparently more difficult concept, 'complexity,' we have a much more compelling reason to follow this strategy; we should start with a typical example of complex systems. Then, it must be the most natural conclusion that we must scrutinize 'organisms.' The developmental process of an organism may not be a simple process, but it seems to occur with ease without any new input. However, the developmental process itself and its outcome are not easy to understand. Thus, we come up with the following intuitive idea:[177]

(*) If a system is easier (for us) to 'make' than to 'understand,' then it is a complex system.

To make this slogan precise, we must specify unambiguously the words 'make' and 'understand,' but already at the intuitive level, the content is not empty. For example, (mimicking organisms) to make an artificial organism chemically must be easier than to understand why the system is alive. Or, we could make a conscious brain mimicking the developmental process, but to understand how the constructed brain has consciousness must be far more difficult. We could train a neural net to predict the tertiary structure of proteins from their primary sequences, but to understand how the neural net performs the task must be much harder.

To see that (*) above is not empty even logically, we could make a computation-theoretical caricature. Then, its fundamental defect, which may already be obvious to the reader, becomes clear as well.

Let us model 'to understand a set' by 'to be able to compute its indicator' in the sense of the theory of computation. If we cannot tell whether an element belongs to the set or not, we can hardly say that we understand the set, so this modeling is not very unnatural. Let us model 'to make a set' by

---

[175] In the Japanese counterpart of this book, *The Tale of Genji*, the world's first novel (11th century), is taken as a concrete example.

[176] The length of any masterpiece is finite, so some readers might say it is nonsense to use it to discuss asymptotic properties of long sequences. Here, we assume the length of the masterpiece is sufficiently long.

[177] According to T. Tsujishita, von Neumann had a similar idea.

'to have a Turing machine that produces the set' (here we consider only a subset of positive integers as sets). As is well known, there is a set that can be produced by a Turing machine, but its indicator is not computable (a recursively enumerable but not recursive set; henceforth, we abbreviate this as a RENR set) (See Appendix 2.14A). That is, there is a set that can be produced but is not understandable.

Note 2.9. Idealization and asymptotics
The asymmetry we considered above need not be qualitative as 'recursive' vs. 'recursively enumerable.' The asymmetry in the case of RENR sets is an absolute difference, but the asymmetry may be the difference in computational complexity. In reality, a gap that can never be overcome and a very wide gap are indistinguishable. Therefore, no absolute gap is required for (*).

However, one more remark is in order. The existence of a gap is clearly recognized only asymptotically. The concept of recursive set does not make sense for a finite set. We must recognize that clear understanding is possible only through idealization. To understand the world clearly is to idealize it. Without infinity clear distinction of many concepts become impossible. Examples include the distinction between rational and irrational numbers, the distinction between chaotic and periodic orbits, the distinction between ice and liquid water, etc. Thus, it should be recognized that the distinction is only possible with limiting or asymptotic concepts. It seems to be an empirical fact that the world becomes easier to understand if we use concepts that become clear asymptotically.[178] If the reader reads the next chapter, she will feel that this is related to the fact that we cannot be microscopic beings. The world we recognize is an asymptotic world from the microscopic point of view.

In reality, no observation can be infinitely precise, so everything can be treated as a rational number. Therefore, some people blame the idealization above as unrealistic and meaningless. Even more extremely, there are people who denounce introducing irrational numbers and infinity itself as a very evidence of not confronting the world without prejudice. However, it is likely that these people have never thought sufficiently seriously about the human activity called 'understanding.' We could perhaps say that a role of mathematics is to guarantee that idealization if properly performed does not lead to contradictions.

The developmental process may look spontaneous and proceeds smoothly, but to make a system that can develop spontaneously must be hard. A chick does not emerge from a beaten egg. It was said that a RENR set can be produced easily, but it is so, only when the needed Turing machine is already given. Thus, it is obvious that (*) captures only one small feature of complex systems. What was meant by 'to make' there is almost equivalent to 'to grow.' A complex system is generated smoothly if we utilize something already organized.

In short, an important feature of a complex system is that it is very hard to produce the organization (prerequisite for development). It has turned out that the above attempt to characterize 'complexity' captures only its mere

[178] The understanding need not be the really true understanding; we are made to feel that we understand things if we use asymptotic concepts. Our nervous system has been selected in this way.

relatively unimportant aspect. In the last chapter it will be stressed that complex systems are such systems that require a huge amount of prerequisite conditions and materials to produce. The reader is a complex system without doubt, because she has her parents.

Appendix 2.14A Recursively enumerable set and recursive set

We only consider number-theoretical functions.

Recursive set: A set whose indicator is computable (is a recursive function) is called a *recursive set*. If a set is recursive, there is an algorithm (that can be completed with finite steps) to judge whether a given non-negative integer belongs to it or not. In other words, a recursive set may be characterized irrespective of the method by which it is generated. In terms of the Turing machine, we may say that if a set is a recursive set, there is a Turing machine that produces 1 after a finite steps of computation, if the input belongs to the set, and 0, otherwise.

Recursively enumerable set: If a set is the range of a partial recursive function, it is called a *recursively enumerable set*. If a set A is recursively enumerable, there is a programmable function g such that $A = \{g(n)\}$ $(n \in \boldsymbol{N})$.[179] Notice that 'programmable' does not imply that there is any guarantee that computation is completed within a finite number of steps. If the program does not stop for a particular input n, we understand that $g(n)$ is not defined; g need not be a total function.

Theorem 2.12. *There is a set that is recursively enumerable but not recursive (a RENR set exists).*

This is a very important theorem.[180] It can be shown as follows. Let $\phi_x(y)$ be the output of the Turing machine whose Gödel number is x, when the input is y, where $x, y \in \boldsymbol{N}$. Construct the following set:

$$K \equiv \{\, x : \phi_x(x) \text{ is defined}\}. \tag{2.69}$$

K is a set of all the numbers x such that the Turing machine specified by the Gödel number x halts after a finite number of steps with input x. This set is recursively enumerable, since each step to compute $\phi_x(x)$ is precisely programmable; we can write a program to print x when the computation is completed with input x (although we do not care whether the computation actually halts after a finite number of steps or not). Since there is no guarantee for the computation to halt, in order to judge whether x is in K or not, we have no other means but to actually run the program. Therefore, it is intuitively obvious that K cannot be recursive.

To be more precise, the argument goes as follows. Define a function f:

---

179 It is known that this function can be chosen such that its output produces all the elements of A without repetition. See A. K. Zvonkin and L. A. Levine, "The complexity of finite objects and the development of the concepts of information and randomness by means of the theory of algorithms," Russ. Math. Surveys **25**(6), 83 (1970), Theorem 0.4.

180 For example, see M. B. Pour-El and J. I. Richards, *Computability in analysis and physics* (Springer, 1989).

$$f(x) \equiv \begin{cases} \phi_x(x)+1, & \text{if } \phi_x(x) \text{ is defined,} \\ 0, & \text{otherwise.} \end{cases} \tag{2.70}$$

That is,

$$f(x) \equiv (\phi_x(x)+1)\,\chi_K(x), \tag{2.71}$$

where χ_K is the indicator of K. If K is recursive, χ_K is computable, so there must be a Turing machine that computes f (f has an algorithm). That is, for any input z there must be X such that $f(z) = \phi_X(z)$. This must hold for any z, so $f(X) = \phi_X(X)$ must hold. Since $\phi_X(X)$ must be defined, $X \in K$, implying $\phi_X(X) = \phi_X(X) + 1$. A contradiction. Therefore, χ_K cannot be computable (this is an example of the famous *diagonal argument*).[181]

It is a good occasion to mention the following theorem:

Theorem 2.13. *A necessary and sufficient condition for a set Q to be recursive is that Q and its complement Q^c are both recursively enumerable.*

This can be shown as follows. Prepare the Turing machines T generating Q and T_c generating Q^c and run both simultaneously with an input x. At least one of them eventually halts, so if neither has halted yet, we must be patient. Thus, within a finite number of steps the membership of x is decided, so Q is recursive.

References

Adamchik V, Wagon, S (1996) π: a 2000-year search changes direction. Mathematica in Education and Research, 5(1):11-19

Adler RL, Weiss B (1970) Similarity of automorphisms of the torus. Memoir Am Math Soc 98

Alekseev VM (1968-9) Quasirandom dynamical systems I, II, III. Math. USSR Sbornik 5:73-128, 6:505-560, 7:1-43

Aref H, Balachandar S (1986) Chaotic advection in a Stokes flow. Phys Fluids 29: 3515-3521

Baba Y, Kubo I, Takahashi Y (1996) Li-Yorke's scrambled sets have measure 0. Nonlinear Analysis 26:1611-1613

Bailey D, Borwein P, Plouffe S (1997) On the rapid computation of various polylogarithmic constants. Math Computat 66:903-913

Bailey DH, Borwein JM, Borwein PB, Plouffe S (1997) The quest for pi. Math Intelligencer 19(1):50-56

Banks J, Brooks J, Cairns G, Davis G, Stacey P (1992) On Devaney's definition of chaos. Am Math Month 99:332-334

Bayley PJ, Frascino JC, Squire LR (2005) Robust habit learning in the absence of awareness and independent of the medial temporal lobe. Nature 436:550-553

Billingsley P (1960) Ergodic theory and information. Wiley

181 The demonstration above is almost the same as the proof of the nonexistence of the algorithm to decide whether a Turing machine halts or not for a given input. There are many undecidability problems that are close to ordinary mathematics. See, for example, the introductory part of D. T. Stallworth and F. W. Roush, "An undecidable property of definite integrals," Proc. Amer. Math. Soc. **125**, 2147 (1997). For Hilbert's 10th problem see B. Poonen, "Undecidability in number theory," Notices Am. Math. Soc. **55**, 344 (2008). A short movie *Julia Robinson and Hilbert's Tenth Problem* (A. K. Peters/CRC Press, 2008) is recommended.

Block L (1978) Mappings of the interval with finitely many periodic points have zero entropy. Proc Amer Math Soc 67:357-360
Block L (1978) Homoclinic points of mappings of the interval. Proc Amer Math Soc 72:576-580
Bowen R (1975) Equilibrium states and the ergodic theory of Anosov diffeomorphism. Lecture Notes in Math 470
Bridges DS (1994) Computability, a mathematical sketchbook (Graduate Texts in Mathematics 146). Springer
Brin M, Katok A (1983) On local entropy. Lecture Notes Math 1007:30-38
Brzeźniak Z, Zastawniak T (1998) Basic stochastic processes, a course through exercises. Springer
Cartwright ML, Littlewood JE (1945) On non-linear differential equations of the second order: I the equation $\ddot{y} + k(1 - y^2)\dot{y} + y = \lambda \cos(\lambda t + a)$, k large. J London Math Soc 20:180-189
Cencini M, Falcioni M, Olbrich E, Kantz E, Vulpiani A (2000) Chaos or noise: difficulties of a distinction. Phys Rev E 62:427-437
Chapter N, Tenenbaum JB, Yuille A (2006) Probabilistic models of cognition: Conceptual foundations. Trends Cognitive Sci 10:287-291
Cornfeld IP, Fomin SV, Sinai YaG (1982) Ergodic theory. Springer
Cover TM, King RC (1978) A converging gambling estimate of the entropy of English. IEEE Trans Inf Theor IT-24:413-421
Cover TM, Thomas JA (1991) Elements of information theory. Wiley
Cox RT (1946) Probability, frequency and reasonable expectation. Am J Phys 14:1-13
Crow JF (2001) Shannon's brief foray into genetics. Genetics 159: 915-917
Crutchfield J (1994) The calculi of emergence: computation, dynamics and induction. Physica D 75:11-54
Davis M (1982) Computability and unsolvability. Dover
Davis PCW, Brown JR (1986) The ghost in the atom. Cambridge University Press
Dawkins, R (2006) The God delusion. Houghton Mifflin
Durrett R (1991) Probability, theory and examples. Wadsworth & Brooks
Frege G (1971) On the foundations of geometry and formal theories of arithmetic (translated and edited by E H Kluge). Yale University Press, New Haven
Gandy R (1988) The confluence of ideas in 1936. In: Herken R (ed) The universal Turing machine, a half-century survey, Oxford University Press, Oxford
Gillis D (2000) Philosophical theories of probability. Routledge
Golomb SW et al (2002) Claude Elwood Shannon (1916-2002). Notices Amer Math Soc 49:8-16
Grassberger P (1989) Problems in quantifying self-generated complexity. Helv Phys Acta 62:489-511
Heisenberg W (1971) Physics and beyond (translated by Pomerans AJ). Harper & Row
Hellmuth T, Walther H, Zajoc A, Schleich W (1987) Delayed-choice experiments in quantum interference. Phys Rev A 35:2532-2541
Hintikka J (1996) The principles of mathematics revisited. Cambridge University Press
Hirsch MW (1984) The dynamical systems approach to differential equations. Bull Amer Math Soc 11:1-64
Hopcroft JE, Ullman JD (1979) Introduction to automata theory, languages and computation. Addison Wesley
Ito K, Oono Y, Yamazaki H, Hirakawa K (1980) Chaotic behavior in great

earthquakes—coupled relaxation oscillator model, billiard model and electronic circuit model—. J Phys Soc Jpn 49:43-52
Ito Sh, Tanaka S, Nakada H (1979) On unimodal linear transformations and chaos I, II. Tokyo J Math 2:221-239, 241-259
Kan I (1984) A chaotic function possessing a scrambled set with positive Lebesgue measure. Proc Amer Math Soc 92:45-49
Katok A, Hasselblat B (1996) Introduction to the modern theory of dynamical systems. Cambridge University Press
Kitagawa T (1958) Statistical inference (Suisoku-Tokei Gaku). Iwanami
Kline M (1982) Mathematics; the loss of certainty. Oxford University Press
Kolmogorov AN (1957) Foundations of the theory of probability (2nd edition). Chelsea
Kolmogorov AN (1958) New metric invariant of transitive dynamical systems and automorphisms of Lebesgue spaces. Dokl Acad Nauk SSSR 119:861-864
Krieger W (1970) On entropy and generators of measure-preserving transformations. Trans Amer Math Soc 149:453-464; (1972) corrections *ibid* 168: 519.
Kuhn T (1996) The Structure of Scientific Revolutions (3rd ed). University of Chicago Press
Langton CG (1990) Computation at the edge of chaos: phase transition and emergent computation. Physica D 42:12-37
Ledrappier F (1981) Some properties of absolutely continuous invariant measures on an interval. Ergodic Theor Dynam Syst 1:77-93
Levinson N (1949) A second order differential equation with singular solutions. Ann Math 50:127-153
Li M, Vitànyi P (1993) An introduction to Kolmogorov complexity and its applications. Springer
Li S-H (1992) Dynamical properties of the shift maps on the inverse limit space. Ergodic Theor Dynam Syst 12:95-108
Li S-H (1993) ω-chaos and topological entropy. Trans Amer Math Soc 339:243-249
Li T-Y, Yorke JA (1975) Period three implies chaos. Am Math Month 82:985-992
Lissauer JJ (1999) Chaotic motion in the solar system. Rev Mod Phys 71:835-845
Lorenz EN (1963) Deterministic nonperiodic flow. J Atmospheric Sci 20:130-141
Lorenz EN (1964) The problem of deducing the climate from the governing equation. Tellus 16:1-11
Matheson C, Kirchhoff E (1997) Chaos and literature. Philosophy and Literature 21:28-45
Matsumoto K, Tsuda I (1983) Noise-induced order. J Stat Mech 31:87-106
McCulloch WS, Pitts WA (1943) A logical calculus of the ideas imminent in nervous activity. Bull Math Biophys 5:115-133
McLaughlin JB, Martin PC (1975) Transition to turbulence in a statically stressed fluid system. Phys Rev A 12:186-203
McMurran S, Tattersall J (1999) Mary Cartwright (1900-1998). Notices Amer Math Soc 46:214-220
Meshalkin LD (1959) A case of Bernoulli scheme isomorphism. Dokl Acad Sci USSR 128(1):41-44
Milani A, Nobili AN (1992) An example of stable chaos in the Solar System. Nature 357:569-571
Miyataka O, Nakayama T (1960) Monte Carlo method. Nikkan-Kogyo
Moser J (1973) Stable and random motions in dynamical systems. Princeton

University Press
Nagashima H, Baba Y (1999) Introduction to chaos, physics and mathematics of chaotic phenomena. Institute of Physics Publishing
Nusse HE, Yorke JA (1996) Basin of attraction. Science 271:1376-1380
Oono Y (1978) Period $\neq 2^n$ implies chaos. Prog Theor Phys 59:1029-1030
Oono Y (1978) Kolmogorov-Sinai entropy as disorder parameter for chaos. Prog Theor Phys 60:1944-1945
Oono Y, Osikawa M (1980) Chaos in nonlinear difference equations I. Prog Theor Phys 64:54-67
Ornstein DS (1970) Bernoulli shifts with the same entropy are isomorphic. Adv Math 4: 337-352
Ornstein DS (1974) Ergodic theory, randomness, and dynamical systems. Yale University Press
Ornstein DS (1995) In what sense can a deterministic system be random? Chaos, Solitons & Fractals 5:139-141
Ornstein DS, Weiss B (1991) Statistical properties of chaotic systems. Bull Amer Math Soc 24:11-116
Osikawa M, Oono Y (1981) Chaos in C^0-diffeomorphism of interval. Publ RIMS 17:165-177
Parry, W (1969) Entropy and generators in ergodic theory. Benjamin
Pesin YaB (1977) Characteristic Lyapunov exponents and smooth ergodic theory. Russ Math Surveys 32(4):55-114
Petzold C (2008) The annotated Turing. Wiley
Poonen B (2008) Undecidability in number theory. Notices Amer Math Soc 55:344-350
Pour-El MB, Richards JI (1989) Computability in analysis and physics. Springer
Rohlin VA (1967) Lectures on the entropy theory of transformations with an invariant measure. Russ Math Surveys 22(5):1-22
Ruelle D (1978) An inequality for the entropy of differentiable maps. Bol Soc Brasil Mat 9:83-87
Ruelle D (1989) Elements of differentiable dynamics and bifurcation theory. Academic Press
Ruelle D (1993) Chance and chaos. Princeton University Press
Ruelle D, Takens F (1971) On the nature of turbulence. Comm Math Phys 20:167-192
Rukhin AL (2001) Testing randomness: a suite of statistical procedures. Theory Probab Appl 45:111-137
Shafer G, Vovk V (2001) Probability and finance: its only a game! Wiley-Interscience
Shermer M (2011) The believing brain. Times Books Henry Holt and Co.
Sieg W (1997) Step by recursive step: Church's analysis of effective calculability. Bull Symbolic Logic 3:154-180
Sinai YaG (2010) Chaos theory yesterday, today and tomorrow. J Stat Phys 138:2-7
Singer IM, Thorpe JA (1976) Lecture notes on elementary topology and geometry (Undergraduate Texts in Mathematics). Springer (original 1967)
Smale S (1967) Differentiable dynamical systems. Bull Amer Math Soc 73:747-817
Smital J (1984) A chaotic function with a scrambled set of positive Lebesgue measure. Proc Amer Math Soc 92:50-54
Stallworth DT, Roush FW (1997) An undecidable property of definite integrals. Proc Amer Math Soc 125:2147-2148
Sugiura S, Shimizu A (2012) Thermal pure quantum states at finite temperature. Phys Rev Lett 108:240401

Tsuda I (2009) Hypotheses on the functional roles of chaotic transitory dynamics. Chaos 19:015113 (1-10)
Umeno K (1997) Method of constructing exactly solvable chaos. Phys Rev E 55:5280-5284
Valone TJ (2006) Are animals capable of Bayesian updating? An empirical review. Oikos 112:252-259
van Lambalgen, M (1990) "The axiomatization of randomness. J Symbolic Logic 55:1143-1167
van Lambalgen M (1992) Independence, randomness and the axiom of choice. J Symbolic Logic 57:1274-1304
van Lambalgen M (1996) Independence structures in set theory. Chapter 12 in Hodges W, Hyland M, Truss J (ed) Logic: from foundations to applications, European logic colloquium. Clarendon Press
Viana M (2000) What's new on Lorenz strange attractor. Math Intelligencer 22(3):6-19
Wackerbauer R, Witt A, Atmanspacher H, Kurths J, Scheingraber H (1994) A comparative classification of complexity measures. Chaos, Solitons & Fractals 4:133-173
Wagon S (1993) The Banach-Tarski paradox. Cambridge University Press
Walters P (1982) An introduction to ergodic theory. Springer
Weisstein EW Lorenz attractor. From MathWorld—A Wolfram Web Resource, http://mathworld.wolfram.com/LorenzAttractor.html
Williams RF (1998) The universal templates of Ghrist. Bull Amer Math Soc 35:145-156
Wittgenstein L (2001) Philosophical investigations. Blackwell (original 1953)
Zvonkin AK, Levine LA (1970) The complexity of finite objects and the development of the concepts of information and randomness by means of the theory of algorithms. Russ Math Surveys 25(6):83-124

Chapter 3
Phenomenology

—Renormalization and Asymptotic Analysis—

As chaos simply shows us, due to scale interference the phenomena on the human scale and those on the disparate scales cannot generally be decoupled; the world we directly observe is not self-contained. Then, the 'nonlinear world' we observe could be extremely disordered. And, indeed, there seem to be some who claim that this is just the case: "many phenomena are unpredictable and reason is not very important. This recognition is causing the collapse of secular rationalist humanism."[1]

Isn't it an empirical fact, however, that the world is not so lawless as can be expected from the studies of chaos and randomness? Isn't our existence as intellectual creatures evidence of the lawfulness to some extent at least of the world?[2] Intelligence is employed to simulate the near future based on the information available up to the present. If the world is totally lawless and unpredictable, to have a large brain would not only be a waste of resources but could be a liability, attracting parasites and pathogens.[3]

---

[1] See, e.g., "Chaos theory seeps into ecology debate, stirring up a tempest," by D. Farney in The Wall Street Journal vol. LXXV, no. 188 (July 11, 1997).

[2] ⟪**Even vague regularity makes a difference**⟫ It is said that a flap of a butterfly's wing could cause a hurricane later somewhere, but the wind pattern in the tropics is regulated by the surface temperature of the ocean, so it cannot depend sensitively on the initial condition. See J. Shukla, "Predictability in the midst of chaos: a scientific basis for climate forecasting," Science **282**, 728 (1998). Even if the weather may be chaotic, the average behavior such as climate is certainly predictable. See C. F. Keller, "Climate, modeling, and predictability," Physica D **133**, 296 (1999). Needless to say, we should not expect that really accurate quantitative prediction becomes possible, but even a vague regularity makes intellectual capability meaningful.

[3] ⟪**Positive devolution**⟫ Actually, there are examples in which devolution of unused organs/functions is a result of selection (i.e., decays can be not due to neutral fluctuations because of disuse): M. Protas, M. Conrad, J. B. Gross, C. Tabin, and R. Borowsky, "Regressive Evolution in the Mexican Cave Tetra, *Astyanax mexicanus*," Curr. Biol. **17**, 452 (2007). According to the study using bacteria, there are various modes of decay, depending on the selection pressure: A. R. Hall and N. Colegrave, "Decay of unused characters by selection and drift," J. Evol. Biol. **21**, 610 (2008).

Why do we feel that the world is to some extent lawful despite being full of unknowable noises? An important ingredient of the answer to this question seems that the world tolerates *phenomenological understanding* (or *phenomenology*, in short), which was briefly mentioned in Chapter 1. There are indeed extensive and diverse effects of noises from the small scales (more generally, from the unknowable scales) to the phenomena we experience on our scale. However, these effects do not pop up haphazardly, but show up rather systematically at restricted places. The phenomenological way to see the world is to see it exploiting this special feature of the world. Why is such a way of seeing the world effective? Because we live in such a world.[4] Why is the world in that way? This is a difficult question. However, it is likely that if the world is not like that, no intelligent organisms would have evolved.

The idea of 'renormalization' outlined in this chapter is a way to extract phenomenology (a way to understand the world), utilizing the fact that often the effects of unknowable scales show up only at restricted places. 'The unknowable' may be separated often in the asymptotic limit of, e.g., the observation scale being quite removed from the unknowable scales. Thus, a point of view also emerges that renormalization is a fairly general approach to asymptotic analysis. Furthermore, by interpreting renormalization as a general way of observing the world—to look at the world through structurally stable observables, the range of application of renormalization may be expanded. This is the reason why renormalization is explained.

The above point of view is quite different from the traditional point of view of the 'rationality' of the world. Isn't it the conventional point of view that the world is rational, because its fundamental laws are logical and reliable (because God presides over the universe or thanks to God's benevolence)? Even if the logical and fundamental laws may be solid, this does not mean much to support the rationality of the world we actually experience; this is the objection that scale interference due to nonlinearity implies. Indeed, the serious effect of scale interference illustrated by chaos is an antithesis to traditional garden-variety rationalism. The reason for the rational or lawful nature of the world is not because of the lawfulness of the microscopic world due to

---

[4] ⟪**Why phenomenological?**⟫ A crude analogy may be of some help. Suppose the details of the microscopic world show up only in colors of the (macro)objects. Colors change incessantly and bewilderingly, so it is not wise to pay attention to colors when we wish to do something in the macroscopic world. But suppose the shapes of the objects are stable. Then, a general cognitive capability of shapes would be highly significant for surviving in the macroscopic world. Consequently, we would be selected to pay attention predominantly to shapes, and, furthermore, we would evolve (we would be selected) to value shape-sensitive cognition. 'Shapes' are important not because we pay attention to them, but because they are highly relevant to our survival. Analogously, phenomenological structures are important not because we wish to see the world in a phenomenological way. We have been selected to see them, because this world abounds with such structures. It is a natural idea that phenomenology and the existence of slowly varying variables are closely connected.

'fundamental laws.' This is the point of view the renormalization theoretical worldview offers. Such a point of view is, however, not novel. There has been a consensus that the effects of the 'unknown scales' are random, so they must obey the law of large numbers, and consequently lawfulness prevails in the world. The triumph of information technology amply vindicates this view.

Let us begin with the question: what is phenomenology?

3.1 What is phenomenology?

We can say 'we understand something,' or 'we know something,' when we know common features of a set described by this 'something,' if this 'something' is a common noun. The assertion that we know 'dog' implies that we know something general about dogs instead of particular knowledge such as 'the dog in the next house is affable'. It also implies that we can tell that Chihuahua and Borzoi are both dogs and we can distinguish them from wolves.[5] When we understand something, some sort of generalization (or abstraction) is always involved.[6,7] This is the core of phenomenological understanding. To

---

[5] The following paper tells us that apparently trivial behavioral feature can have a profound consequence: A. Miklósi, E. Kubinyi, J. Topál, M. Gáacsi, Z. Virányi, and V. Csányi, "A simple reason for a big difference: wolves do not look back at humans, but dogs do," Curr. Biol. **13**, 763 (2003).

[6] We must think why we have common nouns.

[7] ⟨⟨**Priority of essential intuition**⟩⟩ It has long been argued that individual facts and their generalized essences are incommensurate, and that however many facts we may collect, no (mathematical) essence can be extracted from them. Therefore, it has been concluded that we must suppose 'essential intuition' (*Wesensschau*) or categorical cognitive capability prior to any experiences. From this line of thoughts comes naturally the idea that essential existence absolutely predominates actual existence as asserted by Plato and, subsequently, the idea disagreeing with the former by Aristotle. These ideas later led to the idea that, if we reflect on our cognitive capability without any prejudice, these two existences are inseparable; metaphysics is born when we forget this inseparability (e.g., Heidegger).

Aristotelian empiricism appears to respect facts, but lacks sufficient introspection. The point of view of Heidegger sounds like an opinion of a sage knowing all the possible positions, but it commits the folly of taking the midpoint of the view points with and without deep introspection. The following quotations that clearly demonstrate an understanding of the problem are from T. Izutsu, *Consciousness and Essence—quest of the spiritual Orient* (Iwanami paperback, 1991): "If we reflect on our own consciousness, we realize that our ordinary function of consciousness is most often built upon cognition of 'essence' of various things and phenomena. It is said that consciousness is intrinsically 'the consciousness of $\cdots$.' This intrinsic intentionality of consciousness is not realized without somehow grasping the 'essence' of "$\cdots$" (X) beforehand to which consciousness is directed out of itself."(p8) "The world we usually recognize is already the one dissected into the beings that are recognized as already pre-formed beings by omitting the primary and original process of recognizing their 'essences' —or without recognizing the recognition process. We exist as subjects in

understand the world phenomenologically is to dissect the world into a set of phenomena each of which can be understood by a particular phenomenological framework (although we need not consciously recognize this).

Let us try to make slightly clearer the word 'phenomenology' (or 'phenomenological'), which we have used so far without explanation. What is a phenomenological description (*phenomenology*) of a class of phenomena? Instead of pursuing its definition, let us observe typical examples.[8] If quantitative nature is demanded, unfortunately, good examples are restricted to physical or physical-chemical ones. Some examples will be quoted in the subsequent chapters.

Example 3.1. Not fast flow dynamics of fluids
Let us consider a not fast flow of an 'ordinary' fluid. It is governed by the Navier-Stokes equation:[9,10]

$$\rho \left[\frac{\partial \boldsymbol{v}}{\partial t} + (\boldsymbol{v} \cdot \nabla) \right] \boldsymbol{v} = \eta \Delta \boldsymbol{v} - \nabla P, \tag{3.1}$$

where P is the pressure, $\boldsymbol{v}$ is the flow velocity field with an incompressibility condition $\nabla \cdot \boldsymbol{v} = 0$, ρ is the density of the fluid, and η is the shear viscosity of the fluid. Here, ρ and η are the '*phenomenological parameters*' different for each system. A remarkable fact is that without changing the structure of the equation we can describe the slow flow of diverse fluids (e.g., air, wa-

---

such a world that is constructed by these beings, and recognize those surrounding us as objects. Then, our consciousness naturally takes the form of 'consciousness of $\cdots$' without almost any awareness of the 'essence' recognition already hidden in "$\cdots$"."(p13-14).

8 ⟪**Cognitive significance of examples**⟫ As the reader has already realized, repeatedly in this book, instead of pursuing general definitions, we study supposedly representative examples. As already stated in footnote 48 of Chapter 2, the reason for this lies in the author's total trust in the intrinsic cognitive capability of human beings. In other words, the author believes that the important structures of the world we live in are woven into our cognitive capability that is a product of evolution. This is quite natural, because without this it is hardly thinkable that we can recognize the world. Therefore, as to an important matter, even if it transcends the expressive capability of our language, it is natural to expect that our natural intelligence is equipped with the capability to recognize it directly (capability of essential intuition; see the preceding footnote). Showing a suitable array of examples must be a powerful means to stimulate natural intelligence (and its fundamental component, emotion) *without being obstructed* by language.

9 Whether this equation has a unique solution for not very slow flow is still unclear. Its mathematical status may be summarized in O. A. Ladyzhenskaya, "Sixth problem of the millennium: Navier-Stokes equations, existence and smoothness," Russ. Math. Surveys **58**, 251 (2003). This is one of the million-dollar prize problems: `http://www.claymath.org/prizeproblems/navierstokes.pdf`.

10 About Navier (1785-1836), see M. Cannone and S. Friedlander, "Navier: Blow-up and Collapse," Notices AMS **50**, 7 (2003). Navier wrote down the complete Navier-Stokes equation, but the meaning of η and its measurement method were due to Stokes (1819-1903).

ter, molasses) only by changing the phenomenological parameters ρ and η. Notice that the Navier-Stokes equation has (so far) never been derived from the atomistic picture of fluids (microscopic many-body classical mechanical model).[11] The phenomenological parameters in the equation are not mere adjustable or fitting parameters. The methods to measure them are unambiguous and if they are determined once for a particular fluid, the values cannot be further adjusted to explain various experiments on the fluid. □

Example 3.2. Critical phenomenon of binary mixture
Critical phenomena gave physicists great motivation to think about phenomenology consciously.[12] Take a mixture of liquid methanol and n-hexane. If the temperature is sufficiently high, these two substances mix well and make a single transparent liquid phase. However, if the temperature is low enough, these two liquids cannot mix well, and we observe two coexisting liquid phases (phases tend to separate; a schematic *phase diagram* is Fig. 3.1). As the system is cooled down, small domains with excess methanol (or n-hexane) become likely to form. That is, the restoring force against concentration fluctuations weakens. Still, if the composition of the mixture is arbitrary, this restoring force cannot be very small. Since the total amounts of both methanol and n-hexane are fixed (the numbers of molecules of the both chemical species are constant), to make a domain containing methanol more than its average concentration, a domain with excess n-hexane must be created around it. That is, to make concentration fluctuation, an excess component must be squeezed out from the average composition background. Consequently, if a system contains more methanol than n-hexane, it is not easy to squeeze out n-hexane further. This means that in such a solution concentration fluctuations are subjected to a considerable (thermodynamic) restoring force.

However, if the composition and temperature are just right (under a given pressure), the restoring force for the concentration fluctuation can be very weak. The point on the phase diagram around which this happens is called the *critical point.* Close to this point, since the restoring force for concentration fluctuation is weak (precisely speaking, the linear term vanishes; fluctuation becomes 'linearly neutral'), large fluctuation domains can be generated. The representative size of the domain where the correlation is significant is called

---

[11] The state of the art may be seen in H.-T. Yau, "Asymptotic solutions to dynamics of many-body systems and classical continuum equations," in *Current Developments in Mathematics, 1998* (International Press, 1998). Within the framework of linear irreversible thermodynamics (and, consequently, within Onsager's fluctuation theory), the equation is natural. See K. Miyazaki, K. Kitahara and D. Bedeaux, "Nonequilibrium thermodynamics of multicomponent systems," Physica A **230**, 600 (1996). There are numerous papers deriving the Navier-Stokes equation from the Boltzmann equation , but since the Boltzmann equation holds only for dilute gases, this approach is an insufficient approach to justify the Navier-Stokes equation from the microscopic point of view.

[12] H. E. Stanley, *Introduction to Phase Transition and Critical Phenomena* (Oxford UP, 1971) is still an excellent book summarizing the phenomenology.

the *correlation length*. Just at the critical point, this length diverges. Since

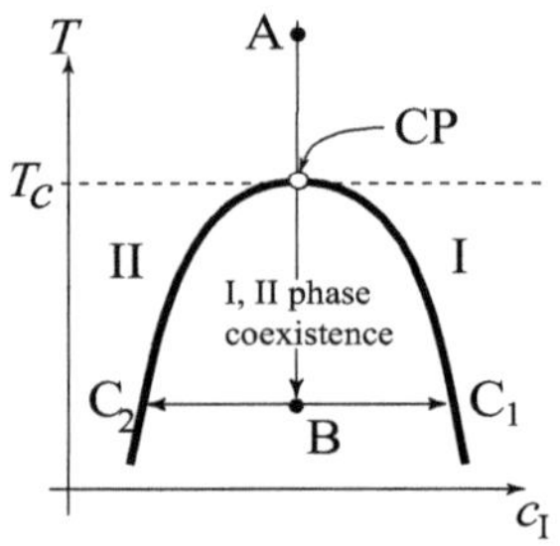

Fig. 3.1 A schematic phase diagram of a binary fluid. The vertical axis denotes the temperature T, and the horizontal axis the concentration c_I of component I. The point (c_I, T) on the diagram denotes the system at temperature T whose overall composition is given by c_I (the average concentration of I over the system, if the system is not uniform). The domain denoted as I (respectively, II) is a single-phase state with more component I than II (respectively, with more component II than I). At high temperatures components I and II mix well (single-phase state) irrespective of the overall composition. If the temperature is lowered (below T_c, the critical point), and if the system whose overall composition is below the parabola-like curve (called the phase coexistence curve), the single-phase state is no more in equilibrium; in equilibrium the system separates into two phases with different compositions. When phase separation occurs, the resultant coexisting phases are on the coexistence curve; the two phases with the concentrations on the coexistence curve at the same temperature coexist. In Example 3.4 we will consider the spinodal decomposition process in which the unstable uniform phase at B obtained by rapid temperature quench from a high temperature uniform state A decomposes into two phases, C_1 and C_2.

fluctuations cannot decay easily, correlation lasts long in time as well. That is, the space-time correlation of the concentration becomes very large near the critical point.

We may restate the above more quantitatively as follows. The spatial concentration fluctuation can be described by the field of '*order parameter*' $\psi(\boldsymbol{r})$. $\psi(\boldsymbol{r})$ for the present example is the quantity proportional to the local concentration deviation of (say) methanol from its average concentration, normalized appropriately to be of order unity (it is a mere number in this example). This quantity can describe the deviation of the composition from the well-stirred (randomly stirred) average state, so it is called the order parameter. The spatial *correlation function* of the order parameter is defined as

$$G(\boldsymbol{r}) \equiv \langle \psi(\boldsymbol{r})\psi(\boldsymbol{0}) \rangle, \tag{3.2}$$

where $\langle\ \rangle$ is the average over the equilibrium state. In the so-called disordered state at temperatures higher than the critical point, the correlation decays exponentially as[13]

$$G(\boldsymbol{r}) \sim \mathrm{e}^{-|\boldsymbol{r}|/\xi}. \tag{3.3}$$

[13] In the following equation $\sim$ implies the ratio of the logarithms of the both sides is globally asymptotically unity.

In this way, the *correlation length* ξ is introduced. The correlation function can be measured by (light) scattering experiments.

If the concentration is fixed to be the same as the critical point, it is experimentally established that the correlation length depends on T as

$$\xi \simeq \xi_0 |T - T_c|^{-\nu} \tag{3.4}$$

near the critical point, where T_c is the critical temperature and ξ_0 is a system-dependent constant (these two are phenomenological parameters). The exponent ν (> 0) (one of the examples of *critical indices*) is the same for any binary fluid. That is, the divergence has a universal form. At the critical point, the correlation length diverges, so the correlation function cannot decay exponentially. It is known that generally it is algebraic: $G(\boldsymbol{r}) \sim 1/|\boldsymbol{r}|^{d-2+\eta}$, where d is the spatial dimensionality, and η is another critical index. Critical indices are related to anomalous dimensions we will encounter later. □

Example 3.3. Polymer semidilute solution
A polymer is a long molecule formed by connecting (low molecular weight molecules called) monomers.[14] For example, polyethylene is a polymer linearly connecting the chemical repeating unit -(CH_2-CH_2)- (which is called the monomer unit; industrially, it is made of ethylene C_2H_4) successively as CH_3-(CH_2-CH_2)$_{N-1}$-CH_3. The number N is called the degree of polymerization. Here, we consider only flexible polymers. In contrast to low molecular weight solute solutions, for polymer solutions we can consider the *semidilute limit*.[15] Even if the solution is very dilute, since each polymer spreads spatially, polymers can have extensive overlaps among them. Therefore, we cannot simply ignore the polymer-polymer interaction even in the dilute limit; numerous degrees of freedom of polymer chains still entangle each other.

Let c be the number density of polymers (the number of centers of mass of polymers in the unit volume). The domain where a single polymer spreads is roughly a sphere of radius $R \sim N^\nu$ (in 3-space $\nu \simeq 3/5$; ν is a universal exponent).[16] Therefore, the number of polymers overlapping with a single

---

[14] ⟨⟨**Staudinger and Heidegger**⟩⟩ The concept of the polymer was established by Staudinger (1881-1965). See Y. Furukawa, *Inventing Polymer Science — Staudinger, Carothers, and the emergence of macromolecular chemistry* — (University of Pennsylvania Press, 1998). Page 162-166 of this book tells us how Heidegger was eager to please the Nazi officials. Heidegger was the President of the University of Freiburg, succeeding Möllendoff, who was dismissed for forbidding anti-Semitic placards. Staudinger was a professor there. "In September 1933, Heidegger secretly reported details about Staudinger's political past to a Nazi official. ... According to Heidegger, during WWI Staudinger had publicized his anti-war ideas and opposed German militarism.

The information was confirmed by the Gestapo, which asked Heidegger to act quickly: Heidegger dashed off a reply to the Baden Minister of Culture and Education, "... Dismissal rather than retirement may be suitable. Heil Hitler!"."

[15] This concept was introduced by Jacque des Cloizeaux, "The Lagrangian theory of polymer solutions at intermediate concentrations," J. Phys. (France) **36**, 281 (1975).

[16] Mathematically, it is related to the critical exponent ν in Example 3.2.

polymer chain may be estimated as $cR^d \sim cN^{d\nu}$. On the other hand, the number density of the repeating units (= monomer number density) is $\sim cN$. Since $d\nu > 1$, we can take the monomer density zero limit while keeping $cN^{d\nu}$ constant; we have only to take an appropriate simultaneous limit $c \to 0$ and $N \to \infty$. This limit is called the semidilute limit. In this limit, the osmotic pressure of the polymer solution has the following form:

$$\pi = ck_BTf(c/c^*), \tag{3.5}$$

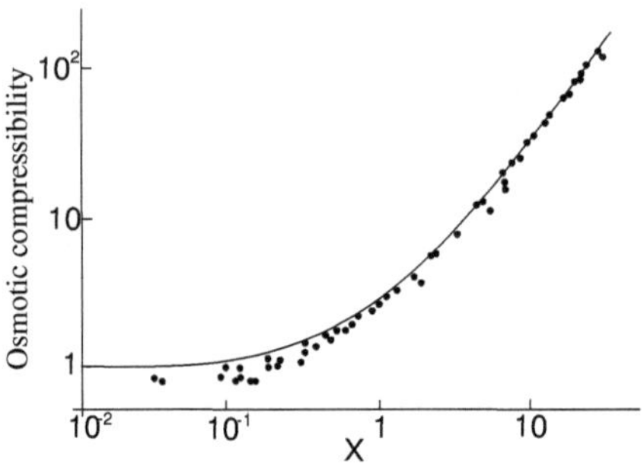

Fig. 3.2 An example of the renormalization group calculation. The osmotic pressure (more precisely, the osmotic compressibility) of the semidilute solution of the polymer solution. This plot does not depend on any adjustable parameter. Osmotic compressibility is $\partial\pi/\partial c$. The solid curve is the result of renormalization group theory based on T. Ohta and A. Nakanishi, J. Phys. A **16**, 4155 (1983). The points are experiments due to I. Noda, N. Kato, T. Kitano and M. Nagasawa, Macromolecules **14**, 668 (1981).

where k_B is the Boltzmann constant, and T is the absolute temperature. The function f is a universal function independent of the details of the solution (polymers and solvents). c^* is a phenomenological parameter determined by the details of the system. This universal function was calculated in detail by Takao Ohta and his collaborators, and is a representative example of the (quantitative) success of the renormalization group theory (Fig. 3.2).[17] This example was adopted by Michael Fisher to illustrate that quantum mechanics is 'almost irrelevant': "Quantum mechanics was not needed to do the cutting-edge research, so one feels that if some of the giants of the past, like Boltzmann or Gibbs or Rayleigh, were able to rejoin us today, they would be able to engage in research at the cutting edges of condensed matter physics..."[18] □

[17] T. Ohta and A. Nakanishi, "Theory of semi-dilute polymer solutions: I. Static properties in a good solvent," J. Phys. A **16**, 4155 (1983); T. Ohta and Y. Oono, "Conformational space renormalization theory of semidilute polymer solutions", Phys. Lett. **89A**, 460 (1982). There is a fairly kind exposition of the work in *Stealing the Gold: A Celebration of the Pioneering Physics of Sam Edwards* (P. M. Goldbart, N. Goldenfeld, and D. Sherrington editors, Oxford University Press, 2005).

[18] M. E. Fisher, "Condensed Matter Physics: Does Quantum Mechanics Matter?" in *Niels Bohr: Physics and the World* (Proceedings of the Niels Bohr Centennial

Example 3.4. Spinodal decomposition of binary fluid
Prepare a mixture (say, the critical mixture; see Example 3.2) of two fluids in its disordered state, i.e., the single-phase state (the high-temperature state A in Fig. 3.1). What happens if the temperature is suddenly lowered to one far below the critical temperature T_c (e.g., B in Fig. 3.1)? Just after this sudden cooling, nothing significant should happen; the order parameter field exhibits only small fluctuations around 0. However, the effect of thermal agitation decreases, and demixing should ensue (note that in equilibrium at this temperature, the system consists of two separate phases C_1 and C_2). Therefore, if the order parameter is slightly positive in a small region,[19] it grows more positive. Subsequently, a region with a positive order parameter tries to grow at the expense of regions with negative order parameter, and the same applies to a region with a negative order parameters. In this way, the size ℓ of a single-phase (e.g., the C_1 phase) domain formed by segregation increases gradually. But this coarsening process (called *spinodal decomposition*) is a tortuous process because molecules that wish to join their own kind cannot traverse the domains dominated by the other kinds of molecules. Thus, the process produces a complicated pattern (see Fig. 4.4) and takes a long time. We will consider this phenomenon as a modeling example in Chapter 4.

The system can be studied by scattering experiments. The Fourier transform of the spatial correlation function, called the *form factor* (or the static scattering function), $S(k,t)$ at time t is empirically known to have the following form:

$$S(k,t) = \langle k \rangle_t^{-3} F(k/\langle k \rangle_t), \tag{3.6}$$

where $\langle k \rangle_t$ is the peak position of the form factor at time t, which behaves, after a sufficiently long time, as $\langle k \rangle_t = at^{-1}$ (for solid systems $\langle k \rangle_t = at^{-1/3}$), where a is a phenomenological constant depending on the idiosyncrasy of the mixture, but F seems a universal function (at least they take the shape indistinguishable between low molecular weight fluid mixtures and polymer mixtures). As we will see in the next chapter, there is a good reason to believe in universality.

The phase transition occurring here is a first-order phase transition (the average value of the order parameter jumps upon phase transition), and the actually generated pattern is determined by the initial microscopic fluctuations. Every time we reheat and equilibrate the system and then repeat the quenching process, different patterns emerge. History never repeats in the phase transition dynamics (but (3.6) always holds).[20] □

Symposium), edited by H. Feshbach, T. Matsui, and A. Oleson (Harwood Academic Publishers, Chur, 1988) p65.

[19] Its size is about the correlation length at the initial high temperature.

[20] ⟪**Intermediate asymptotics**⟫ As the reader must have realized, if k^{-1} becomes close to the system size, (3.6) cannot be true. This relation is a typical '*intermediate asymptotics*' (see Appendix 3.5B and Section 4.6). It is not the true asymptotics that must be, in the present case, the coexisting two bulk phases with a simple boundary surface between them (a dead-end equilibrium state without any further

The above are examples of quantitative (or we could even say rigorous) phenomenologies, but many examples of less quantitative phenomenologies have been known for a long time.

Example 3.5. van der Waals gas
A typical example of an equation of state of imperfect gases is the *van der Waals equation of state*:

$$\left(P + \frac{aN^2}{V^2}\right)(V - Nb) = Nk_BT, \tag{3.7}$$

where P, T, V, N are the pressure, the temperature, the volume, and the number of molecules, respectively, k_B is the Boltzmann constant, and a and b are positive constants. a has a dimension of energy ($\times$ volume), and b that of volume (for the dimension see Appendix 3.5A). This equation also consists of the universal structure and phenomenological parameters that are peculiar to each gas.

It is empirically well known that the equations of state of various gases can be superposed fairly well on a single master curve in terms of reduced quantities that are scaled variables with their values at the critical point: $P_r = P/P_c$, $V_r = V/V_c$, and $T_r = T/T_c$ (the *law of corresponding states*, also proposed by van der Waals). In this case, the master curve is the universal structure, and critical parameters, P_c, T_c, V_c, are materials constants, only two of which are independent: $P_cV_c/Nk_BT_c = 3/8$.

Let us consider the problem of computing the equation of state when a molecular interaction potential is given. Suppose the molecular interaction potential contains only two parameters, the length scale of the molecule (or its core size) and the energy scale (e.g., the depth of the attractive potential) (notice that the majority of explicit models of molecular interaction potentials such as the Lennard-Jones potential has this form). Then, the existence of the master curve (i.e., the law of corresponding states) is naturally understood with the aid of dimensional analysis. In short, when the imperfect gas is modeled in terms of a molecular interaction potential with two parameters, the universality of the potential function is the key to the universality of the equation of states. However, in this case the universality must remain approximate, because the intermolecular interactions are not completely determined by the energy scale and the volume scale.[21]

Near the critical point, there is a true universality of critical fluid, although

---

time evolution). The relation (3.6) holds when the pattern size is much larger than the molecular scale, but much smaller than the system size. However, if we use a big enough system, (3.6) holds for a very long time. That is why it is said to describe an *intermediate asymptotic* state. The actually realized patterns we can observe during the intermediate asymptotics in this example are very sensitive to initial conditions.

[21] For example, electrical multipole interactions cannot be ignored. See T. Kihara, *Intermolecular forces* (Wiley, 1978).

it is qualitatively different from the van der Waals equation (for example, the critical indices are different). Away from the critical point, it incorporates the approximate universality due to the general form of the binary interaction potentials. Thus, the law of corresponding states approximately holds thanks to the two kinds of universalities. □

Example 3.6. Phonon specific heat
The Debye theory of the low-temperature specific heat of (non-conducting) solids (the phonon specific heat) is a good example of the approximate phenomenology. According to this theory the heat capacity at constant volume reads[22]

$$C_V = 3Nk_B f\left(\frac{T}{\Theta_D}\right), \tag{3.8}$$

where N is the number of atoms (actually, the number of lattice unit cells) in the solid, Θ_D is a materials constant (= a phenomenological constant) called the Debye temperature[23] and f is a universal function called the Debye function. If x is roughly less than 0.1, we may approximate it as $f(x) \simeq 4\pi^4x^3/5$. This gives the famous T^3 law for low-temperature specific heat of insulating solids. This asymptotic result is rigorously universal.[24]

There are many other similar examples such as the $C_V \propto T$ in the $T \to 0$ limit of the electron specific heat. □

As we have seen above there are two kinds of phenomenology: one is with the universality that is the outcome of complicated entanglement of numerous degrees of freedom (clear examples are Examples 3.2 and 3.3) (let us call it 'true universality') and the other with the universality emerging in the limit of simplicity (e.g., Example 3.6 and the ideal gas law) (let us call it 'elementary universality'). Elementary universalities are observed in, say, the low-temperature limit or low-density limit in systems consisting of noninteracting entities (atoms, molecules, elementary excitations), and the observed universality is ascribed to the universality of the idealized elementary elements themselves. In contrast, the universality seen in the limit of complicated interactions seems often due to many-body effects and the observed universality makes sense only in the macroscopic limit. These two kinds of universality exhibit quite different responses to perturbations. In the case of elementary universality the effect of perturbation to the system sensitively reflects the details of the perturbation. For example, when we add a small attractive interactions between molecules, it matters whether we add it as a part of the van der Waals interaction (as a term proportional to $1/r^6$) or as an

---

22 Any statistical mechanics textbook explains this. For an elementary introduction, the author likes F. Mandle, *Statistical Physics* 2nd edition (Wiley, 1988).

23 Precisely speaking, it is a materials-specific function of temperature, but does not depend on it sensitively and reaches a constant at sufficiently low temperatures.

24 It depends on the spatial dimensionality and the fact that phonons may be treated formally as noninteracting bosons with zero chemical potential.

electrical quadrupole interaction. That is, the effect of attractive interactions depends on their functional forms. In contrast, in the case of true universality, although there are effects of perturbations, they can be described in terms of a small number of phenomenological control parameters. That is, even if the space of perturbations is infinite dimensional, the space of responses can often be finite (even very low) dimensional. For example, in the case of polymer semidilute solutions, we can change solvents and polymers (infinite dimensional perturbations), but in the phenomenology, only c^* is perturbed (one-dimensional response).

True universalities are stable universalities that give us stable phenomenological descriptions. The phenomenological relations in the limit of simplicity ('elementary universality') are 'macroscopically inflated' microscopic laws, and their universalities are the universalities of the microscopic laws being exposed thanks to the uncorrelatedness of non-interacting microscopic degrees of freedom, so they are not worth discussing here (elementary statistical mechanics is boring because it discusses only these).

In this book the term *phenomenology*[25] or *phenomenological description* should imply the following: the phenomenological description of a particular system (or a phenomenon) in a certain class (= a set of systems or phenomena) consists of the universal structure common to the class and phenomenological parameters peculiar to each system. In short, symbolically,

$$\text{'Phenomenology'} = \text{'universal structure'} + \text{'phenomenological parameters'}. \tag{3.9}$$

Needless to say, unless the characterization of a class (the class sharing a universal structure) is fairly clear, the phenomenology cannot be said to be a good one. The values of the phenomenological parameters cannot be determined by the phenomenology, and must be supplied 'from outside' (e.g., experimentally fixed). The modern science started with the dissection of motion into the general framework, the laws of mechanics, and factors peculiar to the individual phenomenon, the auxiliary conditions (see Section 5.3). Phenomenology explained in this book may be understood as a sort of generalization of this structure of the modern science.

The existence of mathematical structures we can phenomenologically recognize in the world is a prerequisite of the existence of intelligent beings.[26] We could even say that a 'phenomenon realized in a certain system' is a representation of (often mathematical and abstract) 'universal structure' in terms of the concrete system. Organisms with only primitive sensory organs live in a very abstract world. Long ago, all the organisms were like that. This implies

[25] The relation of our phenomenology and phenomenology in philosophy is discussed briefly in Note 3.2 in Section 3.2.

[26] "Nature, I was certain, is made to be understood, or, rather, our thought is made to understand nature." is found in W. Heisenberg, *Physics and Beyond* (translated by A. J. Pomerans, Harper & Row, 1971) 8. p101.

that the more abstract the concepts are, the older they are phylogenetically. Such ability to recognize (respond to) abstract features is at the foundation of our cognition, so the world in which sophisticated cognitive capability can evolve must have abstract lawfulness. This is an important viewpoint when we study complex systems.

3.2 Phenomenology too universal to be recognized

The universalities discussed in the preceding section are not so important universality classes (a *universality class* is a collection of systems sharing the same universality). We must not forget much more important universal structures of the world or universality classes in the world. These classes are so big that those who are immersed in them cannot recognize them (just as air) (and cannot escape from them just as Sun Wukong could not escape from Buddha's palm in *Journey to the West*).

Example 3.7. Thermodynamics
Let us take a uniform object of the 'human' scale (= macroscopic size). For the macroscopic properties of its equilibrium state, equilibrium thermodynamics holds. It must hold if the molecular interactions (microscopic interactions) are short-ranged (i.e., not like gravity; not long-ranged without any shielding effects) and if the total energy of the system is bounded from below by a quantity proportional to the number of particles in the system (i.e., if the molecules have sufficiently hard cores and the system does not collapse due to attractive interactions).[27] That is, the mathematical structure of equilibrium thermodynamics is a universal property shared by any ordinary substance in equilibrium. In thermodynamics, the equation of state describing the individual substance corresponds to phenomenological parameters. For systems slightly deviated from equilibrium linear nonequilibrium thermodynamics holds. This is also quite universal. The fundamental nature of thermodynamics is clearly seen by the fact that while statistical mechanics had to be revised, thermodynamics was never shaken even slightly, when quantum mechanics emerged; rather, it prepared the quantum revolution.[28]
□

---

[27] See the beginning portion of D. Ruelle, *Statistical Mechanics, rigorous results* (Benjamin, 1969, NY).

[28] ⟪**Textbooks of thermodynamics**⟫ Perhaps, practically, H. B. Callen, *Thermodynamics* (Interscience Publ., New York, 1960) is still a good textbook of thermodynamics. However, if the reader is serious about thermodynamics, she must read E. Lieb and J. Yngvason, *The physics and mathematics of the second law of thermodynamics*, Phys. Rep., **340**, 1-96 (1999); an expository article by the same authors for mathematicians is also recommended: "A guide to entropy and the second law of thermodynamics," Notices AMS **45**, 571 (1998). R. Kubo, *Thermodynamics. An advanced course with problems and solutions* (North-Holland Pub. Co., 1968) is a good problem book with full solutions.

Example 3.8. Classical mechanics
The Newtonian mechanics can be interpreted as a phenomenological description of classical motions in general, and makes a large universality class. In this case, the masses and potentials for each system may be understood as phenomenological parameters. That is, in Newton's second law, $md\boldsymbol{v}/dt = \boldsymbol{F}$, $\boldsymbol{F}$ and m are, for mechanics, extrinsic and contingent factors (just so is the equation of state for thermodynamics); as a law of physics Newton's second law is an incomplete law just as thermodynamic laws are. That the equation of motion is a second-order ordinary differential equation is the universal structure as an expression of the general empirical fact: the *Newton-Laplace determinacy* (= that the initial position and velocity determine the future of the system). That there is no first-order derivative term is an expression of the law of inertia. To understand the world phenomenologically is to recognize such general mathematical structures behind various phenomena. However, no mathematical knowledge is required to recognize them. Even dogs and cats understand that classical mechanics is governed by a second-order differential equation, so they rarely fail to eat or are rarely eaten. Of course, to express consciously what one knows (or what is embodied) is difficult, so there is no differential equation in the dog's brain. However, most human beings are not different from dogs in this respect. Mathematics is built into our body (embodied). The law of inertia must have been holding long before dogs and cats and we emerged in the world, so it is a universality that exists apart from us (it exists irrespective of our existence/non-existence). What is universal in the world is not determined by 'our condition' but by the intrinsic structure of the world. □

Discussion 3.1. If we do not exist, then the existence of others cannot be observed, so isn't the statement 'exists apart from us' meaningless? How can you counter this objection? □

Example 3.9. Mechanical motion of a particular system
For both classical and quantum mechanics the motion of each system can be dissected into the equation of motion and the initial (and boundary) conditions (*auxiliary conditions*). For each system the equation of motion is the universal structure and the auxiliary conditions correspond to phenomenological parameters. That is, the large universality class of systems obeying mechanics contains numerous systems, and each system can in turn be considered as a universality class that may be dissected into the universal aspect and the rest. Actually, recognizing this separation was an important aspect of the Newtonian revolution. As we will see later, auxiliary conditions are crucial for complex systems.

Needless to say, in this and previous examples we could discuss quantum mechanics quite parallelly as classical mechanics. □

Sometimes universality is stressed for its significance (as in critical phenomena) and sometimes even the universal nature of quite fundamental universalities has not been stressed. This difference seems to depend to a large extent on whether the recognition of universality leads us to further understanding of the world. For example, in the case of critical phenomena, people were surprised to see that there were universal aspects in the phenomena in which they expected only individual specific understandings. In this case the existence of universality was felt highly nontrivial due to microscopic diversity supporting the same macrophenomena (i.e., the universality was not an elementary one in this case), so universality was regarded worth emphasizing. Or, as in Example 3.9, all the motions allowed to a system can be dissected into a single equation of motion and fortuitous initial conditions, a crucial core of the Newtonian revolution. The universality of equations of motion for celestial bodies made a profound impression on contemporary people.

In contrast, from the beginning, thermodynamics captures the universal features of this world. Since there is only one world, the recognition of this universality does not allow us to classify various worlds and to understand them systematically. Therefore, thermodynamics as the universal structure of the world does not seem to have been stressed. Perhaps, there is no merit in emphasizing thermodynamics from the universality point of view.[29] However, thermodynamics might be looked at from a different point of view through the lens of renormalization, which is not only a means to extract universal features but also has the possibility of giving insight into the reason for the universality. Furthermore, although there is only one thermodynamics for our only one world, this applies only to equilibrium states. It should not be useless *a priori* to expect some sort of universality classes for nonequilibrium phenomena.

Note 3.1. The role of phenomenology in constructing statistical theory
The basic principle of equilibrium statistical mechanics is the principle of equal probability, but to what set of microscopic states we may apply the principle cannot be specified without phenomenology. The principle of equal probability is assumed to hold for a set of microstates specified by the set $\{E, X_i\}$ of macroscopically controllable variables, energy E and work coordinates X_i.[30] The set of macrovariables is selected so that the first and the second laws of thermodynamics hold with these variables. Therefore, it is obvious that the framework of statistical mechanics does not make sense apart from thermodynamics.

Suppose the basic quantities for a theory are extensive quantities (= the quantities proportional to the system size). If this is the case, it is likely that remote volumes in a single system are statistically almost uncorrelated, and the law of asymptotic equipartition (2.51) is likely to hold, which suggests that the statistical framework behind the theory is based on the set of microstates on which the principle of equal probability is assumed.

For example, if there were a statistical mechanical theory of nonequilibrium steady state, it would be natural to expect that the framework of the

[29] However, E. Lieb and J. Yngvason, *The physics and mathematics of the second law of thermodynamics*, Phys. Rep., **340**, 1(1999) describes a certain basic part of thermodynamics as a general mathematical theory of the equivalence of the existence of a certain ordering relation and the existence of consistent 'height function.' This might be an important suggestion on quantification of complexity.

[30] *Work coordinates* are extensive variables that can be specified by mechanics and electrodynamics. Volume and magnetization are examples.

theory takes the form just discussed above. Its implication is that there must be a thermodynamics of steady states. The first law that corresponds to the conservation law may not be problematic. Is there a law corresponding to the second law? Even if a state is in nonequilibrium, there must be processes that cannot be performed purely mechanically. Therefore, the key question must be how to find a set of extensive work coordinates consistent with the first and second laws suitably generalized to steady states. The hardest part seems to be the establishment of the phenomenology. Struggling at this stage is the steady state thermodynamics.[31]

The existence of a suitable phenomenology is a prerequisite for a set of phenomena or systems to be considered as a whole, because the understanding of the universal part in the schema (3.9) is the understanding of the set as a whole. As we will see in the next chapter, this is quite crucial in modeling as well, because a model lacking details (maybe analogized with an impressionistic painting) can be a meaningful model of a phenomenon, if and only if the model captures the universal part of its phenomenological description. If we wish to understand a complicated phenomenon, we should first attempt to extract a phenomenological framework to describe it. If this is impossible, one must give up its general understanding. Therefore, if we wish to have a general understanding of the complex world, we must first try to understand it phenomenologically.

The universal part of a phenomenology is inevitably abstract and mathematical. As has already been stated earlier at various places, the author believes that the existence of phenomena with universal features in the world is a prerequisite for the existence of the intelligent being. The phenomenological understanding of the world may be a viewpoint that emphasizes 'essence' more than 'existence.' As has already been written, phylogenetically speaking, an abstract grasp of the world precedes cognition of concrete and individual objects in the world; the former is much more primitive than the latter. Since we have refined sensory organs, and, for example, can see an object clearly, recognizing abstract features is (at least consciously) harder than recognizing individual concrete features. Therefore, we tend to believe many abstract concepts are more advanced, higher-level concepts than concrete concepts (e.g., topology is a more advanced study than the Euclidean geometry). However, organisms with much more primitive perceptive and processing capacities, without any abstraction, directly perceive abstract features of the world. In cognition abstract concepts phylogenetically precedes concrete concepts, so it is quite possible that this ordering is still preserved in our (unconscious) cognition.[32] Thus, the phenomenological point of view is more basic than the ordinary human recognition of the world. Abstract and mathematical

---

[31] Attempts are Y. Oono and M. Paniconi, "Steady state thermodynamics," Prog. Theor. Phys., Supplement **130**, 29 (1998); S. Sasa and H. Tasaki, " Steady state thermodynamics," J. Stat. Phys. **125**, 125 (2006).

[32] ⟪**More abstract features may be perceived more easily**⟫ Lin Chen has been asserting for more than 20 years that topological features (existence/no-existence, in/out distinction, presence/absence of a hole, etc.) are the basic visual stimuli.

concepts are biologically more primitive than concrete concepts.[33] Abstract concepts are, as long as they are natural, biologically more concrete than what we usually recognize as concrete. The statement, "There is a beautiful flower; the beauty of the flower does not exist." may sound philosophically deep, but it is a mere illusion or is trivially false. Such abstract concepts as 'beauty' that are directly connected to our emotion can be the primary concepts firmly recognized by us. Only where there is the beauty of the flower can there emerge a beautiful flower.

The examples of phenomenology or phenomenological summary shown above are universal laws in 'quantitative' natural science, and no one would object to recognizing them as such. However, there are many other general observations in the world. For example, all the fundamental laws of physics are *variational principles.* Or, we may restate them as *forbidding principles* dictating that such and such is impossible.[34] What is the use of such general observations? In high-energy physics, variational principles are used as a guiding principle to construct new theories, so there must be a deep meaning.

If the biological world is comprehensible to us, at least such general observations must be possible and they should have deep meaning. For example, such statements that a cell must be produced by another cell ("Omnis cellula e cellula"), and that life comes only from another life sound banal, but they are significant statements about complex systems. As will be stated at the beginning of Chapter 5, this is an important observation at least to the extent that the so-called complex systems studies have been off the mark by ignoring it.

Another quite banal statement may be, "if killed, an organism cannot be resurrected (if a cell is destroyed, that is the end)." However, notice that

("Topological structure in visual perception," Science **218**, 699 (1982)). A recent work of his group (B. Wang, T. G. Zhou and L. Chen, "Global topological dominance in the left hemisphere," Proc. Natl., Acad. Sci. **104**, 21014 (2007)) demonstrates that the left hemisphere (for right-handed persons) is more sensitive to topological differences than the right hemisphere. They think this is because the basic cognition of objects is the topological cognition. Furthermore, they showed that topological differences can be recognized within a shorter time than other geometrical distinctions.

[33] For example, the concept of 'sensory organ' may be considered as a man-made secondary concept in contrast to the 'God-given' eyes and ears, but if we look at the gene regulation of the developmental process of sensory organs of *Drosophila*, there is a gene *atonal* that regulates all the imaginal sensory organs (eye, Johnston's organ and chordotonal organs; the latter two are sensors of mechanical stimuli). The famous *ey* (a homolog of *Pax 6*) works under this. See N. Niwa, Y. Hiromi and M. Okabe, "A conserved developmental program for sensory organ formation in *Drosophila melanogaster*," Nature Gen. **36**, 293 (2004). This work discusses the existence of the developmental program of the universal protosensory organ.

[34] As a forbidding principle, the second law of thermodynamic is famous, but any variational principle may be reformulated as a forbidding principle. This is what Carathéodory taught us.

cells modeled in terms of detailed biochemical networks studied by system biology do not die, or more precisely speaking, it is not hard to resurrect them. This strongly suggests that something fundamental is missing. The lack of spatial structures is suspected as the chief factor. The people working in exact sciences may well say that such observations are too vague to found any respectable theory, but they have grave (logical) consequences. We must recall that thermodynamic laws are similar in nature.

Note 3.2. Phenomenology in philosophy and in physics
In philosophy the term 'phenomenology' is said to have appeared first[35] in *Novum Organum* (1764[36]) by Lambert (1728-1777).[37] He gave this name to the 'final step of the study of the means to find truth.' Kant (1724-1804), who respected Lambert, seems to have thought that the 'study to decide the appropriateness and limitation of knowledge based on senses' should be called 'general phenomenology.' He also called this the 'critique of pure reason.'

The modern usage of the word 'phenomenology' is to imply 'phenomenology' initiated by Husserl (1859-1938). His motto was 'to phenomenon itself.' As he himself admits, his usage of this word was based on Mach's (1838-1916) quite anti-metaphysical viewpoint[38] (the attitude to study Nature based only on the comparison and descriptions of observable phenomena) (this name was originally due to Boltzmann and others to indicate Mach's (philosophical) standpoint with contempt). 'Phenomenology' implies, as can be seen from its history, that in contrast to atomism it is confined to superficial description of phenomena, never to try to go beyond them. Consequently, 'phenomenology' does not have a good connotation in physics, or rather, the word has an obvious reactionary nuance. Husserl's early usage of this word meant the 'purely descriptive research of various phenomena.'[39]

As Husserl writes in *Ideas of Phenomenology*: "The playground of these obscure and contradictory theories, as well as the endless controversies associated with them, is *epistemology* and *metaphysics* — for metaphysics is bound up with epistemology both historically and by the way of the subject matter."[40] His anti-metaphysical stance is clear. Following Descartes, Husserl takes as the starting point of critical epistemology the observation that 'the being of *cogitatio*, of experience as it occurs and is simply being reflected upon, cannot be

[35] in the sense now being used; this word appeared first in 1736, invented by Christoph Friedrich Oetinger (1702-1782, German theosophist) as the name of a discipline to study the divine relationships among various things at the surface of the world we can observe.

[36] [1764: Hargreaves invented the Spinning Jenny, Cao Xueqin (of *Dream of the Red Chamber*) died.]

[37] A Swiss mathematician, physicist, astronomer contemporary of Euler and the Bernoullis. He had many works including the Lambert-Beer law for the attenuation of light and Lambert function $f(x) = x \exp x$.

[38] The word 'metaphysics' (meta = transcend, physica = nature) is used as a discourse of transcendental concepts beyond experience. The original meaning had no deep implication; it was simply named so, because it was arranged 'after' Physica in Aristotle's work.

[39] These summaries are based on *Dictionary of philosophy and thoughts* (Iwanami, 1998) and G. Kida, *Phenomenology* (Iwanami, 1970) Introduction.

[40] E. Husserl, *The Idea of Phenomenology*, Lecture 1 (translated by L. Hardy, Kluwer Academic Publishers, Dordrecht, 1999) p18.

doubted.'[41] However, he critically scrutinizes the experience itself: "only the *pure phenomenon*, the reduced phenomenon (is really an absolute givenness). The experiencing ego, the object, the human being in world-time, the thing among things, etc., are not absolute givenness, and therefore experience as this being's experience is not an absolute givenness either."[42] Here, the word 'reduce' used by Husserl is not at all the word used in the context of reductionism, but to imply the removal of transcendental elements (i.e., elements coming from outside experiences only) through introspection (this is called the phenomenological reduction).[43] That is, phenomenological reduction means exactly the opposite of physicists' reduction; it is to confine oneself rigorously to what can be obtained by experiences alone. Thus, Husserl was an arch anti-reductionist in physicists' sense. We may say the hallmark of Husserl's phenomenology is characterized by the reflection on what is observable (can be experienced) and by the scientific attitude to place only what we can observe without metaphysics as the basis of any philosophy.

Husserl had serious influences on Gödel and Weyl (1885-1955); we should remember that Gödel was very sympathetic to this philosopher.[44]

Eventually, Husserl came to realize (at last?) that the transcendental reduction can be performed because there was a live human being who could perform it, and that this fact was fundamentally important. Although he consistently rejected the use of evolutionary biological considerations, the author feels that, since he was a fundamentalist thinker, eventually he would not have avoided taking the basis of our cognition and consequently phylogenetic learning seriously into account. Evolutionary epistemology seems to be the inescapable foundation on which any serious philosophy must be built.

3.3 How to obtain phenomenology—relation to renormalization

One reason why this world, where the unknowable could come to the fore due to scale interference, is (felt to be) still understandable is that at various 'strategic junctures' phenomenological understanding is possible. This is because (albeit approximate) universal structures exist. Such is a summary of the main line of the story in this chapter up to this point. We have examined what phenomenology is through examples. As can be seen from the usual pejorative phrase, "it's a mere phenomenology," phenomenology has a connotation that it is, compared with the microscopic theories, approximate and crude secondary theories (in the first place the word was derogatory in physics. See Note 3.2). Also phenomenology tends to give an impression that it is a mere summary of observed results without pursuing their 'essence'

---

[41] E. Husserl, *ibid*, The train of thought in the lectures, p62.

[42] E. Husserl, *ibid.*, p64.

[43] "to abandon our belief in the existence of the world, and to try to see how such belief has been generated from the conscious experiences alone that are directly given to us" is the transcendental reduction. (G. Kida, *ibid.* p45)

[44] R. Tieszen, "Gödel's path from the incompleteness theorems (1931) to phenomenology (1961)," Bull. Symbolic Logic **4**, 181 (1998) is recommended.

(just as curve fitting to empirical data). However, as we have seen in the preceding section, a quantitative phenomenology cannot be regarded as a crude approximation of a more refined theory; thermodynamics is not an approximation of any other theory.[45]

Then, how can we see the backbone of a phenomenology, i.e., the universal core of the mathematical structure of a set of phenomena?

Large scales lie far above the cloud, so to speak, if we climb up the ladder of length scale from the microscopic world. Then, isn't perceiving some sort of asymptotic behaviors equivalent to grasping the universal aspect of the phenomenology? Doesn't capturing structurally stable features insensitive to microscopic details correspond to extracting the universal (mathematical) structure in the macroscopic world?

Note 3.3. Structural stability
This concept was introduced by Andronov (1901-1952) and Pontrjagin (1908-1988) in 1937[46] for dynamical systems. Differential equations appearing in science and engineering often have parameters that can be determined only within some error bounds. That is, dynamical systems in reality can be determined only with some uncertainty (for example, it is hard to empirically check that there is no double Laplacian term, $\propto \Delta^2$, in the Navier-Stokes equation). If the outcome of the model in terms of a dynamical system is very sensitive to such uncertainty, then such a model is almost useless as a model of an actual phenomenon. Therefore, in natural science useful dynamical systems should be stable against *structural perturbations* (= small changes of the dynamical system itself).[47] Thus, Andronov and Pontryagin introduced the concept of *structural stability*. To formulate this idea as mathematics we must clarify what we mean by "perturbation is small," or "the solution does not change very much." They require topology and distance concepts for dynamical systems, so we will not discuss this topic further here.

We will see that if a system is structurally stable, it is likely that its reduced system can capture qualitative features of the original full system (Section 3.8).

Needless to say, there are many structurally unstable systems as well. Therefore, precisely speaking, we must say that a model of a reproducible phenomenon must be structurally stable. It may also be a modeling strategy that the model of irreproducible phenomena must be structurally unstable.

Discussion 3.2. Is the Navier-Stokes equation structurally stable?[48] □

---

[45] This statement is due to H. Tasaki.

[46] [1937: Guernica, Rape of Nanking]

[47] Notice that *structural perturbation* is quite different from the perturbation we usually consider to study the stability of solutions of a given dynamical system. In the latter case we perturb the solutions (or initial conditions) and do not modify the system itself. Thus, typical chaotic systems exhibit unstable trajectories, but the system themselves are often (C^1-) structurally stable; chaos does not disappear even if the system is modified slightly. Typical properties related to chaos are often open generic.

[48] Interestingly, if the viscosity is velocity-gradient-dependent or if the biharmonic term proportional to $\Delta^2 \boldsymbol{v}$ is added, then we can readily show the unique existence of the global solution even in 3-space for any Reynolds' number. That is, at least functional theoretically, its character changes drastically however small such pertur-

To find universal macroscopic features of a system, let us consider the limit ζ = (macroscale)/(microscale) $\to \infty$. To take this limit is to discuss what we can observe on very large scales. For example, if we take the example of polymers, we discuss the world where we feel the monomer size to be extremely small.[49] The crux of what we wish to do here is to change microscopic details and to see how it affects macroscopic observables. In practice, to change ζ we can change the observation scale or we can change monomers. That is, to change ζ may symbolically imply varying microscopic structures (or observing different substances). Notice that this is methodologically exactly the same as the chief weapon of natural history, *comparative studies.* Natural history does not simply compare systems different in microscopic scales; we must first consider very carefully what perturbation is suitable for a particular comparative study, so natural history is much harder than physics.

If an observable f of a system being studied depends on the observation scale, it may be investigated as a function $f(\zeta)$ of ζ. What can happen in the $\zeta \to \infty$ limit? If the limit exists as $\lim f(\zeta) = c$, then the observable has a definite value in the macroscopic world. This means that such an observable is 'uninteresting' if ζ is sufficiently large. We may conclude that for this observable the details of the microscopic world do not matter; you can imagine any microscopic world to explain it. An interesting case is that there is no limit for f. That an observable does not have any definite value in the $\zeta \to \infty$ limit implies that the microscopic effects remain to be observed on macroscopic scales. In other words, such a quantity sensitively depends on microscopic details. In reality, ζ is finite,[50] so $f(\zeta)$ must have a definitive value for each system. However, since there is no limit, we must conclude that this value depends sensitively on ζ (if it is large; see the schematic figure Fig. 3.3) or on the idiosyncrasy of each microscopic description. As we know the world is full of diverse substances, this should be the rule.[51]

If, however, divergent quantities can be separated out from the observable in the $\zeta \to \infty$ limit, then the remainder should be universal (i.e., insensitive to microscopic details). The separated divergent quantities must be sensitive to the individual microscopic details of the system, so they should be

bations are. See, e.g., the appendix to O. A. Ladyzhenskaya, *Mathematical theory of incompressible fluids* (Gordon & Breach, 1963).

[49] **⟪How small molecules are⟫** It may be a good occasion to appreciate how small molecules are. The number of water molecules in a tablespoonful of water ($= 5 \times 10^{23}$ molecules) is about five times as large as the amount of the total ocean water on the Earth measured in tablespoons ($= 9 \times 10^{22}$ tablespoons).

[50] We never know all the microscopic details. Therefore, for any microscopic theory or model its description of the system under study remains always incomplete beyond some scales (or beyond some energy scales, because to probe small scales we need short-wave or high-energy particles). ζ is always finite.

[51] A more careful argument will be given in Section 3.5. Here, we proceed in a simple fashion. However, even if there is a divergence, the asymptotic form of f can be a part of the universal structure, because, although ζ is dependent on individual examples, f is common to the class of phenomena under study.

regarded as adjustable parameters, that is, as phenomenological parameters within the phenomenological framework. However, as remarked in conjunction with the Navier-Stokes equation, for a particular system after measuring the parameters, their values cannot be adjusted any more.

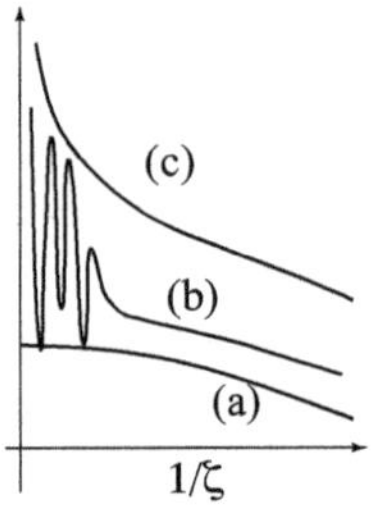

Fig. 3.3 Is there divergence in the macroscopic limit? $1/\zeta=$ microscale/macroscale. If there is no divergence as (a) in the $1/\zeta \to 0$ limit, we may assume anything we wish at the microscopic level, because it is a constant for any system macroscopically. In cases (b) and (c) the effects of microscopic details cannot be ignored because they persist even in the macroscopic limit for the macroscopic observable described by the vertical axis. In the actual world, $1/\zeta$ is a small nonzero value, but if we change it slightly, the value of this macroscopic observable changes drastically.

To separate out the sensitive dependence on microscopic details[52] (i.e., divergence due to microscopic details) into phenomenological parameters is called *renormalization*. If this program can be accomplished, we say the system under study is *renormalizable.*[53] In this way, we can extract the structure of the phenomenology: phenomenology = universal structure + phenomenological parameters. The effects of microscopic details may be quite large, but they do not always show up haphazardly. This is the reason why we can have a phenomenological description. If an observable is renormalizable, the parts sensitively affected by changes of microscopic details can be absorbed into several phenomenological parameters, and the rest ceases to be sensitively dependent on microscopic details (it settles down to a definite limit in the macroscopic limit). Thus, the universal structure (for a set of phenomena under study) has been extracted as the remaining invariant part (structurally stable part). A renormalizable system may be described by a phenomenology. That is, the slogan "phenomenology = universal structure + phenomenological parameters" is realized. From the examples in the next section, we will see that it is a good strategy to bring universal features out through identifying microscopic-detail sensitive parts. Conversely, a set of systems with a phenomenology must be described by a renormalizable model.

Whether a phenomenon is renormalizable or not is a property of the phe-

[52] More generally, these are the details on the scale we cannot observe, the unknowable scale.

[53] This rather informal usage of the term does not exactly overlap with that in high-energy physics.

nomenon, and is not up to us. Whether a phenomenon can be understood by us in general terms (not in terms of individual specific theory)—whether there is a phenomenological framework for the phenomenon—is a property of the phenomenon. If there is a phenomenological framework to understand a collection of phenomena, its universal structure should be captured by various microscopic models, because it is insensitive to microscopic details. To each phenomenology corresponds a compatible set of microscopic systems (or models), which is an equivalence class of microscopic systems with respect to sharing the universal structure of the phenomenology. This set is called a *universality class*. If we wish to know the phenomenological framework, we have only to study the 'simplest model' (*minimal model*) in the class. It is impossible to give a precise definition of the minimal model, because we often pay aesthetic attention to it, but at least it is a model containing exactly the same number of parameters appearing in the phenomenology. Now, it should become clear that renormalization, phenomenology, and modeling are intimately related.

For example, to describe the equilibrium statistical properties of polymer solutions in Example 3.3 two ingredients are needed phenomenologically: the degree of polymerization (polymer molecular weight) and the quality of the solvent. Their microscopic-model counterparts are: (1) the molecule is something like a chain of length N (a polymer is a sort of long, flexible chain; here N is proportional to the molecular weight); (2) monomers mutually exclude each other and their effective (excluded) volume v reflects the nature of the solvent (if a solvent dissolves the polymer well, in other words, if the solvent is 'good,' v is large). The polymer solution phenomenology is expected to be reproducible with a 'minimal model' having two parameters. Indeed, this looks empirically true. Obviously, however, such a minimal model cannot explain why there are only two phenomenological parameters. This question must be considered with the aid of more general models.[54]

In Chapter 1 the word 'system' was explained, and we proceeded, assuming that we understand it practically: a 'system' is a part of the world with structural stability against the changes outside the system (unless some particular crucial factors are not changed; for example, if the system is in thermal contact with its outside, this contact should not be severed). A part of the world for which the external effect may be described by a small number of (control) parameters, despite its inseparable (and strong) interactions with the external world, is deservingly called a system. It is not required that its relation to 'the outside' be weak. We need not disregard the inseparability of the world. However, we should clearly recognize that unless we adopt the point of view of phenomenology discussed above, that is, without understanding the word

[54] The reader would say that is obvious and is a rather stupid comment. Yes, it is, indeed, but this question was seriously discussed in the polymer physics community based on the so-called two-parameter theory containing only two parameters. The reader may well be appalled, but we should not be shocked by such elementary logical errors in a field apart from mathematics.

'system' in the 'correct sense', or in the sense endorsed by renormalization, the concept of 'system' is narrow and its usefulness is limited.[55]

3.4 Two approaches to renormalization

Phenomenology has, roughly speaking, the structure: 'phenomenology' = 'universal structure' + 'phenomenological parameters.'

The universal structure (if any) should be found through searching for results that remain invariant under changing microscopic details, that is, through pursuing 'structurally stable' results by modifying the 'unobservable details.'[56] There are at least two strategies to pursue invariance under microscopic changes.
(1) Observe the system from an increasing distance with a fixed resolution (resolving power) and find the properties that persist to be observed.
(2) Find the structure insensitive to the change of microscopic parameters (a strategy faithful to the idea explained in the preceding section).

The renormalization strategy based on (1) is called the *Wilson-Kadanoff renormalization procedure* and that based on (2) is called the *Stückelberg-Petermann renormalization procedure.*

Suppose we wish to make a phenomenology for an observable (say, the heat capacity near the critical point of a magnet).

To realize (2) we need a fairly explicit form of the quantity as a function of microscopic parameters, so (approximate) methods to compute it are required (e.g., perturbation calculation). Often minimal models are studied.

Approach (1) is equivalent to *coarse-graining* the system while shrinking its size. We could perform this transformation (*renormalization group transformation*) step by step to watch how observables change. If the system is sufficiently large and if its boundaries need not be taken into account, then the global features must keep being observed. On the other hand, small-scale details that have no global effects would become increasingly invisible by the procedure. Therefore, the properties that survive the numerous applications of the renormalization group transformation (= coarse-graining + shrinking) must govern the stable structure we are pursuing.

Against the research program advocated in this chapter (and in the next chapter that carries on its philosophy) there will be harsh criticism from the *ab-initio*-side people who assert everything should be obtained from *ab initio* quantum mechanical calculations. What we wish to understand is, however,

---

[55] The individuality of organisms should also be reconsidered from this point of view.

[56] These details need not be unobservable; they could be the factors we do not wish to consider, although whether we are allowed to ignore them or not cannot be up to us to decide.

a question such as: why are both a piece of chalk and flour white? The answer to this question should not require any details of atoms and molecules. The author does not claim that the atomistic approach is wrong,[57] but it is a simple fact that there are many phenomena for which pursuit of microscopic details does not lead us to their understanding.[58] The idea of the contemporary computational physics is reductionism through and through; if you went to the ultimate microscopic description of a system, you would get the truth, nothing but the truth. The reason why this strategy is often regarded as the frontier of science is that we do not know what is important and what not for most phenomena. We do not know where we need not pay close attention. Rigorous solutions are needed in mathematics if we do not have much insight into the problem. Or, rigorous proofs are needed because our 'math sense' is feeble.[59] Therefore, our goal must be that we can have insight without any unnecessary computation and must not be that we can perform any computation.

3.5 ABC of renormalization

The purpose of this section is to use the supposedly simplest example to have a hands-on experience of renormalization methods. The example, though simple, exhausts all the key points.

Self-similar geometrical objects like the von Koch curve (Fig. 3.4)[60] appear often in simple systems without representative scales. For example, for chaos, the microscopic scale is expanded indefinitely, so there is no small fundamental scale. If a trajectory cannot cover a region densely, it is often the case that we have such a self-similar structure transversal to the orbits (as in the Lorenz system in Section 2.1). The system producing chaos is simple, so

---

[57] Even if one admits that materials consist of atoms and molecules, the assertion that they simultaneously obey mechanics is a transcendental assertion. Of course, there is an (overwhelming majority) opinion that even if this assertion cannot be directly verified, if the idea always works well, we may conclude that the assertion is correct. However, when it is asserted that something always works, it is not rare that 'always' is actually due to biased sampling; it is often the case that we see only things we wish to see. Where there is a logical hole, there is very often a real hole.

[58] Here, 'whiteness' is a metaphor of an interesting property we wish to understand. There can be an opinion that atoms and molecules look irrelevant simply because we pay attention to such an uninteresting secondary property. However, the existence of highly nontrivial phenomenologies almost free from material details (as illustrated in Section 3.1) should be enough to silence this opinion.

[59] according to Kunihiko Kodaira

[60] H. von Koch (1870-1924) proved, for example, that if the Riemann hypothesis holds, the discrepancy between the true distribution of prime numbers (not exceeding X) and the Gauss's distribution $\int_2^X dx/\log x$ is bounded by a number proportional to $\sqrt{X}\log X$.

repetition occurs.

The length of the von Koch curve depends on how we measure it (on what scale we observe the curve). Our problem is a simplified version of how to measure the length of a coastal line. Such self-similar curves are often called *fractal curves*.[61] The problem of measuring the coast length has a deeper meaning than is understood by the fractal aficionados, because the length is always fluctuating due to waves, and in the limit of accuracy, the coast itself becomes meaningless (non-existent). This illustrates that the existence of objects in the microscopic world and that on our scale can be fairly different.

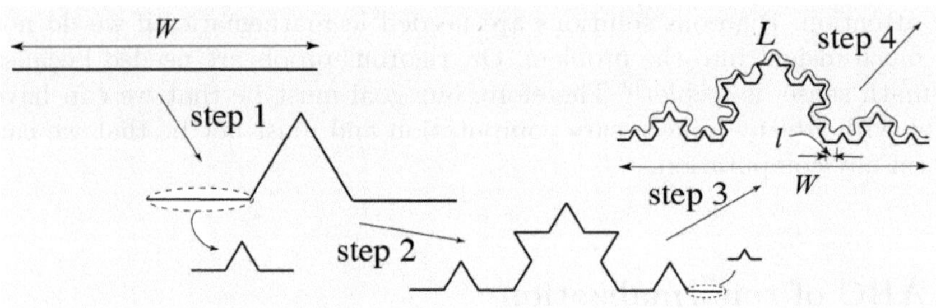

Fig. 3.4 How to construct the von Koch curve. First, take a line segment of length W and make a segment of length 1/3 of the former. Prepare four such segments and make a piecewise linear figure with a triangular mound at the center. This is Step 1. Next, each linear segment of this figure (a representative is encircled with a broken ellipse) is replaced by a 1/3 shrunk copy of the whole figure. This is Step 2. The resultant figure at the bottom of the figure is made of unit segments (monomer units) of length $\ell = W/9$. These length $W/9$ monomer units are all replaced by the small copy of the figure constructed in Step 1 as shown just below the Step 3 arrow. After Step 3 the length of the 'monomer unit' is $W/3^3$. Now, we repeat this procedure *ad infinitum* beyond Step 4, and we will obtain the von Koch curve. In this section, we regard the length of the 'monomer unit' to be finite, so we repeat these steps only finite times. However, since the number of repetition n is large, the length of the monomer unit $\ell = W/3^n$ is invisibly small.

Let ℓ be the length of the smallest unit of the curve ('the atom size'), L be the total length of the curve, and W be, as in Fig. 3.4, the span of the curve. From these lengths, we can make two independent dimensionless quantities: L/W and W/ℓ, so dimensional analytically (see Appendix 3.5A) we can conclude

$$\frac{L}{W} = f\left(\frac{W}{\ell}\right), \tag{3.10}$$

[61] About fractals, see the book by the originator of the concept: B. B. Mandelbrot, *Fractal Geometry of Nature* (W. H. Freeman, San Francisco, 1983). If attractors of chaotic systems are visualized with the aid of computer graphics, numerous fractal-like figures can be seen. An unbiased first impression would be that there are no such rather monotonous repetitive patterns are in Nature. Those who are impressed by computer-graphic fractals are likely to be the people who have never been overwhelmed by the diversity and complexity of a forest. One must feel the distinction between 'complex systems' and not complex systems as fractal curves.

where f is an appropriate function. As is explained in Appendix 3.5A, this is required by the invariance of the relation under changing the unit of length. According to the ordinary common-sense dimensional analysis instruction, we may ignore the dimensionless quantities whose values are too large or too close to zero. If f in (3.10) does not diverge in the $\ell \to 0$ limit, then $L \propto W$. That is, the conventional instruction works. For example, the circumference L of the regular n-gon with the edge length of ℓ inscribed in the circle of diameter W reads

$$L = \frac{\pi \ell}{\text{Arcsin}(\ell/W)} = W \frac{\pi(\ell/W)}{\text{Arcsin}(\ell/W)}. \tag{3.11}$$

Therefore, indeed this has a structure of (3.10) and $f(x) = \pi/[x\text{Arcsin}(1/x)]$ converges to π in the $x \to \infty$ limit. Thus, the ordinary instruction is very reasonable that recommends ignoring W/ℓ if it is extremely large.

However, for the von Koch curve, we can easily see that this is not the case. Although the 'hidden length' ℓ is disparate from W, we cannot ignore it (there is a more systematic explanation in Appendix 3.5B). This is exactly the scenario of interference between disparate scales. The nonlinearity in this example is in the relation between the number of units (on a certain scale) and the total span of the curve W. However, we will see that the effect of this hidden scale (or the effect of the divergence in the $\ell \to 0$ limit) does not show up in the result haphazardly. That is, the length of the von Koch curve has the structure of phenomenology.

Let us check the above assertion beforehand without using renormalization. To this end f is determined in an elementary fashion. Applying n-times the procedure explained in Fig. 3.4 to construct the von Koch curve, we realize that the monomer unit length is $\ell = W/3^n$, and that the total length of the curve is $L = \left(\frac{4}{3}\right)^n W$. Since

$$n = \log_3(W/\ell), \tag{3.12}$$

we obtain

$$L = W^{\log 4/\log 3} \ell^{1-\log 4/\log 3}. \tag{3.13}$$

That is, $f(x) = x^{\log 4/\log 3 - 1}$. If we collect von Koch curves with various values of ℓ (i.e., the curves with different microscopic structures), their 'true' length L as a function of W is always proportional to $W^{\log 4/\log 3}$ (the universal mathematical structure). However, the proportionality constant between $W^{\log 4/\log 3}$ and L varies (the phenomenological adjustable parameter), since it is sensitive to ℓ; although ℓ varies, for each example of the von Koch curve, it is a definite constant. Notice that the 'phenomenological parameter' $\ell^{1-\log 4/\log 3}$ diverges in the $\ell \to 0$ limit. That is, it is sensitive to ℓ.

This is an elementary example, but the structure of phenomenology stated in Section 3.1 appears clearly in the result. Although the value of the phenomenological parameter depends on the detail ℓ, notice that the functional form $\ell^{1-\log 4/\log 3}$ is also universal to all the von Koch curves. If we pay attention only to macroscopically easily observable $L \propto W^{\log 4/\log 3}$, since both L and W have the dimension of length, the result seems to suggest that the dimension of L deviates from the simple dimensional analytical result $L \propto W$ by $\log 4/\log 3 - 1$. This deviation is called the *anomalous dimension*. This is a telltale sign of a hidden dimensional quantity that cannot be ignored.

Let us first perform strategy (1), the Wilson-Kadanoff renormalization: "To observe the system far away with a constant eyesight." Each step of this procedure is (see Fig. 3.5):

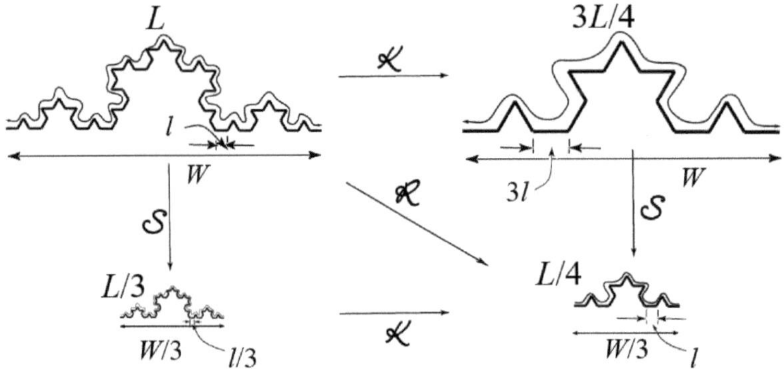

Fig. 3.5 Wilson-Kadanoff renormalization illustrated with the aid of the von Koch curve. $\mathcal{K}$ is the coarse-graining (the Kadanoff transformation) and $\mathcal{S}$ is the scaling transformation (1/3 shrinking). In this case, they commute. Their composition is the renormalization group transformation $\mathcal{R}$. If W is infinitely large, then it preserves the figure. Therefore, the features surviving the transformation must be global features (the features that can be observed from a distance). The procedure (i) in the text is the transformation from the upper left to the right, and (ii) is the one from the upper right downward.

(i) Perform coarse-graining (this procedure is called the Kadanoff transformation $\mathcal{K}$[62]); During this procedure, the size of the figure is intact: $W = \mathcal{K}(W)$. However, the size of the smallest unit (the monomer unit) becomes larger: $\ell \to \mathcal{K}(\ell) = 3\ell$. Therefore, the total length of the curve becomes $L \to \mathcal{K}(L) = (3/4)L$. Next,
(ii) Apply a scaling transformation $\mathcal{S}$ that shrinks the whole figure uniformly so that the size of the smallest unit returns to that of the original figure

[62] What is the reason that Professor Kadanoff has not been awarded a Nobel Prize? It is likely because the judges have not experienced creative breakthrough. It was Kadanoff who saw the mathematical core of the procedure. There are people who say that he should have done one more step (at least in retrospect), but if one is convinced (perhaps unconsciously) that he has grasped the crux of the matter, it is all too natural to relax at that point; in any case checkmate is inevitable several moves away. In these days actually demonstrating every step to checkmate seems excessively appreciated (because everyone can understand what is going on). This is amply proved by the appallingly late Nobel Prize for Y. Nambu.

(scale the whole figure so that the monomer size becomes the same as the original figure): $\mathcal{S}(\ell) = \ell/3$ (indeed, $\mathcal{S}(\mathcal{K}(\ell)) = \ell$). Obviously, $\mathcal{S}(L) = L/3$ and $\mathcal{S}(W) = W/3$ hold.

The composition of the above two transformations $\mathcal{R} \equiv \mathcal{SK}$ is called the *renormalization group transformation.* By construction, obviously, it preserves the microscopic structure of the von Koch curve. $\mathcal{R}^2$ is also a renormalization transformation, so we say the totality of such transformations makes a renormalization group (RG).[63] Since $\mathcal{R}(L) = L/4$, we obtain

$$\mathcal{R}^n(L) \equiv L(n) = L/4^n. \tag{3.14}$$

Here, $\mathcal{R}(W) = W/3$. To count the number of renormalization group transformations applied, let us introduce the *dilation parameter* n with the aid of $W(n) = \mathcal{R}^n(W) = W/3^n$ as $n = \ln\{W/W(n)\}/\ln 3$. Then, (3.14) may be rewritten as

$$L = 4^n L(n) = \left(\frac{W}{W(n)}\right)^{\ln 4/\ln 3} L(n). \tag{3.15}$$

If we choose n appropriately to make $L(n)$ and $W(n)$ of order 1, we obtain essentially (3.13).

Next, let us discuss procedure (2), the Stückelberg-Petermann renormalization procedure: "To pursue invariants under changing microscopic details." The basic idea is: shake the system and remove the most shaken parts, and then the rest should be universal. In the current example $L/\ell^{1-\log 4/\log 3}$ is the invariant that is not shaken even if ℓ is shaken. That is, if we could find such a quantity, we could separate out the universal structure in the phenomenological description.

The quantities a macroscopic observer knows about the von Koch curve are W, the scale of observation (resolution) λ, and the length $\tilde{L}$ measured on this scale. The true length L is not observable, but this and $\tilde{L}$ should be proportional (if the observation scale is fixed):

$$\tilde{L} = ZL. \tag{3.16}$$

This may be called "the condition of realism." If this relation is not assumed, $\tilde{L}$ does not tell us anything about the object.[64] The divergence in the limit of small $\ell \to 0$ (the 'smallness' of a quantity is meaningless unless it is compared with something else, so, correctly speaking, in the limit of $\ell/\lambda \to 0$) cannot be observed as long as our resolution is λ. Consequently,

---

[63] Actually, the inverse transformation is not defined, so it is not a group. However, since it has the unit element (that corresponds to doing nothing), the totality of these transformations makes a monoid which is slightly more structured than a semigroup.

[64] Things we can never observe may be assumed to be non-existent, and it is conceivable to assume only $\tilde{L}$ is real. However, this is not the way we do physics at present.

Z must cancel the divergence in L so that it cannot be seen in $\tilde{L}$. Such a constant that absorbs divergence is called a *renormalization constant.* If we perform a single step in the construction of the von Koch curve in Fig. 3.4 (ℓ is shrunk by factor 3 to $\ell/3$), the total length of the curve is multiplied by 4/3, so the divergence of L in the $\ell \to 0$ limit should behave as $(4/3)^{-\log_3 \ell} = \ell^{1-\log 4/\log 3}$.[65] Therefore, the renormalization constant Z should be selected to remove this divergence $\ell^{1-\log 4/\log 3}$ (i.e., ZL does not have any divergence): $Z(\lambda/\ell) \propto (\lambda/\ell)^{1-\log 4/\log 3}$; Z is dimensionless, so it must be a function of the dimensionless quantity ℓ/λ (if L/ℓ is chosen instead, we lose the proportionality between L and $\tilde{L}$, so no other choice is possible).

λ is the quantity the observer has introduced, and has nothing to do with the system (the von Koch curve) itself. Therefore, the 'true' length L cannot depend on λ (this is a consequence of our belief that the external world exists).[66] That is, if a system (model) is fixed (i.e., ℓ and W are fixed), then even if λ is altered, L cannot be affected:

$$\lambda \frac{\partial L}{\partial \lambda} = 0. \tag{3.17}$$

Here, the multiplying extra λ has no deep meaning; it is only to make the subsequent formulas streamlined. Its essence is $\partial L/\partial \lambda = 0$: "The world exists irrespective of our existence."[67] This condition can also be said as 'the condition of objectivity.'

The quantity the macroscopic observer knows are W, the actually observed $\tilde{L}$, and the observation scale λ, so the result of her dimensional analysis reads:

$$\frac{\tilde{L}}{\lambda} = f\left(\frac{W}{\lambda}\right). \tag{3.18}$$

f here is also a well-behaved function. From (3.16) and this, we have

$$L = Z^{-1}\lambda f\left(\frac{W}{\lambda}\right). \tag{3.19}$$

---

65 If we write $L = g(\ell)$, $g(x/3) = (4/3)g(x)$ is required. Thus, $g(3^n) = (4/3)^{-n}g(1)$. Therefore, although there are many different solutions, $g(x) \propto (4/3)^{-\log_3 x} = x^{1-\log 4/\log 3}$ is a solution.

66 Needless to say, no one can prove the existence of the world beyond our cognition. The belief here is, precisely speaking, the belief in the notion that there is no contradiction due to assuming the existence of the external world. Phenomenologically speaking (or if Husserl's *Ideas of Phenomenology* is mimicked), it is the belief in the notion that our cognition certainly imparted to us can hit what is not given to us or in the notion that there is a reason for the 'undoubtedness' of the external world.

67 But, as discussed in a previous footnote, the microscopic world need not exist definitively in the macroscopic sense. Just as with Monet's waterlily, there need not be any internal structure when closely observed. The meaning of 'existence' is subtle, and need not be the same as our daily sense.

Introducing this into (3.17), we obtain (it is wise to compute the logarithmic derivative $\partial \log L/\partial \log \lambda$)

$$f(x) - \alpha f(x) - x f'(x) = 0, \tag{3.20}$$

where

$$\alpha \equiv \partial \log Z/\partial \log \lambda. \tag{3.21}$$

Such equations as (3.17) or its consequence (3.20) are called *renormalization group (RG) equations*. Especially if α converges to a constant in the $\ell \to 0$ limit (for the current example, it converges to $\alpha = 1 - \log 4/\log 3$), (3.20) becomes an equation governing the universal asymptotic relation independent of the microdetails.

Solving (3.20), we obtain

$$f(x) \propto x^{1-\alpha}, \tag{3.22}$$

or

$$\tilde{L} \propto W^{1-\alpha}\lambda^{\alpha} \propto W^{\log 4/\log 3}. \tag{3.23}$$

Thus, the universal part of the phenomenological result obtained already has been reproduced.

The renormalization group equation is derived from an almost trivial condition, but combining this and the requirement that the λ/ℓ-dependence of the observable can be absorbed into the renormalization constant can produce nontrivial results. The requirement that the divergence can be absorbed in the renormalization coefficient is nothing but the requirement of the existence of a phenomenological description (of the type discussed in Section 3.1). This requirement may not always be fulfilled. However, to require renormalizability is a natural attitude for empirical scientists, when empirically there is a phenomenology. Eventually, the renormalizability must be demonstrated mathematically, but this is a highly nontrivial task, so it is a productive strategy to assume renormalizability and pursue the consequences. The requirement of renormalizability imposes strong constraints on the structure of the observable quantities. Therefore, with other partial information, such as the result of perturbation calculation, we can arrive at fairly nontrivial results. The reader must clearly recognize, however, that perturbation is a mere tool and has nothing intrinsically to do with renormalization itself.

To illustrate the preceding paragraph, let us study the von Koch curve from a perturbative point of view. This example can be studied exactly, but such examples are extremely rare, so its perturbative study is not useless. For the actual von Koch curve, when one step of its construction illustrated in Fig. 3.4 is performed, the microscopic unit length is changed from ℓ to $\ell/3$, and the total length (true length) L to $4L/3$. 4/3 is fairly different from 1, but let us write this expansion ratio as $e^{\varepsilon} \simeq 1 + \varepsilon$ and assume ε to be small. After n construction steps, to the first order in ε, we obtain

$$W \to L = (1 + n\varepsilon)W. \tag{3.24}$$

This equation is reliable only when $\varepsilon n \ll 1$. In other words, this perturbation cannot be used uniformly in n.[68] If n is expressed in terms of ℓ, we get

$$L = \{1 + \varepsilon \log_3 (W/\ell)\} W. \tag{3.25}$$

Expanding the renormalization coefficient as $Z = 1 + \varepsilon A + \cdots$, we choose A to remove the divergence in the $\ell \to 0$ limit. However, it is convenient to rewrite (3.25) as follows by introducing λ:

$$L = \{1 + \varepsilon [\log_3 (W/\lambda) + \log_3 (\lambda/\ell)]\} W. \tag{3.26}$$

Now, the equation corresponding to (3.16) can be expanded as

$$\tilde{L} = ZL = \{1 + \varepsilon [A + \log_3 (\lambda/\ell)] + \varepsilon \log_3 (W/\lambda)\} W. \tag{3.27}$$

Therefore, if we choose $A = -\log_3 (\lambda/\ell)$, the divergence is absorbed into Z. The resultant formula

$$\tilde{L} = \{1 + \varepsilon \log_3 (W/\lambda)\} W \tag{3.28}$$

is called the renormalized perturbation series (although here we have only two terms). From this, it is natural to guess

$$\tilde{L} \sim W^{1+\varepsilon/\log 3}, \tag{3.29}$$

but this exponentiation is required by the renormalization group equation (3.20). The only difference from what we did before is that Z is determined perturbatively here:

$$Z = 1 - \varepsilon \frac{1}{\log 3} \log(\lambda/\ell). \tag{3.30}$$

Introducing this into the definition of α (3.21), we obtain $\alpha = -\varepsilon/\log 3$, so (3.23) concludes the above 'natural result' $\tilde{L} \propto W^{1+\varepsilon/\log 3}$. We wish to set $\varepsilon = \log 4 - \log 3$; in the present case, the result happens to agree with the exact answer.

The explanation up to this point simplified the actual situation by assuming that disparate scales interfere without any restriction and that microscopic effects have indefinitely large effects on macroscopic observables. Generally speaking, it is rare that the materials parameters change their values when we change our observation scale. For example, the density and viscosity in the Navier-Stokes equation observed at 1mm and at 1m are the same. If an observable settles down to a particular value in the $\zeta = \lambda/\ell \to \infty$

68 In this sense, the term proportional to n corresponds to the *secular term* in *singular perturbation* theory (see Section 3.6).

limit, this means that the effect of the microscopic parameters drops out from macroobservables. However, this contradicts the materialistic diversity of the world. This implies that $\ell \to 0$ and $\lambda \to \infty$ are not the same thing in contrast to the treatment above. Therefore, we must conclude that there is more than one dimensionless quantity (at least one more other than ζ) constructed from lengths. Since the properties of materials are determined by microscopic details, between our observation scale λ and the atomic scale ℓ must be some length scale ξ, and materials constants are determined by the world of the scales up to ξ (they depend on the dimensionless quantity ξ/ℓ). For example, the correlation length is such a length, which decouples λ and ℓ, so the classical instruction for the dimensional analysis often works.

To extract a macroscopic phenomenology from a microscopic description of a system is, in short, to derive the description of the system on the scale of ξ.[69] Then, what determines the correlation length ξ? It is not determined solely by microscopic parameters (say, the size of molecules). As can be seen from statistical mechanical calculation, it depends on the energy (temperature) of the system and other fields (say, magnetic field). At the critical point, ξ diverges, so our observation scale directly confronts the microscopic scale ($\xi > \lambda$), but away from criticality, ξ is not at all large.

How much is the idea of RG useful to compute the property of the system on the ξ scale? It is in principle possible to study how observables change when ξ/ℓ is varied, but as has been seen in the above simple example, the renormalization group idea is really effective, when asymptotic analysis is possible. Whether asymptotic analytical results are useful or not depends on the individual details of the system, because ξ/ℓ is not actually very large. There is no general theory that can tell us when asymptotic results become meaningful.

This concludes the explanation of the ABC of renormalization group theory with a bit of general comments. Immediately after this should be detailed illustrations of renormalization procedures for the actual examples mentioned in Section 3.1, but there is no new concept nor new idea needed. Here, some supposedly useful references to learn established methods for critical phenomena and polymer solutions are mentioned in the footnote.[70] The accom-

---

[69] If the correlation length is small, to suppress the effect of thermal fluctuations, we need a larger scale so that the law of large numbers holds.

[70] **⟨⟨Self-study guide to low-energy-physics renormalization group⟩⟩** To practice the Wilson-Kadanoff approach in critical phenomena the Ising model on a triangular lattice is the best. This is discussed in most statistical mechanics textbooks. An excellent explanation of the Stückelberg-Petermann approach (the so-called field-theoretical renormalization group theory) for a minimal model called the Ginzburg-Landau model (See Chapter 4) can be found in M. Le Bellac, *Quantum and Statistical Field Theory* (Oxford 1991). For 2D XY model see Chapter 9 of P. M. Chaikin and T. C. Lubensky, *Principles of Condensed Matter Physics* (Cambridge University Press, 2000). This book also gives a good explanation of renormalization group theory. A newer book, I. Herbut, *A modern approach to critical phenomena* (Cambridge Univer-

panying supplementary pages[71] contain relevant lecture notes with detailed calculated examples.

The rest of this chapter is devoted to an introductory exposition of renormalized perturbation approach to singular perturbation theory and system reduction.

Appendix 3.5A. Introduction to dimensional analysis

Usually, a certain physical quantity, e.g., mass or length, is expressed by numbers indicating how many certain units the quantity corresponds to. The statement, "the length of this stick is 3," does not make sense; we must say, for example, "The length of this stick is 3 m," with an appropriate unit. The number itself appearing with a unit has no meaning of its own ($\underline{3}$ m and $\underline{9.8425\cdots}$ ft are identical). That is, we can freely scale them (by changing the units). However, the statement, "the length ratio of this and that sticks is 4" makes sense irrespective of the adopted unit of the length. The quantity whose numerical expression does not depend on the choice of units is called a *dimensionless quantity*. The number 4 above is an example. In contrast to the number 3 above, which depends not only on the property of the stick alone but also on how we observe it (how we measure it), this 4 does not depend on our observation mode.

An expression of a quantitative relationship among physical observables is a relationship among several numbers. If the relationship exists independent of us (e.g., does not depend on how we look at it), then the truth value of the relationship should at least be free from the choice of units. For example, a relationship that is correct only when the length is measured in meters is not pleasant (it does not capture an essential relationship; at least it is inconvenient).

If we alter the units of a quantity, its numerical value is scaled; the numbers using the same unit must be scaled in the same way (if one does not wish to write units meticulously everywhere, the length must be declared to be measured, say, in inches, and everywhere the convention must be applied). The numerical values corresponding to various quantities measured in different (independent) units may be scaled differently. If the numerical values expressing two quantities always scale identically when the units are changed, we say these two quantities have the same *dimension*. In other words, scientists and engineers express "independently scalable" as "having distinct dimensions." Objectively meaningful quantitative relationships should remain true irrespective of the choice of the units. Therefore, such relationships must be expressible in terms of dimensionless quantities (quantities that are not affected by unit changes). To analyze problems according to this requirement is called *dimensional analysis*.

Let us look at a simple example, the 1D diffusion equation considered in Chapter 1:

$$\frac{\partial \psi}{\partial t} = D\frac{\partial^2 \psi}{\partial x^2}, \tag{3.31}$$

---

sity Press, 2007), is also excellent. For polymer solutions, Y. Oono, "Statistical physics of polymer solutions. Conformational-space renormalization group approach," Adv. Chem. Phys., **61**, Chapter 5 (i.e., 301-437) (1985) is reasonably kind.

[71] http://www.yoono.org/NonlinearWorldSpringer/Nonlinear_World_Supplements.html

where t is the time, x is the position coordinate, ψ is, e.g., the temperature, and D is the diffusion constant. The integral Q of $\psi(x,t)$ over the whole space must be a time-independent constant.

There are three distinct physical units: time, length, and the unit of the quantity Q in this equation. Let us write these dimensions as T, L, and M, respectively: the usual convention is: $[X]$ = the dimension of X, so we write $[t] = T$, $[x] = L$, and $[Q] = M$. ψ is the density of the quantity Q, so in 1-space $[\psi] = M/L$. Differentiation is essentially division, so $[\partial\psi/\partial t] = M/LT$. We wish this equation to hold irrespective of our choice of the units. Then, the both sides of the equation should have the same dimension (we say the equation is dimensionally *homogeneous*[72] Then, $M/LT = [D]M/L^3$ is required, so $[D] = L^2/T$.[73] We can assert that objectively meaningful relationships can always be expressed in a homogeneous form.

To perform the dimensional analysis of the diffusion equation (3.31), we must first construct dimensionless quantities (scaling-invariant quantities). Combinations whose powers of T, L, M are all zero are such quantities: tD/x^2 and $\psi x/Q$ or $\psi\sqrt{tD}/Q$.[74] For example, we can check $[tD/x^2] = T\cdot(L^2/T)/L^2 = 1$. Dimensionless quantities may always be considered as functions of dimensionless quantities (this assertion is easy to understand if we imagine what happens otherwise), so the solution to (3.31) must have the following form:

$$\psi = \frac{Q}{\sqrt{Dt}} f\left(\frac{x^2}{Dt}\right), \tag{3.32}$$

where f is a well-behaved function.[75] Thus, (3.31) becomes an ordinary differential equation for f.

Here, a remark is in order. Whether we may independently scale two different physical quantities or not can be a problem. For example, in ordinary engineering problems, energy E and mass m have distinct units, but for relativistic phenomena, it is much more natural to regard both as having the same unit. The reason why there is a 'conversion factor' c^2 in Einstein's relation $E = mc^2$ is simply the non-relativistic convention; the speed of light c is an absolute constant, so c must be regarded as a dimensionless quantity. This concludes that T and L should not be distinguished. In the non-relativistic world, however, this identification is not convenient; we should regard time and length in different categories. (We may perhaps say that c is so large that it cannot be distinguished from infinity, so it drops out of the theory.)

Dimensional analysis is often fundamental. It is clear dimension-analytically that classical physics cannot explain the atomic structure. In a hydrogen atom

---

72 It is said that the importance of dimensional homogeneity was pointed out first by Fourier.).

73 The unit of the diffusion constant in SI is indeed m^2/s.

74 ⟨⟨**Number of independent dimensionless quantities**⟩⟩ There is no other independent combination; other dimensionless quantities are written in terms of the products of some powers of these quantities. The general theory counting the number of independent dimensionless quantities is called *Bridgeman's Π-theorem*, but such a general statement may not be needed; there are three independent dimensions T, L, M and 5 quantities Q, ψ, D, t, x, so $5 - 3 = 2$ independent dimensionless quantities can be constructed. As seen here, the Π-theorem is a restatement of an elementary theorem in linear algebra.

75 The meaning of "well-behaved" depends on the context, but usually, it means sufficient smoothness (differentiability) and boundedness in the domain under study.

an electron is bound to a proton by the Coulomb interaction. Therefore, its Newton's equation of motion reads

$$m\frac{\mathrm{d}^2}{\mathrm{d}t^2}\boldsymbol{r} = -\frac{e^2\boldsymbol{r}}{4\pi\varepsilon_0 r^3}, \tag{3.33}$$

where e is the electron charge, and m is the electron mass. $4\pi\varepsilon_0$ always appears with e^2, so there are only two basic quantities: m and $e^2/4\pi\varepsilon_0$. Their dimensions are $[m] = M$, $[e^2/4\pi\varepsilon_0] = ML^3/T^2$ (this is required by the dimensional homogeneity of the equation of motion). There is no way to construct a quantity with the unit of length from them. However, if Planck's constant h, whose dimension is $[h] = ML^2/T$ (recall that h times frequency is the photon energy), is relevant as Bohr thought, then, since

$$[e^2/m\varepsilon_0] = L^3T^{-2}, \quad [h/m] = L^2T^{-1}, \tag{3.34}$$

we can solve L as $(h/m)^2/(e^2/m\varepsilon_0)$:

$$[(h/m)^2/(e^2/m\varepsilon_0)] = (L^2T^{-1})^2/(L^3T^{-2}) = L. \tag{3.35}$$

$(\varepsilon_0 h^2/me^2)/\pi$ is the Bohr radius $= 0.53$ Å $= 0.053$ nm (nanometer; 1 nm $= 10^{-9}$ m), which is roughly the size of the hydrogen atom in its ground state. Actually, this argument convinced Bohr that h is necessary.

Discussion 3.3. Derive Kepler's Third Law dimensional analytically. □

Discussion 3.4. A wonderful book by Migdal[76] begins with a dimensional analytical 'proof' of Pythagoras' theorem. The argument relies on the expression of the area $S = a^2 f(\alpha)$ of an orthogonal triangle whose smallest angle is α with the length of its hypotenuse being a (Fig. 3.6). As can be seen from the figure $a^2 + b^2 = c^2$. (Can we prove $S = a^2 f(\alpha)$ within the Euclidean geometry?[77]) □

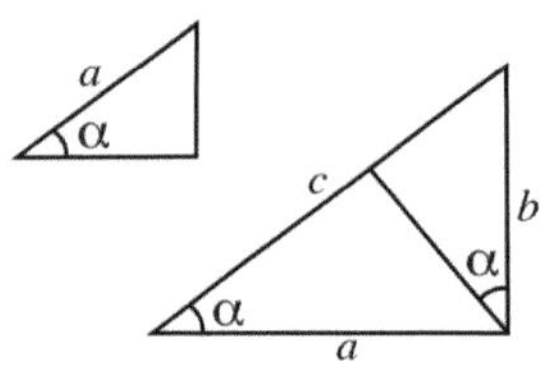

Fig. 3.6 Pythagoras' theorem. This dimensional analytic result does not contradict the result of Euclidean geometry. However, is this a respectable proof of the theorem? How can you demonstrate $S = a^2 f(\alpha)$? Is Migdal logical?

Appendix 3.5B. Dimensional analysis and renormalization

Any relationship meaningful in physics may be written as a relationship among

[76] A. B. Migdal, *Qualitative methods in quantum theory* (translated by A J. Leggett) (Westview Press, 2000).

[77] The argument here only implies that the key issue of the proof is, logically speaking, to demonstrate $S = a^2 f(\alpha)$. This requires far more subtle theories than the usual elementary proof(s) of the theorem (we need at least Book 5 (ratios and proportions) and the next volume about similar figures).

dimensionless quantities (the quantities indifferent to the choice of units, see the preceding Appendix 3.5A). Such relationships generally have the following form:

$$\Pi = f(\Pi_0, \Pi_1, \cdots, \Pi_n), \tag{3.36}$$

where Π and Π_i $(i = 0, 1, \cdots, n)$ are dimensionless quantities and f is a certain function. In the following, we review in a more general fashion what we have seen in this section. If we read a standard explanation of dimensional analysis, it is usually stated that among the dimensionless quantities "extremely large ones and ones too close to zero may be ignored." This recommendation may be formalized as follows. Suppose Π_0 is very small. "The following limit

$$f(0, \Pi_1, \cdots, \Pi_n) = \lim_{\Pi_0 \to 0} f(\Pi_0, \Pi_1, \cdots, \Pi_n) \tag{3.37}$$

'exists,' so we may consider Π without Π_0." This procedure is the assertion that we may ignore any interference between disparate scales.

The key effect of nonlinearity arises when this assertion is false. If the limit in the right-hand side of (3.37) does not exist, essentially nonlinear phenomena can be observed. Let us assume that all the dimensionless quantities in (3.36) include a quantity that cannot be observed directly (quantities that cannot be operationally defined; the unit segment length ℓ of the von Koch curve or the monomer unit of polymers are examples). In the $\Pi_0 \to 0$ limit, these dimensionless quantities may all diverge (going to zero is also a kind of divergence in the present context). Let us separate Π_0 into the part dependent on the observation condition $\Pi_O(\lambda)$ and the part that goes to zero $\Pi_{NO}(\lambda)$ (recall the separation performed in (3.26)), where λ is the parameter specifying the scale of observation. If we describe a phenomenon that can be directly observed, its phenomenology must be written in terms of renormalized quantities. Let us introduce renormalized quantities denoted by the suffix R and the renormalization constant Z_i as follows:

$$\Pi_R = Z\Pi, \quad \Pi_{Ri} = Z_i \Pi_i. \tag{3.38}$$

The renormalization constants are chosen to absorb divergences in the $\Pi_{NO} \to 0$ limit. Introducing the above definitions into (3.37), we may convert the relation to the relation among renormalized quantities. Suppose the result reads:

$$\Pi_R = F(\Pi_{R1}, \cdots, \Pi_{Rn}). \tag{3.39}$$

The renormalization group equation comes from the fact that λ has nothing to do with the system being observed,

$$\lambda \frac{\partial \Pi}{\partial \lambda} = 0. \tag{3.40}$$

Rewriting (3.39) as

$$\Pi = Z^{-1} F(Z_1 \Pi_1, \cdots, Z_n \Pi_n) \tag{3.41}$$

and introducing this into the renormalization group equation (3.40), we obtain with the aid of the chain rule

$$-\alpha \Pi_R + \sum_i \alpha_i \Pi_{Ri} \frac{\partial \Pi_R}{\partial \Pi_{Ri}} = 0. \tag{3.42}$$

Here,

$$\alpha \equiv \frac{\partial \log Z}{\partial \log \lambda}, \quad \alpha_i \equiv \frac{\partial \log Z_i}{\partial \log \lambda}. \tag{3.43}$$

In the $\Pi_{NO} \to 0$ limit, let us assume that α, α_i have limits.

The renormalization group equation (3.42) may be solved with the aid of the *method of characteristics* (see Note 3.4 at the end of this note). The *characteristic equation* becomes

$$\frac{\mathrm{d}\Pi_{Ri}}{\alpha_i \Pi_{Ri}} = \frac{\mathrm{d}\Pi_R}{\alpha \Pi_R} = \frac{\mathrm{d}\rho}{\rho} \tag{3.44}$$

with the introduction of the *dilation parameter* ρ, which is introduced to parameterize the characteristic curve. To solve this, we first separate it as

$$\frac{\mathrm{d}\Pi_{Ri}}{\mathrm{d}\rho} = \alpha_i \frac{\Pi_{Ri}}{\rho}, \tag{3.45}$$

$$\frac{\mathrm{d}\Pi_R}{\mathrm{d}\rho} = \alpha \frac{\Pi_R}{\rho}, \tag{3.46}$$

and then solve each as follows:

$$\Pi_{Ri} = C_i \rho^{\alpha_i}, \quad \Pi_R = C\rho^{\alpha}. \tag{3.47}$$

Using the fact that the general relation among the integral constants C, C_i becomes the general solution, we obtain the general solution to (3.44) as

$$\Pi_R = \rho^{\alpha} \Phi(\Pi_{R1}\rho^{-\alpha_1}, \cdots, \Pi_{Rn}\rho^{-\alpha_n}), \tag{3.48}$$

where Φ is a well-behaved function. This relation must hold for any (positive) ρ. Therefore, we may set, for example, $\rho = \Pi_{R1}^{1/\alpha_1}$. This gives the following relation:

$$\Pi_R = \Pi_{R1}^{\alpha/\alpha_1} g(\Pi_{R2}^{\alpha_1}/\Pi_{R1}^{\alpha_2}, \cdots, \Pi_{Rn}^{\alpha_1}/\Pi_{R1}^{\alpha_n}). \tag{3.49}$$

Here, g is also a well-behaved function and exponents α, α_i cannot be determined by dimensional analysis (since all the quantities are already dimensionless).

Even if the limit (3.37) does not exist, Barenblatt realized that still some sort of limit can be taken, and demonstrated that solving problems assuming the form of the solution as (3.49) is often successful.[78] He called this limit the *intermediate asymptotics* (of the second kind). The key ingredient of this concept is the relation that holds for a long time before the system reaches its really longtime asymptotic state, which is, for many systems, a sort of dead state (see Example 3.4, footnote 20, and the next chapter). In the actual calculation based on the intermediate asymptotic limit theory the form (3.49) is assumed and the exponents are selected (as a nonlinear eigenvalue problem) so that there is a solution. In any case, this theory does not ask the question why the form (3.49) is natural; it is an assumption.[79]

---

78 G. I. Barenblatt, *Similarity, Self-Similarity, and Intermediate Asymptotics* (Cambridge University Press, New York, 1996).

79 ⟪**History of renormalization group theory for differential equations**⟫ The author learned the problem of intermediate asymptotics in conjunction to the *Barenblatt equation*: $\partial_t u = [1+\varepsilon\Theta(-\partial_t u)]\partial_x^2 u/2$, where Θ is Heaviside's step function and ε is a (small) positive constant, from his longtime collaborator Nigel Goldenfeld. Goldenfeld had been interested in intermediate asymptotics for some years and expected the relevance of renormalization (as Barenblatt himself), but since the author,

It may not be necessary anymore, but let us review the von Koch curve. As before, let us write the actual length of the curve as L, the width as W, and the length of the smallest segment as ℓ. There are two dimensionless quantities L/W and ℓ/W, so dimensional analysis tells us that

$$L/W = f(\ell/W). \tag{3.50}$$

The right-hand side diverges in the $\ell \to 0$ limit, so the equation corresponding to (3.49) is

$$L/W \propto (\ell/W)^{-\alpha/\alpha_1}. \tag{3.51}$$

That is, we obtained $L \propto W^{1+\alpha/\alpha_1}$ again, naturally. However, we cannot determine the exponents within this formalism.[80]

Note 3.4. Method of characteristic equation
The idea of the method is used to solve the general first-order partial differential equations, but here we confine ourselves to *quasilinear first-order partial differential equations* on a domain U in an n-dimensional space:

$$\sum_{i=1}^{n} b_i(u, \boldsymbol{x}) \frac{\partial u}{\partial x_i} = c(u, \boldsymbol{x}). \tag{3.52}$$

A solution to this equation $u = f(\boldsymbol{x})$ describes a surface floating on U (Fig. 3.7; it is a hypersurface in the $(n+1)$-space). If we draw a curve $\boldsymbol{x} = \boldsymbol{x}(s)$ in U parameterized by s, a curve $z(s) = u(\boldsymbol{x}(s))$ is drawn on the solution surface (see Fig. 3.7; the curve is actually given by the position vector $(\boldsymbol{x}(s), z(s))$.). Conversely, if this curve z is known, we can construct a solution surface. Since

$$\frac{du}{ds} = \sum_i \frac{dx_i}{ds}\frac{\partial u}{\partial x_i}, \tag{3.53}$$

ignorant about the theory of intermediate asymptotics, did not pay much attention. However, on one day in 1987 or '88, his student gave a seminar on the Barenblatt equation, and the author knew the intermediate asymptotics problem. In the case of the Barenblatt equation the structure of the problem became clear as soon as perturbation calculation was performed (however, the perturbation term is so singular that one needs desperation to summon courage); the similarity of the divergence structure to that in polymer solutions (see Example 3.3) immediately clarified the situation. N. Goldenfeld, *Lectures on phase transitions and renormalization group* (Addison Wesley, 1992) Chapter 10 explains this [the original paper is N. D. Goldenfeld, O. Martin and Y. Oono, "Intermediate asymptotics and renormalization group theory," J. Scientific Comp. **4**, 355 (1989)].

The next important progress came when the Burgers equation was recognized as a renormalization group equation [L.-Y. Chen, N. Goldenfeld, Y. Oono, and G. Paquette, "Selection, stability and renormalization," Physica A **204**, 111 (1993)]. This implied that *singular perturbation theory* and *reductive perturbation theory* could be unified within the renormalization group theory. See Sections 3.7 and 3.9. After all, the relation to reductive perturbation was the essence of the renormalization group theory of differential equations. This view was perfectly vindicated by the next breakthrough, Hayato Chiba's qualitative theory of renormalization group equations [H. Chiba, "C^1 approximation of vector fields based on the renormalization group method," SIAM J. Appl. Dym. Syst. **7**, 895 (2008)]. See Section 3.8.

[80] That is, the theory of intermediate asymptotics corresponds to the scaling theory in critical phenomenon.

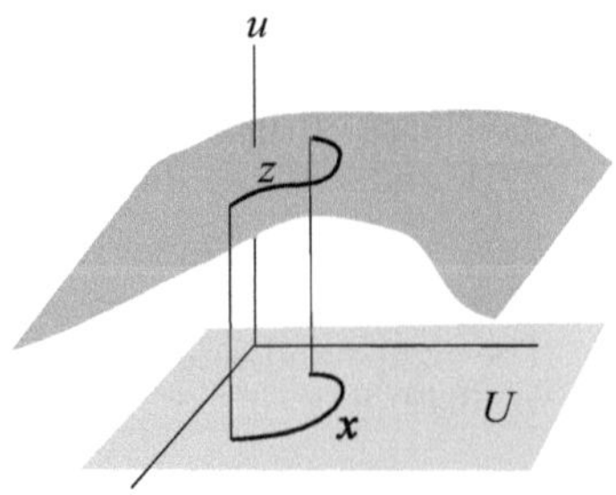

Fig. 3.7 The gray curved surface is a solution surface. The curve $\boldsymbol{x}$ on the plane is $\boldsymbol{x} = \boldsymbol{x}(s)$, and the curve above it in the solution surface is $z(s) = u(\boldsymbol{x}(s))$. [Actually, the trajectory of the vector $(\boldsymbol{x}(s), z(s))$.]

comparing this with (3.52) and choosing a curve $\boldsymbol{x}$ satisfying

$$\frac{dx_i}{ds} = b_i(u, \boldsymbol{x}), \tag{3.54}$$

we can write (3.53) as

$$\frac{du}{ds} = c(u, \boldsymbol{x}). \tag{3.55}$$

Thus obtained differential equations (3.54) and (3.55) are called *characteristic equations*. We can write them as

$$\frac{dx_1}{b_1} = \cdots = \frac{dx_n}{b_n} = \frac{du}{c} \left(= \frac{d\rho}{\rho} \right), \tag{3.56}$$

where $\rho = e^s$ is the *dilation parameter*. The differential equation (3.44) is an example. From this we obtain n equations. Integrating them gives the solution: $F_1(\boldsymbol{x}, u) = C_1, \cdots, F_n(\boldsymbol{x}, u) = C_n$, containing n integration constants.

Except for special cases, these equations describe n hypersurfaces mutually in the general position, so the common set of all these surfaces determines a curve floating on U. It is parameterized by $C_1, \cdots, C_n$. Since this curve must be embedded in a solution surface, moving this curve should make a solution surface. However, if we vary all the parameters $C_1, \cdots, C_n$ independently, the curve would fill a part of the $(n+1)$-space. A hypersurface is a subset of codimension 1, so if we specify one relationship among $C_1, \cdots, C_n$, then the curves determined by $C_1, \cdots, C_n$ should as a whole determine a hypersurface. Therefore, choosing a well-behaved function G, we demand

$$G(C_1, \cdots, C_n) = 0. \tag{3.57}$$

In other words,

$$G(F_1(\boldsymbol{x}, u), \cdots, F_n(\boldsymbol{x}, u)) = 0 \tag{3.58}$$

must be the general expression of the hypersurface determined by (3.52).

3.6 Longtime behavior and renormalization: a simple example

What we have clearly seen from such a simple example as the von Koch curve is that renormalization can be used as a tool of asymptotic evaluation. We started this chapter with the point of view that extracting a phenomenology is a sort of asymptotic evaluation, and that renormalization is a strategy for it, so there is no surprise. Let us proceed with the banner, "All the asymptotic evaluation in the world can be done by renormalization."

For the time being, in this section under this banner we will inspect in detail an example that convinces us that a renormalization group method allows us to understand the longtime behavior of differential equations. It is a very simple example, but later we will see it is an essential example.

We consider the following ordinary differential equation:

$$\varepsilon \frac{\mathrm{d}^2 y}{\mathrm{d}t^2} + \frac{\mathrm{d}y}{\mathrm{d}t} + y = 0, \tag{3.59}$$

where $\varepsilon > 0$ is a small constant. If this is zero, the solution decays exponentially ($A\mathrm{e}^{-t}$ is its general solution, where A is a numerical constant). If $\varepsilon > 0$, for a sufficiently small time, the second-order derivative term is important (notice that however small the mass may be, the inertial effect is crucial for a short time at the beginning of the motion). However, after a long time, the system behavior should be similar to the $\varepsilon = 0$ case. Then, why don't we take this term into account through perturbation? This problem is easily solved by hand exactly, but let us pretend that we cannot do so, and perform a perturbative calculation.

Let us expand the solution formally as

$$y = y_0 + \varepsilon y_1 + \cdots \tag{3.60}$$

and then introduce this into the equation. Equating the terms with the same power of ε, we obtain

$$\frac{\mathrm{d}y_0}{\mathrm{d}t} + y_0 = 0, \tag{3.61}$$

$$\frac{\mathrm{d}y_1}{\mathrm{d}t} + y_1 = -\frac{\mathrm{d}^2 y_0}{\mathrm{d}t^2}, \tag{3.62}$$

etc. Let us write the solution to the first equation as $y_0 = A_0 \mathrm{e}^{-t}$, where A_0 is an integration constant. Then, the general solution to the second equation reads

$$y_1 = A_1 \mathrm{e}^{-t} - A_0 t \mathrm{e}^{-t}, \tag{3.63}$$

where A_1 is also an integration constant. Combining these two results, we obtain to order ε

$$y = A_0 \mathrm{e}^{-t} - \varepsilon A_0 t \mathrm{e}^{-t} + O(\varepsilon^2). \tag{3.64}$$

Here, $A_0 + \varepsilon A_1$ is redefined as A_0 (we ignore $\varepsilon^2 A_1$ in the second term; as can be seen from this, in the perturbative expansion we have only to find special solutions).

In this way, we can compute any higher-order terms, but this is usually regarded as a bad solution, because ε appears with t, which increases indefinitely. Thus, the perturbation effect that should be small becomes not small, and the perturbation method breaks down. In other words, the perturbation result may be used only for the time span much shorter than ε^{-1}. Mathematically speaking, the convergence is not uniform in time. The term with multiplicative t is traditionally called a *secular term*.

Doesn't the secular term correspond to the divergence appearing in the renormalization approach? See (3.24). If we could remove such divergences, then naive perturbation series as obtained above could make sense. For the present problem to watch the behavior just in front of us ('at the present time') corresponds to macroscopic observations and the behaviors long ago correspond to microscopic scales.[81] That is, we are interested in the global behavior that does not change very much even if the initial condition is modified. In the ordinary renormalization problem, the microscopic-detail-sensitive responses are separated and renormalized into materials constants. Therefore, in the present problem what must be renormalized is the sensitively dependent behavior on the initial condition, and the place it should be pushed into must be the integration constants (the quantities connecting what we observe now and the initial condition); it is a natural observation, because the integration constants are determined by the initial condition.

Let us review the key points of the perturbative realization of the Stükelberg-Petermann approach applied to the von Koch curve. We first studied the divergence and separated it as (recall (3.26))

$$\log(W/\ell) = \log(W/\lambda) + \log(\lambda/\ell), \tag{3.65}$$

and the second term that diverges in the $\ell \to 0$ limit was pushed into the renormalization constant. The renormalization constant did not show up haphazardly but in the relationship between what we can observe and what we cannot. Since there cannot be any arbitrariness in the relationships among observable quantities, there is no room for renormalization constants for observable relationships. It is crucial to distinguish what we can know and what we cannot. Thus, the following ancient teaching becomes a crucial renormalization instruction: "When you know a thing, recognize that you know it, and when you do not, recognize that you do not."[82]

Based on the above analogy, the procedure for the von Koch curve me-

[81] For ordinary deterministic systems it is hard to obtain the initial condition from the observation result at t (it is asymptotically impossible in the $t \to \infty$ limit). For chaotic systems, however, this is not only always possible, but the estimate of the initial condition becomes more accurate if we observe longer-time asymptotic behaviors (however, we assume there is no noise). That is, in a certain sense, chaos is the antipode of renormalizability.

[82] The Master said, "Yu, shall I teach you what knowledge is? When you know a thing, to recognize that you know it, and when you do not know a thing, to recognize

chanically adapted to the current problem is as follows: Separate the secular divergence as $(t - \tau) + \tau$, and then absorb τ by modifying unobservable A_0 as $A(\tau)$, which is understood as an adjustable parameter to be determined so that the solution agrees with the behavior we directly observe at present, i.e., around time t. After this renormalization, the perturbation series (3.64) reads

$$y = A(\tau)\mathrm{e}^{-t} - \varepsilon(t - \tau)A(\tau)\mathrm{e}^{-t} + O(\varepsilon^2). \tag{3.66}$$

(due to our mechanical procedure, obviously τ corresponds to $\log(\lambda/\ell)$ in the preceding section. As can be seen from this, τ is actually τ − initial time, i.e., the time lapse from the initial time). Such a series is called a *renormalized perturbation series.*

This equation makes sense only when $\varepsilon(t - \tau)$ is small, but it is distinct from the original perturbation series (3.64) we started with (which is often called a 'bare' perturbation series) in that t need not be small, because we can choose τ to be large enough.[83]

Since τ is a parameter not existing in the problem itself, $\partial y/\partial\tau = 0$. This is the *renormalization group equation* for the current problem:

$$\frac{\partial y}{\partial \tau} = \frac{\mathrm{d}A}{\mathrm{d}\tau}\mathrm{e}^{-t} + \varepsilon(t - \tau)\frac{\mathrm{d}A}{\mathrm{d}\tau} + \varepsilon A\mathrm{e}^{-t} + \cdots = 0. \tag{3.67}$$

The equation tells us that $\mathrm{d}A/\mathrm{d}\tau$ must be of order ε (the terms proportional to e^{-t} must cancel each other), so we may discard the second term of (3.67) as a higher-order term. Therefore, the renormalization group equation reads, to order ε,

$$\frac{\mathrm{d}A}{\mathrm{d}\tau} = -\varepsilon A. \tag{3.68}$$

The renormalized perturbation series (3.66) is simplified if we set $\tau = t$ (those who question this procedure should see Note 3.5 (2) below):

$$y = A(t)\mathrm{e}^{-t}. \tag{3.69}$$

(3.68) implies that $A(t)$ obeys the following amplitude equation:

$$\frac{\mathrm{d}A(t)}{\mathrm{d}t} = -\varepsilon A(t). \tag{3.70}$$

The equation indicates that A changes significantly only in the longtime scale of order $t \sim 1/\varepsilon$; only when εt has a visible magnitude can A change

---

that you do not know it. That is knowledge." *Analects* Book 2, 17 [A. Waley, *The Analects of Confucius* (Vintage, 1989)].

83 The above procedure may be modified so that it can be interpreted as solving the equation from the initial time τ with an effective initial condition $A(\tau)$ chosen appropriately to reproduce the actual solution. However, when the secular term is t^2, we often renormalize it as $t^2 - \tau^2$ instead of $(t - \tau)^2$, so this interpretation is not needed. The relation among various renormalization schemes corresponds to the so-called finite renormalization.

significantly. Finally, solving (3.70), we get the following asymptotic behavior,

$$y = Be^{-(1+\varepsilon)t} + O(\varepsilon^2), \tag{3.71}$$

where B is an adjustable parameter. This is our conclusion about asymptotic behavior.[84] Here, notice that the form of $A(t)$ as a function of t is universal in the sense that it does not directly depend on the initial condition ((3.70) is determined by the original differential equation itself) just as the power determining the divergence $\ell^{1-\log 4/\log 3}$ in the von Koch curve example is common to all such curves.

As can be seen from the above example, the core of the renormalization group method for the problems with secular terms is to derive equations that govern the slow systematic motions such as (3.70). This may be interpreted as coarse-graining or reduction of the system behavior.

The problems that produce secular terms under simple-minded perturbation series as (3.60) are generally called *singular perturbation* problems.[85] Since for these problems simple perturbation does not work, tons of techniques collectively called singular perturbation methods have been developed. For example, to solve the above simple problem the multi-time-scale method has been used in which many slow times $\varepsilon t, \varepsilon^2 t, \cdots$ other than the ordinary time t are introduced.

The above simple example suggests the following:
(1) The secular term is a divergence, and renormalization procedure removes this divergence to give the same result singular perturbation methods give. Singular perturbation methods are 'renormalized ordinary perturbations.'
(2) The renormalization group equation is an equation governing slow phenomena. The core of singular perturbation theories is to extract such a slow motion equation, which can be obtained in a unified fashion with the aid of renormalization.

As to (1), as long as the author has experienced, many (almost all?) problems solved by named singular perturbation methods can be solved by renormalization method in a unified fashion without any particular prior knowledge. As to (2) many (all?) famous equations governing phenomenological behaviors (e.g., the nonlinear Schrödinger equation, the Burgers equation, the Boltzmann equation, etc.) can be derived as renormalization group equations. Actually, we will see in the subsequent sections that (2), the reductive use of renormalization group theory (*reductive renormalization*), is, as ex-

---

[84] In terms of the two roots $\lambda_\pm = (-1 \pm \sqrt{1-4\varepsilon})/2\varepsilon$ of the characteristic equation $\varepsilon s^2 + s + 1 = 0$ the analytic solution for (3.59) is $y(t) = Ae^{\lambda_+ t} + Be^{\lambda_- t}$, where for small ε $\lambda_+ = -1 - \varepsilon + O(\varepsilon^2)$, $\lambda_- = -1/\varepsilon + 1 + \varepsilon + O(\varepsilon^2)$, so (3.71) is a uniformly correct order ε solution up to time $\varepsilon t \sim 1$.

[85] The reader may say this is not the standard definition of singular perturbation, but its most reasonable definition must be the perturbation that produces secular terms (or not uniformly convergent series).

pected, the key element of singular perturbation methods. If one masters the renormalization group method, no singular perturbation methods are likely to be necessary to solve ordinary differential equation problems. As we will see later, this conclusion seems to apply to partial differential equations as well. That is, the renormalization group method unifies numerous singular perturbation methods and make the conventional methods unnecessary.[86]

Note 3.5. Technical remarks on the renormalization scheme

(1) To perform calculation more systematically, we introduce the renormalization constant Z as $A = ZA_0$ or more conveniently as (notice that the following Z is the reciprocal of the Z in $A = ZA_0$)

$$A = ZA_R, \tag{3.72}$$

where A is the 'bare' microscopic quantity and A_R the renormalized counterpart (in the current context it is what we observe a long time later). We are performing a perturbation calculation, so we expand $Z = 1 + \varepsilon Z_1 + \cdots$, and the coefficient are determined order by order to remove divergences. In the lowest-order calculation as we have done, there is no danger of making any mistake, so a simple calculation as explained above is admissible, but, as we will see in Appendix 3.7A, a formal expansion helps systematic studies.

(2) In the above calculation putting $\tau = t$ makes everything simple, but there are people who feel that it is a bit too convenient and *ad hoc* a procedure, so let us avoid this procedure. The result of renormalized perturbation series has the following structure:

$$y(t) = f(t; \varepsilon\tau) + \varepsilon(t - \tau)g(t) + O(\varepsilon^2). \tag{3.73}$$

Since f is differentiable with respect to the second variable, with the aid of Taylor's formula we may rewrite it as

$$y(t) = f(t; \varepsilon t) + \varepsilon(\tau - t)\partial_2 f(t, \varepsilon t) + \varepsilon(t - \tau)g(t) + O(\varepsilon^2), \tag{3.74}$$

where ∂_2 denotes the differentiation with respect to the second variable. The second and the third terms must cancel each other, since the original problem does not depend on τ. That is, the procedure to remove the secular term by setting $\tau = t$ is always correct.

(3) To construct an envelope is a renormalization procedure.[87] Suppose a family of curves $\{x = F(t, \alpha)\}$ parameterized with α is given. Its envelope is given by

$$x = F(t, \alpha), \quad \frac{\partial}{\partial \alpha} F(t, \alpha) = 0. \tag{3.75}$$

---

[86] These days, applied mathematicians have organized the results of the renormalization group method as a method of nonlinear variable transformation that requires no idea/philosophy of renormalization. The author has no interest in this direction.

[87] This was pointed out by T. Kunihiro.

T. Wall seems to have been the first to construct a result that can be obtained naturally by a renormalization group method as an envelope of approximate solutions: F. T. Wall, "Theory of random walks with limited order of non-self-intersections used to simulate macromolecules," J. Chem. Phys. **63**, 3713 (1975); F. T. Wall and W. A. Seitz, "The excluded volume effect for self-avoiding random walks," J. Chem. Phys. **70**, 1860 (1979). Later, the same idea (called the coherent anomaly method) was systematically and extensively used by M. Suzuki to study critical phenomena.

The second equation can be interpreted as a renormalization group equation. The envelope curve is such a set of points among the points (x, t) satisfying $x = F(t, \alpha)$ that stay invariant under change of α to $\alpha + \delta\alpha$. Therefore, it must satisfy the second equation describing the condition that the (x, t) relation does not change under perturbation of α. This is exactly the same idea as searching for features that stay invariant even if microscopic details are perturbed. However, renormalization group theory must not be misunderstood as a mere theory of special envelope curves. The theory of (or the procedure to make) envelope curves is meaningful only after a one-parameter family of curves is supplied. Thus, from the envelope point of view, the most crucial point is that renormalization provides a principle to construct the one-parameter family to which the envelope theory may be applied. Needless to say, the key to singular perturbation is this principle and not the envelope interpretation, which may not always be useful, as we see in Appendix 3.7A.

(4) The best reference book (practical book) of singular perturbation is C. M. Bender and S. A. Orszag, *Advanced Mathematical Methods for Scientists and Engineers* (McGraw-Hill, 1978). Numerous examples and numerical confirmation of the results make this book unrivaled. It is a book to be kept at one's side whenever singular perturbation problems are studied. However, the book is wonderfully devoid of any mathematical theory. To have an overview of various singular perturbation methods within a short time, E. J. Hinch, *Perturbation Methods* (Cambridge University Press, 1991) is recommended.

3.7 Resonance and renormalization

A typical case where singular perturbation problems arises is the increase of the order of the differential equation due to perturbation as seen in the example in the preceding section, but another typical case is the existence of *resonance*. If a harmonic oscillator is perturbed by an external perturbation with the frequency identical to the oscillator itself, its amplitude increases indefinitely. For a globally stable nonlinear system, this divergence is checked sooner or later by some nonlinear effect, so there is no genuine divergence. However, if the nonlinear term is treated as perturbation, this effect disappears from the perturbation equations, so singularity due to resonance shows up. Even if the average external force is zero, resonance has a 'secular effect.' That is, there is an effect that accumulates with time. The etymology of 'secular term' lies here.

A typical example illustrating that an ordinary perturbation series is plagued by resonance is the following weak nonlinear oscillator:

$$\frac{\mathrm{d}^2 y}{\mathrm{d}t^2} + y = \varepsilon(1 - y^2)\frac{\mathrm{d}y}{\mathrm{d}t}, \tag{3.76}$$

where ε is a small positive constant (so the nonlinearity is weak). This equation is a famous equation called the van der Pol (1889-1959) equation. Introducing the following expansion

$$y = y_0 + \varepsilon y_1 + O(\varepsilon^2) \tag{3.77}$$

into (3.76), and equating terms with the same power in ε, we obtain

$$\frac{\mathrm{d}^2 y_0}{\mathrm{d}t^2} + y_0 = 0, \tag{3.78}$$

$$\frac{\mathrm{d}^2 y_1}{\mathrm{d}t^2} + y_1 = (1 - y_0^2)\frac{\mathrm{d}y_0}{\mathrm{d}t}, \tag{3.79}$$

etc. The general solution to the first equation may be written as

$$y_0(t) = A\mathrm{e}^{it} + \text{ c.c.}, \tag{3.80}$$

where A is a complex constant and c.c. implies a complex conjugate. Using this in the second equation, we get

$$\frac{\mathrm{d}^2 y_1}{\mathrm{d}t^2} + y_1 = iA(1 - |A|^2)\mathrm{e}^{it} - iA^3\mathrm{e}^{3it} + \text{ c.c.} \tag{3.81}$$

We have only to obtain its special solution. To this end it is the easiest to use *Lagrange's method of varying coefficients.*[88] Thus, we obtain

$$y_1 = \frac{1}{2}A(1 - |A|^2)t\mathrm{e}^{it} + \frac{i}{2}A^3\mathrm{e}^{3it} + \text{ c.c.} \tag{3.82}$$

We have obtained the naive perturbation series as

$$y(t) = A\mathrm{e}^{it} + \varepsilon\left[\frac{1}{2}A(1 - |A|^2)t\mathrm{e}^{it} + \frac{i}{2}A^3\mathrm{e}^{3it}\right] + \text{ c.c.} + O(\varepsilon^2). \tag{3.83}$$

Clearly, there is a secular term. The reason for it is that the right-hand side of the equation for y_1 contains the term proportional to e^{it} that has the same frequency as the harmonic oscillator expressed by the left-hand side. Again, numerous singular perturbation methods have been developed to overcome resonance, but our procedure is exactly the same as in the preceding section.

---

[88] For example, a special solution u to the second-order ordinary differential equation

$$\frac{\mathrm{d}^2 y}{\mathrm{d}x^2} + a\frac{\mathrm{d}y}{\mathrm{d}x} + by = f$$

may be constructed as follows in terms of the fundamental solutions of the corresponding homogeneous equation ϕ_1 and ϕ_2:

$$u = C_1\phi_1 + C_2\phi_2,$$

where the coefficients (functions) C_1 and C_2 are obtained by solving the following equations:

$$\frac{\mathrm{d}C_1}{\mathrm{d}x} = -\frac{f\phi_2}{W}, \quad \frac{\mathrm{d}C_2}{\mathrm{d}x} = \frac{f\phi_1}{W}.$$

Here, W is the Wronskian $W = \phi_1\phi_2' - \phi_2\phi_1'$.

We separate t as $(t-\tau)+\tau$, and then absorb τ into the constant A that depends on the initial condition. The renormalized result is

$$y(t) = A(\tau)e^{it} + \varepsilon\left[\frac{1}{2}A(\tau)(1-|A(\tau)|^2)(t-\tau)e^{it} + \frac{i}{2}A(\tau)^3 e^{3it}\right] + \text{c.c.} + O(\varepsilon^2). \tag{3.84}$$

Since y cannot depend on τ, the renormalization group equation $\partial y/\partial\tau = 0$ becomes

$$\frac{dA}{dt} = \varepsilon\frac{1}{2}A(1-|A|^2) + O(\varepsilon^2), \tag{3.85}$$

where τ has already been replaced with t. This is an equation governing the longtime behavior of the amplitude. Setting $t=\tau$ in (3.84), we get

$$y(t) = A(t)e^{it} + \varepsilon\frac{i}{2}A(t)^3 e^{3it} + \text{ c.c.} + O(\varepsilon^2). \tag{3.86}$$

The key result is the *amplitude equation* (3.85). In this example, the case $\varepsilon = 0$ and the case $\varepsilon > 0$ are qualitatively different. For a harmonic oscillator, any amplitude is allowed. In contrast, too large or too small amplitudes are not stable for (3.76) as can be seen from its right-hand side term: if $(1-y^2) < 0$, then it is a deceleration term that reduces the amplitude; otherwise, it is an acceleration term injecting energy to the oscillator. Indeed, according to the above approximate calculation, the first-order solution slowly converges to a limit cycle expressed by $|A| = 1$.

Notice that the amplitude equation (3.85) contains ε. As expected from the result of the preceding section, the renormalization group equation describes a slow change of the amplitude (the actual motion is a busy rotation of period about 2π). As already suggested in the preceding section, the renormalization group approach supplies a new point of view for singular perturbation: to extract such a slow motion equation is the key point of the singular perturbation problems.

Appendix 3.7A More systematic renormalization approach

This note explains the *proto-renormalization group approach* that considerably simplifies the reductive use of renormalization group theory.[89]

Let us consider an autonomous equation (no explicit t dependence)

$$Ly = \varepsilon N(y), \tag{3.87}$$

where L is a linear differential operator with constant coefficients and N a nonlinear (differential) operator. We assume that the characteristic roots λ_i

[89] Proposed in K. Nozaki and Y. Oono, "Renormalization-group theoretical reduction," Phys. Rev. E **63**, 046101 (2001), but its explanation is hard to understand (the author's fault).

$(i = 1, \cdots, n)$ of L are distinct, so the general solution to the unperturbed equation $Ly_0 = 0$ reads

$$y_0 = \sum_{i=1}^{n} A_i e_i(t), \tag{3.88}$$

where $e_i(t) = e^{\lambda_i t}$, and A_i are constants. We write the formal expansion of the solution to (3.87) as

$$y(t, A) = \sum_{i=1}^{n} A_i e_i(t) + \varepsilon y_1(t, A) + \varepsilon^2 y_2(t, A) + \cdots, \tag{3.89}$$

where the A_i-dependence of the higher-order terms is explicitly written; A denotes collectively $\{A_i\}$. Since each y_k $(k = 1, 2, \cdots)$ has its own secular terms Y_k (that is, $e_i(t)$ times t or its powers; see, e.g., (3.82)), let us write $y_k = \eta_k + Y_k$, where η_k denotes the non-secular terms in y_k. Now, we wish to renormalize this bare perturbation series through $A \to A_R$. That is, we wish to remove Y_k from (3.89), choosing renormalization constants Z^i: $A_i = Z^i A_{Ri}$ (no summation convention implied) appropriately. What we will do in the following is to replace the t in the prefactors of $e_i(t)$ in Y_k with τ (as, e.g., $te^{it} \to \tau e^{it}$; we mark secular time dependence with τ), and then to absorb all these replaced τ into $Z^i(\tau)$. That is, if the above-mentioned replacement $t \to \tau$ converts $y(t, A)$ into $y(t, \tau, A)$, then all the τ-dependence of $y(t, \tau, A)$ is killed by $Z^i(\tau)$; we could write symbolically as $y(t, \tau, A) = y(t, \tau, Z(\tau)A_R(\tau)) = y(t, 0, A_R(\tau))$.

To be systematic, let us expand the renormalization constant in powers of ε:

$$Z^i = 1 + \varepsilon Z_1^i + \varepsilon^2 Z_2^i + \cdots. \tag{3.90}$$

Also we expand $y(t, \tau, A)$ in powers of ε:

$$y(t, \tau, A) = \sum_{i=1}^{n} A_i e_i(t) + \varepsilon y_1(t, \tau, A) + \varepsilon^2 y_2(t, \tau, A) + \cdots. \tag{3.91}$$

Then, we replace A in (3.91) with

$$A_i = A_{Ri} + \varepsilon Z_1^i(\tau) A_{Ri} + \varepsilon^2 Z_2^i(\tau) A_{Ri} + \cdots, \tag{3.92}$$

and order by order the explicit τ-dependence in $y_k(t, \tau, A)$ is absorbed into $Z_k^i(\tau)$ $(k = 1, 2, \cdots)$. While doing this, A_{Ri} is treated as a constant. That is, only the expansion terms $Z_k^i(\tau)$ are regarded τ-dependent, and are chosen to kill all the τ-dependence in $y_k(t, \tau, A)$.

Using A in the form of (3.92), we wish to rewrite (3.91) in terms of A_R, which collectively denotes $\{A_{Ri}\}$. Assuming needed differentiabilities, we have

$$\begin{aligned} y(t, \tau, A) = &\sum_{i=1}^{n} \left(1 + \varepsilon Z_1^i(\tau) + \varepsilon^2 Z_2^i(\tau) + \cdots\right) A_{Ri} e_i(t) + \varepsilon y_1(t, \tau, A_R) \\ &+ \varepsilon^2 [y_2(t, \tau, A_R) + \sum_i \partial_i y_1(t, \tau, A_R) Z_1^i(\tau) A_{Ri}] \\ &+ \varepsilon^3 \Bigg[y_3(t, \tau, A_R) + \sum_i \partial_i y_2(t, \tau, A_R) Z_1^i(\tau) A_{Ri} \\ &+ \sum_i \partial_i y_1(t, \tau, A_R) Z_2^i(\tau) A_{Ri} \end{aligned}$$

$$+\frac{1}{2}\sum_{i,j}\partial_i\partial_j y_1(t,\tau,A_R)Z_1^i(\tau)A_{Ri}Z_1^j(\tau)A_{Rj}\Bigg]$$

$$+\cdots, \tag{3.93}$$

where ∂_i implies the partial derivative with respect to A_i.

To remove the singular terms (secular terms) in y_1 from the first-order result we require

$$\sum_{i=1}^{n} A_{Ri}Z_1^i(\tau)e_i(t) + Y_1(t,\tau,A_R) = 0. \tag{3.94}$$

Note that at this stage A_R are understood as constants. Here, the t in the secular factors has been replaced by τ, so in the secular term Y_1 two time variables t and τ are written explicitly. $Z_2^i(\tau)$ must be determined to absorb all τ-dependences (all the secular terms) from the order ε^2 term:

$$\sum_{i=1}^{n} A_{Ri}Z_2^i(\tau)e_i(t) + y_2(t,\tau,A_R) + \sum_{i=1}^{n}\partial_i y_1(t,\tau,A_R)Z_1^i(\tau)A_{Ri}. \tag{3.95}$$

Thus, the renormalization condition to the second order reads

$$\sum_{i=1}^{n} A_{Ri}Z_2^i(\tau)e_i(t) + Y_2(t,\tau,A_R) + \sum_{i=1}^{n}\partial_i y_1(t,\tau,A_R)Z_1^i(\tau)A_{Ri} = 0. \tag{3.96}$$

Here, notice that in the third term, even the regular (i.e., non-secular) contribution $\partial_i\eta_1(t,A_R)$ cannot be ignored (for now), because it is multiplied by a τ-dependent factor $Z_1^i(\tau)$. Analogously, the third-order renormalization condition requires

$$\sum_{i=1}^{n} A_{Ri}Z_3^i(\tau)e_i(t) + Y_3(t,\tau,A_R)$$

$$+\sum_{i=1}^{n}\partial_i y_2(t,\tau,A_R)Z_1^i(\tau)A_{Ri} + \sum_{i=1}^{n}\partial_i y_1(t,\tau,A_R)Z_2^i(\tau)A_{Ri}$$

$$+\frac{1}{2}\sum_{i,j=1}^{n}\partial_i\partial_j y_1(t,\tau,A_R)Z_1^i(\tau)A_{Ri}Z_1^j(\tau)A_{Rj} = 0. \tag{3.97}$$

Let us study in more detail the structure of the perturbation result. The nth-order result y_n has the following structure

$$y_n(t,\tau,A) = Y_n^{[r]}(t,\tau,A) + Y_n^{[nr]}(t,\tau,A) + \eta_n(t,A), \tag{3.98}$$

where $Y_n^{[r]}(t,\tau,A)$ is the resonant secular term, that is, a linear combination of $e_i(t)$ with polynomial coefficients of τ without a constant term. $Y_n^{[nr]}(t,\tau,A)$ is the non-resonant secular terms, that is, all other secular terms not included in $Y_n^{[r]}(t,\tau,A)$. The secularity of $Y_n^{[nr]}(t,\tau,A)$ is solely due to the lower-order secular terms appearing in the perturbation terms, and as was noted in the original paper, secular terms appearing in them are all killed by lower order

Z_m's already determined by the renormalization of $Y_n^{[r]}(t, \tau, A)$.[90]

We see that the renormalization condition such as (3.97) has the following structure (symbolically, since the third term is not explicitly written out):

$$\sum A_{Ri} Z_n^i(\tau) e_i + Y_n(t, \tau, A_R) + \sum [\text{derivatives wrt } A_R \text{ of } y_k(t, \tau, A_R)] \times \prod Z_m^j(\tau) A_{Rj} = 0, \tag{3.99}$$

where y_k $(k < n)$ in the third term are the results of the lower-order calculations, and Z_m^j $(m < n)$ are the terms already determined by the lower-order renormalization conditions. As noted just above the preceding formula, we have only to consider the resonant part (a linear combination of $e_i(t)$):

$$\sum A_{Ri} Z_n^i(\tau) e_i + Y_n^{[r]}(t, \tau, A_R) + \sum \left[\text{derivatives wrt } A_R \text{ of } Y_k^{[r]}(t, \tau, A_R)\right] \times \prod Z_m^j(\tau) A_{Rj}. \tag{3.100}$$

In order to absorb all the τ in $y(t, \tau, A)$ into $Z^i(\tau)$ when A_R is regarded constant, all the coefficients of $e_i(t)$ appearing in (3.100) must be free of any explicit τ-dependence. Adding all the orders of (3.100), we obtain formally

$$\sum_i [A_i - A_{Ri}] e_i(t) + Y^{[r]}(t, \tau, A) = 0. \tag{3.101}$$

When this is used to determine $Z^i(\tau)$ as above, A_i are interpreted as (3.92) and A_{Ri} are regarded constant. After determining $Z^i(\tau)$ we can determine the τ-dependence of A_R by $Z^i(\tau) A_{Ri}(\tau) = A_i$, which is constant, and rewrite (3.101) as

$$\sum_i [A_i - A_{Ri}(\tau)] e_i(t) + Y^{[r]}(t, \tau, A) = 0, \tag{3.102}$$

where A_i are numerical constants.

Combining (3.99) and these equations determining the renormalization constants order by order, we obtain

$$y(t, \tau, A) = \sum_{i=1}^{n} A_{Ri}(\tau) e_i(t) + \eta(t, A_R(\tau)). \tag{3.103}$$

As expected, the singular terms in the perturbation series are completely removed, and absorbed in the renormalized coefficients A_{Ri}. If we set $\tau = t$, $y(t, t, A)$ is just $y(t, A)$, so we get the renormalized perturbation result.

Notice that

$$\left\{\frac{\mathrm{d}}{\mathrm{d}t}[f(t)g(t)]\right\}_{t \to \tau \text{ in } g \text{ and } g'} = \left(\frac{\partial}{\partial t} + \frac{\partial}{\partial \tau}\right) f(t) g(\tau). \tag{3.104}$$

Since L is a differential operator, we can write it explicitly as $L(\mathrm{d}_t)$, where $\mathrm{d}_t = \mathrm{d}/\mathrm{d}t$. Let us introduce a linear operator $\mathcal{L}$ as

$$\mathcal{L}(\partial_t, \partial_\tau) = L(\partial_t + \partial_\tau) - L(\partial_t), \tag{3.105}$$

[90] This can be demonstrated by mathematical induction. A mathematical proof of renormalizability [H. Chiba, "Extension and Unification of Singular Perturbation Methods," SIAM J. Applied Dynam. Sys. **8**, 1066 (2009)] also implies this.

where $\partial_t = \partial/\partial t$ and $\partial_\tau = \partial/\partial\tau$. Notice that if it is applied to a function that does not contain τ, it gives zero. For an eigenfunction $e_i(t)$ of L with its characteristic root λ_i we define new operator $\mathcal{L}_i$ by

$$\mathcal{L}[f(\tau)e_i(t)] = \{[L(\lambda_i + \mathrm{d}_\tau) - L(\lambda_i)]f(\tau)\}e_i(t) = [\mathcal{L}_i f(\tau)]e_i(t). \quad (3.106)$$

Applying $\mathcal{L}$ to (3.102), we have

$$0 = -\sum_{i=1}^{n}[\mathcal{L}_i A_{Ri}(\tau)]e_i(t) + \mathcal{L}Y^{[r]}(t,\tau,A). \quad (3.107)$$

Since $Y^{[r]}(t,\tau,A)$ has the following structure

$$Y^{[r]}(t,\tau,A) = \varepsilon\sum_i P_1^i(\tau,A)e_i(t) + \varepsilon^2\sum_i P_2^i(\tau,A)e_i(t) + \cdots, \quad (3.108)$$

(3.107) implies

$$\sum_i \mathcal{L}_i\left[-A_{Ri}(\tau) + \varepsilon P_1^i(\tau,A) + \varepsilon^2 P_2^i(\tau,A) + \cdots\right]e_i(t) = 0 \quad (3.109)$$

or

$$\mathcal{L}_i A_{Ri}(\tau) = \mathcal{L}_i P^i(\tau,A), \quad (3.110)$$

where $P^i(\tau,A) = \sum_k \varepsilon^k P_k^i(\tau,A)$. Notice that $A_i = Z^i(\tau)A_{Ri}(\tau)$ on the right-hand side of (3.110) are constants (τ-independent), so when $\mathcal{L}_i$ is applied, they are intact. After this procedure, we replace A with $A_{Ri}(\tau)Z^i(\tau)$s. If the original equation is autonomous, then the renormalization group equation governing A_R should be autonomous as well. Therefore, after this replacement there must not be any τ on the right-hand side; A_R depends on τ, but the equation governing A_R does not explicitly depend on τ. Thus, we can simply set $\tau = 0$ and $Z = 1$; we have reached the *proto-renormalization group equation* (proto-RG equation)

$$\mathcal{L}_i A_{Ri}(\tau) = \mathcal{L}_i P^i(\tau, A_R)|_{\tau=0}. \quad (3.111)$$

Here, on the right-hand side the differential operator $\mathcal{L}_i$ applies only to τ (not on A_R), and after differentiation τ is set to zero. The reason why it is called the proto-RG equation may be clear from the example below.

From the practical point of view, the cleverest method to solve singular perturbation problems is to use the proto-RG equation[91] just discussed.

Let us use the van der Pol equation to illustrate the proto-RG approach. The operator L in this example is given by

$$L(\mathrm{d}_t) = \mathrm{d}_t^2 + 1. \quad (3.112)$$

The eigenfunctions are $e_+(t) = \mathrm{e}^{it}$ and its conjugate $e_-(t) = e_+(t)^*$, so

$$\mathcal{L}(\partial_t, \partial_\tau) = \partial_\tau^2 + 2\partial_t\partial_\tau, \quad \mathcal{L}_+(\mathrm{d}_\tau) = \mathrm{d}_\tau^2 + 2i\mathrm{d}_\tau. \quad (3.113)$$

We know from the perturbation equations (3.81) that

$$\mathcal{L}_+ P_1 = iA(1 - |A|^2). \quad (3.114)$$

---

[91] An error is in the second-order solution to the van der Pol equation in the original Nozaki-Oono paper (the error is the author's) was pointed out by Y. Shiwa as mentioned below.

Thus, the proto-RG equation (3.111) to order ε reads

$$\left(\frac{\mathrm{d}^2}{\mathrm{d}\tau^2}+2i\frac{\mathrm{d}}{\mathrm{d}\tau}\right)A_R = i\varepsilon A_R(1-|A_R|^2). \tag{3.115}$$

Notice that time differentiation raises the power of ε by one, so the second derivative must be a higher-order small quantity. Therefore, to order ε we need not retain the second derivative:

$$\frac{\mathrm{d}}{dt}A_R = \frac{1}{2}\varepsilon A_R(1-|A_R|^2). \tag{3.116}$$

Thus, almost without any calculation we have obtained the RG equation. However, the above approach only gives the reduced equation, although it is the most important piece in the singular perturbation theory from our point of view.

To get the full-order ε result, we need regular (non-secular) terms as well (see (3.103)). The following may be probably the most efficient approach. In the present case, we can easily guess the form of the first-order perturbation result: $y_1 = P_1 \mathrm{e}^{it} + Q_1 \mathrm{e}^{3it} + \text{c.c.}$ The equations that P_1 and Q_1 obey can be obtained by introducing this form into (3.81) and by comparing both sides. We obtain (3.114) and

$$\left(\frac{\mathrm{d}^2}{\mathrm{d}\tau^2}+6i\frac{\mathrm{d}}{\mathrm{d}\tau}-8\right)Q_1 = -iA^3 \tag{3.117}$$

from the first-order perturbation equation. Thus, Q_1 may be a constant (so $Q_1 = iA^3/8$). Now, the renormalized result (3.103) reads

$$y(t) = A_R(t)\mathrm{e}^{it} - \frac{i}{8}A_R^3(t)\mathrm{e}^{3it} + \text{ cc.} \tag{3.118}$$

The correct second-order result is noted in Shiwa's paper.[92]

3.8 How reliable is the renormalization group result?

How reliable is the behavior that the renormalization group method tells us? It is easy to prove that (3.86) stays with the true solution within the error of $O[\varepsilon]$ for the time scale $1/\varepsilon$ by a standard argument with the aid of the *Grönwall inequality*.[93] However, such a result never tells us anything definitive about the long-term behavior of the system. For example, even the existence of a limit cycle cannot be asserted.

---

[92] Y. Shiwa, "Comment on "Renormalization-group theory for the phase-field crystal equation"," Phys. Rev. E **79**, 013601 (2005). Actually, he demonstrated that a straightforward proto-RG equation approach explained above gives the correct result (private communication).

[93] ⟪**Grönwall's inequality**⟫ Inside an interval $I = [a, b]$ let us assume that α is a continuous function and $f'(t) \le \alpha(t)f(t)$ or $f(t) \le f(a) + \int_a^t \alpha(s)f(s)ds$ holds. Then, $f(t) \le f(a)\exp\left(\int_a^t ds\, \alpha(s)\right)$ holds for $t \in I$.

Recent work by H. Chiba[94] considerably clarified this problem. Apart from some technicality, his conclusion is: "the long-time behavior of the system with small ε can be qualitatively inferred from its renormalization group equation." That is, roughly speaking, the invariant manifold of the renormalization group equation is diffeomorphic to that of the original equation. In the resonance example of Section 3.7, it is trivial that the renormalization group equation has a hyperbolic limit cycle, so we can conclude that the original equation also has a hyperbolic limit cycle.

In this section, at least an intuitive explanation of the qualitative reliability of the RG results is attempted. The explanation may oversimplify and may not do justice to the original theory, so those who are seriously interested in the proof should read the original paper. Chiba's argument relies on the following points:
(1) The invariant manifold of the renormalization group equation and that of the equation governing the (truncated) renormalized perturbation series are (very crudely put; see Note 3.6) diffeomorphic.
(2) The differential equation governing the (truncated) renormalized perturbation series is at least C^1-close[95] to the original differential equation.
With (2) and *Fenichel's theorem* (see Note 3.7),
(3) The invariant manifold of the equation governing the (truncated) renormalized perturbation series is diffeomorphic to the invariant manifold of the original equation.

We may expect that the invariant manifolds of C^1-close vector fields are 'close' in some sense. This is, however, a slightly delicate question even under hyperbolicity if we demand C^1-closeness of the manifolds.

At least in the case of diffeomorphisms it is known that *normal hyperbolicity* (see below) is a necessary and sufficient condition for an invariant manifold to persist.[96] Thus, hyperbolicity of invariant manifolds should not be enough to guarantee the qualitative similarity of the renormalization group equation to the original equation.

What Chiba demonstrated is that if the original system has a normally hyperbolic invariant manifold, then the renormalization group equation preserves it. Although for continuous dynamical systems, the relation between the normal hyperbolicity and C^1-structural stability seems not known, probably, we can conjecture that if the original equation is C^1-structurally stable (at least near its invariant manifold), then its renormalization group equation preserves the invariant manifolds of the original system.

[94] H. Chiba, "C^1-approximation of vector fields based on the renormalization group method," SIAM J. Appl. Dyn. Syst. **7**, 895 (2008).

[95] That is, not only the formulas but their derivatives are also close.

[96] M. Hirsch, C. Pugh and M. Shub, "Invariant manifolds," Bull. Amer. Math. Soc. **76**, 1015 (1970), and R. Mañé, "Persistent M manifolds are normally hyperbolic," Trans. Amer. Math. Soc. **246**, 271 (1978). These papers discuss the C^1-closeness of the invariant manifolds.

Note 3.6. Some technical points

The relation between the solution $A_R(t)$ to the renormalization group equation and the renormalized perturbation series solution $y(t, 0, A_R(t))$ of the original equation is given by (here maximally the same notations are used as in Section 3.7) the map α_t defined as $\alpha_t(A) = \sum A_i e_i(t) + \eta(t, A)$ (thus $\alpha_t(A_R(t)) = y(t, 0, A_R(t))$; see (3.103)). Here, we are interested in truncated solutions to some power of ε. Thus, we make a truncated version of α_t by truncating η. The differential equation governing $\alpha_t(A_R(t))$ (both α_t and $A_R(t)$ are truncated) is the equation V' governing the truncated renormalized perturbation series. A technical complication is that the truncated $\alpha_t(A_R(t))$ is generally explicitly time-dependent ($\alpha_t(x)$ is t-dependent), so the invariant set of V' must be considered in the 'space-time' (i.e., (t, y)-space); what (1) asserts is that the invariant set of the truncated original equation V' considered in the (t, y)-space and the direct product of time and the invariant set of the renormalization group equation are diffeomorphic.

(2) should not be a surprise to physicists. The difference between the equation V' governing $y(t, 0, A_R(t))$ (appropriately truncated) and the original equation V must be bounded, since the renormalized result is bounded uniformly. Thus, their closeness should be obvious. Here, we need the closeness of the derivatives as well. The approximate solution and the exact solution are differentiable (actually C^1). Therefore, derivatives needed to calculate the derivatives of the vector fields are all bounded (if the appropriate derivatives of the formula with respect to y exist). Thus, the equation governing $y(t, 0, A_R(t))$ (appropriately truncated) and the original equation are C^1-close.

The final step is (3); since the equation V' governing $y(t, 0, A_R(t))$ (appropriately truncated) and the original equation V are C^1-close, if normal hyperbolicity (see the next note) of the invariant manifold may be assumed, then Fenichel's theorem concludes the demonstration. This is, however, if both V and V' are autonomous, but the equation V' governing $y(t, 0, A_R(t))$ (appropriately truncated) is not generally autonomous. Chiba overcame this problem as in (1); to consider the systems in space-time. Fenichel's theorem can be applied there, and the C^1-closeness of the invariant sets (and the closeness of the diffeomorphisms to identity) is established.

Thus, the invariant manifold of the renormalization group equation and that of the original equation are diffeomorphic.

Note 3.7. Normal hyperbolicity and Fenichel's theorem

Let us consider a (continuous time) dynamical system defined on a subset U of a vector space with a (compact) invariant manifold M. M is assumed to have its unstable and stable manifolds, and the tangent space is decomposed as $T_M U = TM \oplus E^s \oplus E^u$, where E^s is the stable bundle and E^u the unstable bundle. M is *normal hyperbolic*, if the flow along the manifold M is 'slower' than the flows in the stable and unstable manifolds (here, a formal definition is suppressed). The illustration (Fig. 3.8) is basically the same as in Fenichel's original paper.[97]

Theorem 3.1 (Fenichel). *Let X be a C^r-vector field ($r \geq 1$) on $\boldsymbol{R}^n$. Let M be a normally hyperbolic invariant manifold of X without boundary. Then, for any C^r-vector field Y in a certain C^1-neighborhood of X is a Y-invariant manifold C^r-diffeomorphic to M. This diffeomorphism is C^1-close to the identity.*
□

[97] N. Fenichel, "Persistence and smoothness of invariant manifolds for flows," Indiana Univ. Math. J., **21**, 193 (1971).

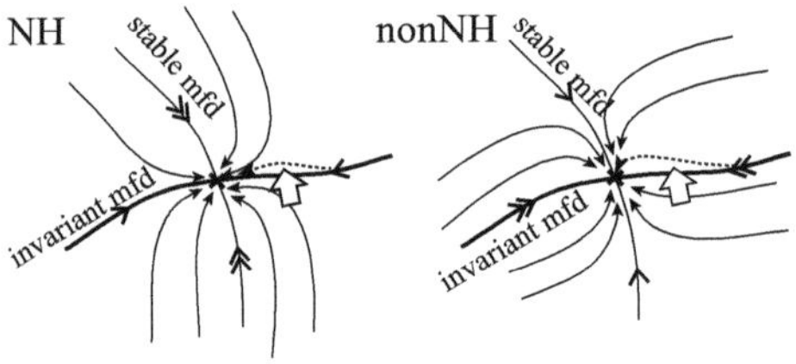

Fig. 3.8 Why normal hyperbolicity is needed. The thick curves denote (a part of) the invariant manifold M and its stable manifold is illustrated. The big white arrow denotes perturbation, and the dotted curves illustrate perturbed flow lines. 'NH' denotes the normal hyperbolic case, where the flow on the stable manifold is 'faster' (double arrows) than the flow on the invariant manifold. That is, normal direction is 'more' hyperbolic. 'nonNH' denotes the hyperbolic but not normally hyperbolic case. In the NH case the perturbation cannot disrupt the smoothness of the vector field near 'x,' but that is not the case for the nonNH case; a cusp could be formed.

3.9 Proto-renormalization group approach to system reduction

We have so far discussed ordinary differential equations. How about partial differential equations? It is very often the case that analytical (and even numerical) studies are hard, and qualitative understanding is crucial. Thus, reducing the equations to relatively easily understandable equations that are qualitatively correct is highly welcomed. For partial differential equations many different singular terms could be produced from a single equation, so minimizing explicit calculations is almost mandatory. Then, the proto-RG equation approaches should be useful. Here, it is pointed out that the proto-RG equations are quite parallel to the ones for the ordinary-differential-equation (ODE) cases at least formally. Thus, this parallelism is explained in Note 3.8, and the results (3.128) and (3.135), which are quite parallel to (3.111), are subsequently applied to the reduction of, e.g., the phase-field crystal equation.

Note 3.8. Proto-RG approach for evolution equations is parallel to the ODE case
We study the following equation whose independent variables are collectively denoted as x, which is often time and space $(t, \boldsymbol{r})$:

$$L(\partial_x)\psi = \varepsilon N[\psi], \tag{3.119}$$

where L is a linear differential operator, and N is a certain nonlinear operator. Here, the notations are so chosen that the parallelism to the exposition in Section 3.7 is maximal. ∂_x implies a set of (partial) differential operators with respect to the independent variables. The solution to the zeroth-order equation

$$L\psi_0 = 0 \tag{3.120}$$

may be written as

$$\psi_0(x) = \sum_k A_k e_k(x), \tag{3.121}$$

where $e_k(x)$ is the basis functions of the null space of L.[98] We may interpret $e_k(x)$ just as corresponding to $e_i(t)$ in the general formula for the ODE in Section 3.7. Thus, we can almost copy the previous results.

The renormalization condition (3.102) now reads with $A_k = Z^k(\chi)A_{Rk}(\chi)$, where χ denotes the variables that mark secular divergences by replacing x in the coefficients of $e_k(x)$ and their products (corresponding to τ in the ordinary-differential-equation cases):

$$\sum_k [A_k - A_{Rk}(\chi)]e_k(x) + Y^{[r]}(x, \chi, A) = 0, \tag{3.122}$$

where $Y^{[r]}(x, \chi, A)$ may be interpreted as the linear combination of $e_k(x)$ with polynomial coefficients of χ without constant terms and A collectively denotes the 'bare' coefficients A_k.

Let us introduce the linear operator $\mathcal{L}$ as

$$\mathcal{L}(\partial_x, \partial_\chi) = L(\partial_x + \partial_\chi) - L(\partial_x). \tag{3.123}$$

We define new operator $\mathcal{L}_k(\partial_\chi)$ by

$$\mathcal{L}[f(\chi)e_k(x)] = [\mathcal{L}_k(\partial_\chi)f(\chi)]e_k(x). \tag{3.124}$$

Applying $\mathcal{L}$ to (3.122), we have

$$0 = -\sum_k [\mathcal{L}_k(\partial_\chi)A_{Rk}(\chi)]e_k(x) + \mathcal{L}Y^{[r]}(x, \chi, A). \tag{3.125}$$

Since $Y^{[r]}(x, \chi, A)$ has the following structure

$$\begin{aligned} Y^{[r]}(x, \chi, A) &= \varepsilon \sum_k P_1^k(\chi, A)e_k(x) + \varepsilon^2 \sum_k P_2^k(\chi, A)e_k(x) + \cdots \\ &\equiv \sum_k P^k(\chi, A)e_k(x), \end{aligned} \tag{3.126}$$

we can write the proto-RG equation as

$$\mathcal{L}_k(\partial_\chi)A_{Rk}(\chi) = \mathcal{L}_k(\partial_\chi)P^k(\chi, A_R)|_{\chi=0}. \tag{3.127}$$

Here, on the right-hand side the differential operator $\mathcal{L}_k(\partial_\chi)$ applies only to χ, and after differentiation χ is set to zero. The right-hand side is obtained from the perturbation equation. For example, the first-order proto-RG equation can be read off immediately from (3.127) and

$$\mathcal{L}(\partial_x, \partial_\chi)Y_1^{[r]}(x, \chi, A)|_{x=\chi} = \mathcal{P}N[\psi_0] \tag{3.128}$$

as illustrated below, where $\mathcal{P}$ denotes projection to the linear space spanned by $\{e_i\}$.

If ψ is a conserved quantity (i.e., ψ integrated over the spatial domain

98 We proceed quite formally here, so we need not worry about the spectrum of L, but if we put the system in a big but compact box, L may be interpreted just as an ordinary differential operator. We assume all the eigenvalues are distinct.

is constant), then the perturbation term $N[\psi]$ often takes the form $\Delta M[\psi]$, where Δ is the Laplacian with respect to $\boldsymbol{r}$ and M is some function(al) of ψ (see Chapter 4). It is impossible to respect this conservation law order by order with the above naive perturbation calculation. Thus, we need a non-perturbative method to impose the conservation law. The most natural approach is to rewrite the original equation as

$$\tilde{L}\psi \equiv \Delta^{-1}L\psi = \varepsilon M[\psi] + H, \tag{3.129}$$

where H is a spatial harmonic function. The solution to the zeroth-order equation $\tilde{L}\psi_0 = H$ may be written as

$$\psi_0(x) = \sum_k A_k e_k(x), \tag{3.130}$$

where $e_k(x)$ is the basis functions of the null space of $L = \Delta\tilde{L}$. The subsequent development is quite parallel to the above 'non-conserved' case. From an analogue of (3.122) we obtain

$$\sum_k \tilde{L}(\partial_x + \partial_\chi)A_{Rk}(\chi)e_k(x) = \sum_k \tilde{L}(\partial_x + \partial_\chi)P^k(\chi, A_R)e_k(x) + H(x). \tag{3.131}$$

Applying $(\partial_x + \partial_\chi)^2$ to the above equation, we obtain

$$\begin{aligned} L(\partial_x + \partial_\chi)A_{Rk}(\chi)e_k(x) &= L(\partial_x + \partial_\chi)P^k(\chi, A_R)e_k(x) \\ &= [\mathcal{L}_k P^k(\chi, A_R)]e_k(x). \end{aligned} \tag{3.132}$$

To the first order

$$L(\partial_x + \partial_\chi)Y_1^{[r]}(x, \chi, A)|_{\chi=x} = \mathcal{P}(\partial_x + \partial_\chi)^2 M\left[\sum_k A_{Rk}(\chi)e_k(x)\right], \tag{3.133}$$

or

$$\mathcal{L}_k P_1^k(\chi, A_R) = \mathcal{P}_k(\partial_x + \partial_\chi)^2 M\left[\sum_k A_{Rk}(\chi)e_k(x)\right], \tag{3.134}$$

where $\mathcal{P}_k$ is a projection operator onto the linear space spanned by $e_k(x)$. Thus,

$$\mathcal{L}_k A_{Rk}(\chi) = \mathcal{P}_k(\partial_x + \partial_\chi)^2 M\left[\sum_k A_{Rk}(\chi)e_k(x)\right]. \tag{3.135}$$

Let us illustrate[99] the proto-RG approach to reduction of evolution equations with the aid of the Swift-Hohenberg equation and the phase-field crystal equation.[100]

---

[99] The following exposition approximately traces Y. Shiwa, "Renormalization-Group for Amplitude Equations in Cellular Pattern Formation with and without Conservation Law," Prog. Theor. Phys. **125**, 871 (2011).

[100] The Swift-Hohenberg equation is proposed in J. Swift and P. C. Hohenberg, "Hydrodynamic fluctuations at the convective instability," Phys. Rev. A **15**, 319 (1977) [cf. M. C. Cross and P. C. Hohenberg, "Pattern formation outside of equilibrium," Rev. Mod. Phys. **65**, 851 (1993)]. The phase-field crystal equation is proposed in

$$\frac{\partial}{\partial t}\psi = (i\partial_{\boldsymbol{r}})^p[\varepsilon\psi - \psi^3 - (\partial_{\boldsymbol{r}}^2 + k_0^2)\psi], \tag{3.136}$$

where ψ is a function of time t and space $\boldsymbol{r}$, and k_0 is a constant expressing the (reciprocal of the) scale of the representative pattern size. Here, $p = 0$ for the Swift-Hohenberg equation and $p = 2$ for the phase-field crystal equation. For $p = 2$ the extra Laplacian in (3.136) makes the order parameter ψ conserved as we will encounter in Section 4.5. These equations are highly nonlinear equations, so analytical solutions are extremely unlikely to obtain. Therefore, reduced equations are highly desirable. As can be seen from the equation, if ε changes its sign from minus to plus, spatially nonuniform solutions become realizable. The representative size of ψ is of order $\sqrt{\varepsilon}$. Thus, it is convenient to scale $\psi \to \sqrt{\varepsilon}\psi$ (i.e., the amplitude ψ is magnified to $O[1]$). Instead of (3.136), we study the scaled equation

$$\frac{\partial}{\partial t}\psi = (i\partial_{\boldsymbol{r}})^p[\varepsilon(\psi - \psi^3) - (\partial_{\boldsymbol{r}}^2 + k_0^2)^2\psi]. \tag{3.137}$$

We must first identify operators appearing in the above general exposition. The correspondences of the variables are $x = (t, \boldsymbol{r})$, and $\chi = (\tau, \boldsymbol{\rho})$ in the following.

First, the $p = 0$ case or the Swift-Hohenberg equation:

$$L(\partial_x) = \partial_t + (\partial_{\boldsymbol{r}}^2 + k_0^2)^2. \tag{3.138}$$

$\partial_{\boldsymbol{r}}^2$ is the Laplacian Δ. From this, we get the bases of the null space as

$$e_{\boldsymbol{k}}(x) = \exp[\omega(\boldsymbol{k})t + i\boldsymbol{k}\cdot\boldsymbol{x}] \tag{3.139}$$

with the dispersion relation $\omega(\boldsymbol{k}) = -(k^2 - k_0^2)^2$, where $k = |\boldsymbol{k}|$. In the following, however, for simplicity, we consider the case of time-independent unperturbed states (frozen patterns in space as the basic patterns), so we assume $k = k_0$. Other operators in the general discussion above read (see (3.123))

$$\mathcal{L}(\partial_x, \partial_\chi) = \partial_\tau + \left\{[(\partial_{\boldsymbol{r}} + \partial_{\boldsymbol{\rho}})^2]^2 - \partial_{\boldsymbol{r}}^4\right\} + 2k_0^2\left[(\partial_{\boldsymbol{r}} + \partial_{\boldsymbol{\rho}})^2 - \partial_{\boldsymbol{r}}^2\right] \tag{3.140}$$

and (see (3.124))

$$\mathcal{L}_{\boldsymbol{k}}(\partial_\chi) = \partial_\tau + \Box_{\boldsymbol{k}}^2(\boldsymbol{\rho}), \tag{3.141}$$

where $\Box_{\boldsymbol{k}}(\boldsymbol{\rho}) = \partial_{\boldsymbol{\rho}}^2 + 2i\boldsymbol{k}\cdot\partial_{\boldsymbol{\rho}}$ ($k = k_0$ is assumed). Therefore, the amplitude for $e_{\boldsymbol{k}}(x)$ obeys the following proto-RG equation (see (3.125) and (3.128)):

K. R. Elder, M. Katakowski, M. Haataja, and M. Grant, "Modeling Elasticity in Crystal Growth," Phys. Rev. Lett. **88**, 245701 (2002); K. R. Elder and M. Grant, "Modeling elastic and plastic deformations in nonequilibrium processing using phase field crystals," Phys. Rev. E **70**, 051605 (2004) [cf. M. Haataja, L. Gránásy and H. Löwen, "Classical density functional theory methods in soft and hard matter," J. Phys.: Cond. Mat. **22**, 360301 (2010)].

$$\mathcal{L}_{\boldsymbol{k}}(\partial_\chi)A_{R\boldsymbol{k}}(\chi) = \varepsilon\mathcal{P}_{\boldsymbol{k}}(\psi_0 - \psi_0^3), \tag{3.142}$$

where $\mathcal{P}_{\boldsymbol{k}}$ is the projection to $e_{\boldsymbol{k}}(x)$. If we assume a striped pattern as the unperturbed pattern, then $\psi_0 = Ae^{i\boldsymbol{k}\cdot\boldsymbol{r}}$. In (3.142) A in ψ_0 must be replaced with $A_R(\chi)$. Therefore, after replacing $\chi \to x$ we obtain

$$\partial_t A_R + \Box^2 A_R = \varepsilon\left(A_R - 3|A_R|^2 A_R\right), \tag{3.143}$$

where $\Box = \Box_{\boldsymbol{k}}(\boldsymbol{r})$. This operator is a rotationally covariant operator.[101]

Now, let us consider the $p = 2$ case, the phase-field crystal equation with

$$L(\partial_x) = \partial_t - \partial_{\boldsymbol{r}}^2(\partial_{\boldsymbol{r}}^2 + k_0^2)^2. \tag{3.144}$$

The crucial difference is that there is a zero-wavevector mode, so even if the unperturbed pattern is a stationary stripe, we have

$$\psi_0(x) = B + A\tilde{e}_{\boldsymbol{k}}(x), \tag{3.145}$$

where B is the amplitude of the zero-wavevector mode and

$$\tilde{e}_{\boldsymbol{k}}(x) = \exp[\omega(\boldsymbol{k})t + i\boldsymbol{k}\cdot\boldsymbol{x}] \tag{3.146}$$

with the dispersion relation $\omega(\boldsymbol{k}) = -k^2(k^2 - k_0^2)^2$ (but again, we set $k = k_0$ here). Other operators in the general discussion above read

$$\mathcal{L}(\partial_x, \partial_\chi) = \partial_\tau - (\partial_{\boldsymbol{r}} + \partial_{\boldsymbol{\rho}})^2\left[(\partial_{\boldsymbol{r}} + \partial_{\boldsymbol{\rho}})^2 + k_0^2\right]^2 + \partial_{\boldsymbol{r}}^2(\partial_{\boldsymbol{r}}^2 + k_0^2)^2, \tag{3.147}$$

and

$$\mathcal{L}_{\boldsymbol{k}}(\partial_\chi) = \partial_\tau - \left(\Box_{\boldsymbol{k}}(\boldsymbol{\rho}) - k_0^2\right)\Box_{\boldsymbol{k}}^2(\boldsymbol{\rho}), \tag{3.148}$$

$$\mathcal{L}_{\boldsymbol{0}}(\partial_\chi) = \partial_\tau - \partial_{\boldsymbol{\rho}}^2(\partial_{\boldsymbol{\rho}}^2 + k_0^2)^2. \tag{3.149}$$

Therefore, the amplitudes obey the following proto-RG equations (see (3.133)-(3.135)):

$$\mathcal{L}_{\boldsymbol{k}}(\partial_\chi)A_R = \varepsilon\mathcal{P}_{\boldsymbol{k}}(\partial_{\boldsymbol{r}} + \partial_{\boldsymbol{\rho}})^2(\psi_0 - \psi_0^3) = \varepsilon(\Box_{\boldsymbol{k}}(\boldsymbol{\rho}) - k_0^2)\mathcal{P}_{\boldsymbol{k}}(\psi_0 - \psi_0^3), \tag{3.150}$$

$$\mathcal{L}_{\boldsymbol{0}}(\partial_\chi)B_R = \varepsilon\partial_{\boldsymbol{\rho}}^2\mathcal{P}_{\boldsymbol{0}}(\psi_0 - \psi_0^3), \tag{3.151}$$

where $\mathcal{P}_{\boldsymbol{k}}$ is the projection to $\tilde{e}_{\boldsymbol{k}}(x)$. Introducing (3.145) into ψ_0, we have

$$\psi_0 - \psi_0^3 = B - B^3 - 6B|A|^2 - (3B^2A + 3|A|^2A - A)e^{i\boldsymbol{k}\cdot\boldsymbol{x}} + \text{c.c.} + \cdots. \tag{3.152}$$

---

[101] introduced by H. Gunaratne, Q. Ouyang and H. L. Swinney, "Pattern formation in the presence of symmetries," Phys. Rev. E **50**, 2802 (1994).

In (3.150) and (3.151) A and B must be replaced with $A_R(\chi)$ and $B_R(\chi)$, respectively. Therefore, (3.150) and (3.151) read, after replacing $\chi \to x$,

$$\partial_t A_R - (\Box - k_0^2)\Box^2 A_R = \varepsilon(\Box - k_0^2)(3B_R^2 A_R + 3|A_R|^2 A_R - A_R), \tag{3.153}$$

$$\partial_t B_R - \Delta(\Delta + k_0^2)^2 B_R = \varepsilon\Delta(B_R^3 - B_R + 6B_R|A_R|^2). \tag{3.154}$$

As the reader may have realized, the above procedure looks quite parallel to the so-called method of variation of constants: simply construct a general solution to the unperturbed equation as (3.130), and then regard the numerical coefficients as functions to absorb the perturbation terms to construct the full solution. A crucial difference is that coefficients are determined to absorb only the 'leading contributions' causing secular terms. This point is the key feature of the proto-RG approach, and that is why it is called by the name related to renormalization group.

The derivation of the proto-RG equation is not the final goal of system reduction; the system reduction is not yet completed. Do not forget that these equations are called *proto*-RG equations, because usually they require further transformations just as we have already experienced in the nonlinear oscillator case (cf. (3.115)-(3.116)).

To proceed further, we must specify what we wish to observe, or more specifically, we must specify on what space-time-scale we wish to study the system. Consistency of the orders of various terms is probably the chief guiding principle to obtain physically meaningful reduced equations. In other words, we should demand that the obtained equation describes a certain intermediate asymptotic processes allowed to the system under study. In practice, if we wish to have a reduction consistent to order ε^q, all the terms lower than or equal to this order must be retained.

For example, if one wishes to describe a diffusive behavior of the order parameter ψ, a natural requirement is $\partial_t \sim \partial_{\boldsymbol{r}}^2 \sim \varepsilon$ (this expression implies that these derivatives are of the order specified (e.g., $\nabla\psi \sim \varepsilon^{1/2}$). If $k_0 \sim 1$ (this is the usual interpretation; we are interested in the global and slow change of the pattern of the basic scale of order 1), the operator $\Box$ consists of two operators of order ε and of order $\sqrt{\varepsilon}$. Thus, to order ε $\Box^2 = (2i\boldsymbol{k}\cdot\nabla)^2$, and (3.143) gives

$$\partial_t A_R - 4(\boldsymbol{k}\cdot\nabla)^2 A_R = \varepsilon\left(A_R - 3|A_R|^2 A_R\right), \tag{3.155}$$

which loses the rotational covariance in the original proto-RG equation. Therefore, if we wish to describe the diffusive space-time dynamics with rotational covariance, since keeping the whole $\Box^2$ requires retention of $O[\varepsilon^2]$ terms, the $O[\varepsilon]$ reduction (3.143) is not enough; we need the next-order reduction (the proto-RG equation to order $O[\varepsilon^2]$ is required).

Instead, if we are interested in the space-time pattern of order 1 (i.e.,

$k_0 \sim 1$) with evolution of the scale $\partial_t \sim \varepsilon$ and $\partial_{\boldsymbol{r}} \sim \varepsilon^{1/4}$, the proto-RG equation for the Swift-Hohenberg equation (3.143) is consistent to order ε; without dropping or adding any term, we may interpret it as a properly reduced equation to $O[\varepsilon]$.

A similar conclusion applies to the proto-RG equations (3.153) and (3.154) for the phase-field crystal equation with $\partial_t \sim \varepsilon^{3/2}$, $\partial_{\boldsymbol{r}} \sim \varepsilon^{1/4}$ and $k_0 \sim 1$ (consistent to order $\varepsilon^{3/2}$; notice that no further contribution comes from the right-hand side). Thus, in these cases we may consistently interpret the proto-RG equations as properly reduced equations.

As can easily be guessed, higher-order reduction requires us to retain numerous terms on the right-hand side of the proto-RG equations; the number of terms is expected to increase exponentially. This implies that, for example, $O[\varepsilon^3]$ reduction is almost impractical, defeating the purpose of system reduction. Although the reliability discussion in Section 3.8 is about ordinary differential equations, as long as we can expect that big but finite systems should not be qualitatively different from the unbounded space-time cases, partial differential equations are essentially finite dimensional ordinary differential equation systems, so the general conclusion of Section 3.8 should apply. We may expect that for sufficiently small ε, the reduced system to order ε should be qualitatively the same as the original system. Use of higher-order reduction results may be tempting (and to order ε^2 they may still be reasonably practical) to understand the cases with slightly larger ε. If there is no new bifurcation for larger ε, the order ε (i.e., the lowest-nontrivial-order) result should be enough. If there is a new bifurcation, then it may be practical to devise the reduction around this new bifurcation point so that the lowest-nontrivial-order result should be used.

There can be an elementary question: can we further reduce renormalization group equations? For example, the Boltzmann equation may be obtained as a renormalization group equation.[102] If we consider the equation on a larger space-time scale, the Navier-Stokes equation can be derived.[103] However, if the observation space-time scale does not change, no further reduction of the renormalization group equation should be possible. For example, the Newell-Whitehead equation derived as a renormalization group equation[104] is intact, if one tries to reduce it by the same method that derived the equation.

The reader might think that the reductive renormalization approach is quite similar to the multiscale expansion approach that has been standard

[102] O. Pashko and Y. Oono, "The Boltzmann equation is a renormalization group equation," Int. J. Mod. Phys. B **14**, 555 (2000).

[103] as demonstrated by Kuramoto, reductive perturbation does this, so obviously it is possible by renormalization (but as noted before, it is not very meaningful as a method to derive the Navier-Stokes equation).

[104] Y. Shiwa, "Renormalization-group theoretical reduction of the Swift-Hohenberg model," Phys. Rev. E **63**, 016119 (2000).

in systematic system reductions. However, there is an important difference. The reductive RG approach, especially the proto-RG approach, does not pay any attention to the space-time scale consistency; the method concentrates on removing secular space-time dependence that ruins any naive perturbative approach (i.e., that ruins qualitative features of the original system). Only after securing the (space-time globally) bounded results obtainable through perturbative approaches can we start paying attention to the scaling consistency.

3.10 Statistics seen from the renormalization group point of view

Renormalization is effective in extracting global behavior and coarse-grained behavior of a system, so it must be natural to expect that it may be utilized to summarize a huge data set into a human-size set of information.

Let us first consider a problem of a dynamical system that is the same as the problem of the law of large numbers and large deviation theory (a problem to derive the Langevin equation). It is a dynamical system with two disparate time scales:

$$\frac{\mathrm{d}x}{\mathrm{d}t} = f(x, y), \quad \frac{\mathrm{d}y}{\mathrm{d}t} = \frac{1}{\varepsilon} g(x, y), \tag{3.156}$$

where x and y are dynamical variables (can be vectors), and f and g are well-behaved functions. Here, ε is a small positive number. We may consider y to be a rapidly changing variable. If the system is observed on a large time scale, rapid changes would look like a noise obeying a certain stochastic process.

Renormalization group approaches to singular perturbation problems study the system on the smallest time scale, so let us stretch the time scale as $\tau = t/\varepsilon$, and rewrite the original equation as (i.e., we watch the system as a slow-motion movie):

$$\frac{\mathrm{d}x}{\mathrm{d}\tau} = \varepsilon f(x, y), \quad \frac{\mathrm{d}y}{\mathrm{d}\tau} = g(x, y). \tag{3.157}$$

If we apply the naive perturbation calculation, we obtain

$$x = A + \varepsilon \int_0^\tau f(A, y_0(A, \sigma)) d\sigma + o[\varepsilon], \tag{3.158}$$

where A is a constant and y_0 is the solution to

$$\frac{dy_0}{d\tau} = g(A, y_0). \tag{3.159}$$

If y_0 is chaotic for all A, it is effectively a random noise for x, so its average behavior can be described in terms of an observable invariant measure μ_A:[105]

$$\frac{1}{\tau}\int_0^\tau f(A, y_0(A,\sigma))d\sigma \simeq \int \mu_A(dy)f(A,y) \equiv \langle f\rangle(A). \tag{3.160}$$

Here, the time average $\langle f\rangle$ is considered as a function of A, so it is explicitly written as $\langle f\rangle(A)$. Consequently, the perturbation result (3.158) can be rewritten in terms of the average as

$$x = A + t\langle f\rangle(A) + o[1]. \tag{3.161}$$

Here, $o[1]$ is with respect to ε. Clearly, there is a secular term. This is renormalized into A as before (that is, $A \to A(t)$). The result is $x = A(t)$, where

$$\frac{dA(t)}{dt} = \langle f\rangle(A(t)). \tag{3.162}$$

This is the first-order result of the renormalized perturbation for (3.157): $x = A(t)$. This is the well-known result of the so-called *averaging method.*

If the correlation time of $y_0(A,\tau)$ is sufficiently short, then the time integral appearing in (3.158) may be understood as a sum of independently and identically distributed random variables. This implies[106]

$$P\left(\frac{1}{\tau}\int_0^\tau f(A, y_0(A,\sigma))d\sigma - \langle f\rangle(A) \sim \xi\right) \sim \exp\left(-\frac{\tau}{2b(A)}\xi^2\right), \tag{3.163}$$

where $b(A)$ is given by

$$b(A) = 2\int_0^\infty d\sigma \int d\mu_A(y_0(A,0))\,[f(A,y_0(A,\sigma)) - \langle f\rangle(A)] \times [f(A,y_0(A,0)) - \langle f\rangle(A)]. \tag{3.164}$$

Consequently, in terms of the Wiener process $B(t)$ we may write it as[107]

---

[105] One may think of it as something like a Sinai-Ruelle-Bowen measure.

[106] The large deviation principle holds for this random sum, so we may approximate it as is written here. For a pedagogical introduction to large deviation, see, e.g., Y. Oono "Large Deviation and Statistical Physics," Prog. Theor. Phys. Suppl. **99**, 165 (1989); H. Tourchette, "The large deviation approach to statistical mechanics," Phys. Rep. **478**, 1 (2009).

[107] ⟪***D*-dimensional Wiener process**⟫ On an infinite D-dimensional cubic lattice consider an infinitely long random walk. Then, apply the Wilson-Kadanoff renormalization to obtain its fixed point. It is the D-dimensional Wiener process, where we normalize the space scale such that the average displacement after t steps is $\sqrt{t}$. That is, if we observe an infinitely long random walk on a lattice from a distance, it looks

$$x(t) = A + t\langle f\rangle(A) + \sqrt{\varepsilon b(A)}B(t) + o[\varepsilon^{1/2}], \tag{3.165}$$

where $\varepsilon B(\tau) = \varepsilon B(t/\varepsilon) = \sqrt{\varepsilon}B(t)$ has been used. This equality is in law. That is, the both sides obey the same statistical law. Now, the fluctuating term proportional to $B(t)$ increases as $\sqrt{t}$, so it is also a secular term and must be removed by renormalization into A. The resultant renormalization group equation is the following *stochastic differential equation* (*Langevin equation*):

$$dA(t) = \langle f\rangle(A(t))dt + \sqrt{\varepsilon b(A(t))}dB. \tag{3.166}$$

The first term corresponds to the (strong) law of large numbers. Therefore, let us make a more explicit correspondence between renormalization and statistics.

Consider, for simplicity, a set of iid stochastic variables $\{X_i\}$. The procedure to make their partial sum $S_n = \sum_{k=1}^{n} X_k$ is, if the number of samples is regarded as time, to solve the following difference equation:

$$S_{n+1} - S_n = \varepsilon x_{n+1}. \tag{3.167}$$

Here, the parameter ε is introduced to emphasize the parallelism between this and the dynamical perturbation problem discussed just above, but it is unity in reality. The general solution reads, according to the law of large numbers and large deviation,[108]

$$S_n = A + \varepsilon \sum_{k=1}^{n} x_k = A + \varepsilon n\langle x_1\rangle + \varepsilon\sqrt{n}\sigma\chi + o[n^{1/2}], \tag{3.168}$$

where A is the integration constant, χ is a stochastic variable obeying $N(0,1)$, σ is the standard deviation of X_i. Absorbing the divergences into A as A_n, we obtain the renormalization group equation (cf. (3.166))

$$A_{n+1} - A_n = \varepsilon\langle x_1\rangle + \varepsilon\sigma\chi. \tag{3.169}$$

---

like a sample path of the D-dimensional Wiener process.

There are many ways to define the one-dimensional Wiener process mathematically. One characterization (definition) is as follows: $B(0) = 0$ and $B(t)$ is a Gaussian process (for any positive integer n at arbitrary n time points $0 < t_1 < \cdots < t_n$, $\{B(t_1), \cdots, B(t_n)\}$ obeys an n-dimensional Gaussian distribution): for $0 \le s < t$ the expectation of $B(t) - B(s)$ is zero and its variance is $t - s$, and the events up to time t and the increment $\mathrm{d}B(t) = B(t + \mathrm{d}t) - B(t)$ ($\mathrm{d}t > 0$) are statistically independent (i.e., non-anticipating). [For example, the expectation of $B(t)B(s)$ is $\min(s,t)$. In particular, the expectation value of $B(t)^2$ is t.]

The D-dimensional Wiener process is a D-vector whose components are all (statistically independent) one-dimensional Wiener processes.

[108] The fluctuation term is usually interpreted as the result of the central limit theorem, but a more proper understanding is in terms of large deviation; if the variance is finite, the results coincide.

If we scrutinize the above renormalization procedure, the way the renormalization constant is determined may be understood as a consequence of the requirement that the result is stable against addition of new samples.[109] For example, the problem of obtaining the first term of the right-hand side of (3.169) (corresponding to the law of large numbers) may be interpreted as the problem of choosing m, requiring

$$\sum_{k=1}^{n}(x_n - m) = o[n]. \tag{3.170}$$

That is, the change due to adding a new sample is $o[1]$. That we can accomplish this with probability one is the *strong law of large numbers.*

The general idea suggested by the above simple observation is that statistical analysis of a huge data set is to extract information that is stable against addition and modifications of data. Notice that this idea supplies a unified point of view to statistics and learning theory.[110] To learn from examples is to extract a certain pattern (hypothesis) from a set of given examples. Therefore, to pursue stable hypotheses against addition of new examples can be a learning strategy. A similar idea has been proposed from the learning theory side as well.[111]

References

Arnold VI (1997) Mathematical methods of classical mechanics. Springer

Barenblatt GI (1996) Similarity, self-similarity, and intermediate asymptotics. Cambridge University Press

Callen HB (1960) Thermodynamics. Interscience Publ.

Cannone M, Friedlander S (2003) Navier: blow-up and collapse. Notices Amer Math Soc 50:7-13

Chaikin M, Lubensky TC (2000) Principles of condensed matter physics. Cambridge University Press

Chen L-Y, Goldenfeld N, Oono Y, Paquette G (1993) Selection, stability and renormalization. Physica A 204:111-133

Chiba H (2008) C^1-approximation of vector fields based on the renormalization group method. SIAM J Applied Dynam Syst 7:895-932

Chiba H (2009) Extension and unification of singular perturbation methods. SIAM J Applied Dynam Sys 8:1066-1115

Cross MC, Hohenberg PC (1993) Pattern formation outside of equilibrium. Rev Mod Phys 65:851-1112

des Cloizeaux J (1975) The Lagrangian theory of polymer solutions at intermediate concentrations. J Phys (France) 36:281-291

---

[109] Actually, we need not be able to compute the perturbation result exactly; we have only to minimize $S_n - A_n$ approximately.

[110] S. Rajaram, Y-h. Taguchi and Y. Oono, "Some implications of renormalization group theoretical ideas to statistics," Physica D **205**, 207 (2005).

[111] T. Poggio, R. Rifkin, S. Mukherjee, P. Niyoki, "General conditions for predictivity in learning theory," Nature **428**, 419 (2004).

Elder KR, Grant M (2004) Modeling elastic and plastic deformations in nonequilibrium processing using phase field crystals. Phys Rev E 70:051605 (1-18)

Elder KR, Katakowski M, Haataja M, Grant M (2002) Modeling elasticity in crystal growth. Phys Rev Lett 88:245701 (1-4)

Fenichel N (1971) Persistence and smoothness of invariant manifolds for flows. Indiana Univ Math J 21:193-226

Fisher ME (1988) Condensed matter physics: does quantum mechanics matter? In: Feshbach H, Matsui T, Oleson A (ed) Niels Bohr: physics and the world (Proceedings of the Niels Bohr Centennial Symposium). Harwood Academic Publishers

Furukawa Y (1998) Inventing polymer science—Staudinger, Carothers, and the emergence of macromolecular chemistry—. University of Pennsylvania Press

Goldenfeld N (1992) Lectures on phase transitions and renormalization group. Addison Wesley

Goldenfeld ND, Martin O, Oono Y (1989) Intermediate asymptotics and renormalization group theory. J Scientific Comp 4:355-372

Gunaratne H, Ouyang Q, Swinney HL (1994) Pattern formation in the presence of symmetries. Phys Rev E 50:2802-2820

Haataja M, Gränäsy L, Löwen H (2010) Classical density functional theory methods in soft and hard matter. J Phys: Cond Mat 22:360301 (1-8)

Hall AR, Colegrave N (2008) Decay of unused characters by selection and drift. J Evol Biol 21:610-617

Heisenberg W (1971) Physics and beyond (translated by A J Pomerans) Harper & Row

Herbut I (2007) A modern approach to critical phenomena. Cambridge University Press

Hirsch M, Pugh C, Shub M (1970) Invariant manifolds. Bull Amer Math Soc 76:1015-1019

Husserl E (1999) The idea of phenomenology (translated by L Hardy). Kluwer Academic Publishers

Izutsu T (1991) Consciousness and essence—quest of the spiritual Orient. Iwanami paper back

Keller CF (1999) Climate, modeling, and predictability. Physica D 133:296-308

Kihara T (1978) Intermolecular forces. Wiley

Kubo R (1968) Thermodynamics. An advanced course with problems and solutions. North-Holland Pub Co.

Ladyzhenskaya OA (1963) Mathematical theory of incompressible fluids. Gordon & Breach

Ladyzhenskaya OA (2003) Sixth problem of the millennium: Navier-Stokes equations, existence and smoothness. Russ Math Surveys 58:251-286

Le Bellac M (1991) Quantum and statistical field theory. Oxford University Press

Lieb E, Yngvason J (1998) A guide to entropy and the second law of thermodynamics. Notices Amer Math Soc 45:571-581

Lieb E, Yngvason J (1999) The physics and mathematics of the second law of thermodynamics. Phys Rep 340:1-96

Mandelbrot BB (1983) Fractal geometry of nature. W H Freeman

Mandle F (1988) Statistical physics (2nd edition). Wiley

Mañé R (1978) Persistent manifolds are normally hyperbolic. Trans Amer Math Soc 246:271-283

Migdal AB (2000) Qualitative methods in quantum theory (translated by Leggett AJ). Westview Press

Miklósi A, Kubinyi E, Topál J, Gáacsi M, Virányi, Z, Csányi V (2003) A simple reason for a big difference: wolves do not look back at humans, but dogs do. Curr Biol 13:763-766

Miyazaki K, Kitahara K, Bedeaux D (1996) Nonequilibrium thermodynamics of multicomponent systems. Physica A 230:600-630

Niwa N, Hiromi Y, Okabe M (2004) A conserved developmental program for sensory organ formation in *Drosophila melanogaster*. Nat Genet 36:293-297

Nozaki K, Oono Y (2001) Renormalization-group theoretical reduction. Phys Rev E 63:046101 (1-18)

Ohta T, Nakanishi A (1983) Theory of semi-dilute polymer solutions: I Static properties in a good solvent. J Phys A 16:4155-4170

Ohta T, Oono Y (1982) Conformational space renormalization theory of semidilute polymer solutions. Phys Lett 89A:460-464

Oono Y (1985) Statistical physics of polymer solutions. Conformational-space renormalization group approach. Adv Chem Phys 61:301-437

Oono Y (1985) Dynamics in polymer solutions — a renormalization-group approach. AIP Conference Proceedings No 137 (edited by Y Rabin) p187-218

Oono Y (1989) Large deviation and statistical physics. Prog Theor Phys Suppl 99:165-205

Oono Y, Ohta T, In Goldbart PM, Goldenfeld N, Sherrington D (ed) Stealing the gold: a celebration of the pioneering physics of Sam Edwards. Oxford University Press

Oono Y, Paniconi M (1998) Steady state thermodynamics. Prog Theor Phys Suppl 130:29-44

Pashko O, Oono Y (2000) The Boltzmann equation is a renormalization group equation. Int J Mod Phys B 14:555

Poggio T, Rifkin R, Mukherjee S, Niyoki P (2004) General conditions for predictivity in learning theory. Nature 428:419-422

Protas M, Conrad M, Gross JB, Tabin C, Borowsky R (2007) Regressive evolution in the Mexican cave tetra, *Astyanax mexicanus*. Curr Biol 17:452-454

Rajaram S, Taguchi Y-h, Oono Y (2005) Some implications of renormalization group theoretical ideas to statistics. Physica D 205:207-214

Ruelle D (1969) Statistical mechanics, rigorous results. Benjamin

Sasa S, Tasaki H (2006) Steady state thermodynamics. J Stat Phys 125:125-224

Shiwa Y (2000) Renormalization-group theoretical reduction of the Swift-Hohenberg model. Phys Rev E 63:016119 (1-7)

Shiwa Y (2005) Comment on "renormalization-group theory for the phase-field crystal equation." Phys Rev E 79:013601 (1-2)

Shiwa Y (2011) Renormalization-group for amplitude equations in cellular pattern formation with and without conservation law. Prog Theor Phys 125: 871-878

Shukla J (1998) Predictability in the midst of chaos: a scientific basis for climate forecasting. Science 282:728-731

Stanley HE (1971) Introduction to phase transition and critical phenomena. Oxford University Press

Swift J, Hohenberg PC (1977) Hydrodynamic fluctuations at the convective instability. Phys Rev A 15:319-328

Tieszen R (1998) Gödel's path from the incompleteness theorems (1931) to phenomenology (1961). Bull Symbolic Logic 4:181-203

Tourchette H (2009) The large deviation approach to statistical mechanics. Phys Rep 478:1-69

Wall FT (1975) Theory of random walks with limited order of non-self-intersections used to simulate macromolecules. J Chem Phys 63:3713-3717

Wall FT, Seitz WA (1979) The excluded volume effect for self-avoiding random walks. J Chem Phys 70:1860-1863

Wang B, Zhou TG, Chen L (2007) Global topological dominance in the left hemisphere. Proc Natl Acad Sci USA 104:21014-21019
Yau H-T (1998) Asymptotic solutions to dynamics of many-body systems and classical continuum equations. In: Current Developments in Mathematics, 1998. International Press

Chapter 4
Modeling

—Description and Understanding of Phenomena—

It has been explained that 'phenomenological understanding' is the basic way of understanding the world, and then the renormalization group philosophy has been explained as a strategy to extract phenomenology. Although renormalization is a basic mode of thinking, if we wish to compute something that can be actually compared with real phenomena, we need models—mathematical expressions of phenomena. We have seen that to understand ordinary universalities for which renormalization works well we need only minimal models. A minimal model is a lean model that just corresponds to the phenomenology of a set of phenomena and represents the state of our understanding of the set of phenomena. Thus, to model a phenomenon is to give an ultimate expression to what we have understood about the universal features of the phenomenon.

If the phenomenon is not too complicated, so-called *abduction*[1] (or abductive reasoning) may well give us such a model. The author interprets abduction as an attempt to understand the world relying confidently on the fact that there are numerous phenomenological structures in the world and on the wisdom embodied in us through phylogenetic learning. It is an attempt to understand the world relying on intuition about the crux of the phenomenon, while respecting the general character of the world (symmetry, conservation laws, etc.) and our aesthetics. We must clearly recognize that to understand a thing is an active process to select the best hypothesis through generating numerous hypotheses. There is no pure induction.

There must be a sensible opinion that many phenomena in the world should not be that 'soft' as can be crunched by abduction. Before trying abduction, efforts to extract patterns intrinsic to the observed data (distillation

[1] This is due to C. S. Peirce (1839-1914). According to him, retroduction should be a better term; "it is a process to try to decipher a phenomenon based on certain hypotheses in order to grasp it as a meaningful and rational whole." (*Dictionary of Philosophy and Thoughts* (Iwanami) p30).

of information) should be useful. Therefore, the operation of constructing a model that represents our state of understanding must be performed together with a distillation of information.

It is important to observe the phenomenon well without any prejudice, but does massive statistical processing of data surpass our capability to detect subtle patterns? Needless to say, comparing huge data sets or reducing the data into easy-to-understand formats, etc., need not be done by us human beings, but beyond such preparatory steps, we will have to rely more or less on abduction. Purely mechanical induction is hardly expected to make a good model. Therefore, in this chapter modeling is illustrated through successful examples of abduction. We cannot avoid the question: what do we mean by 'a model is good'?

Let us return to the basic, and start with a question: what is a model?

4.1 What is a model?

There seem to be roughly three ways to use the word 'model' in science (they are not mutually exclusive).

The first usage may be found in the Heisenberg model of magnets, or the planetary model of atoms. This 'model' means a mathematical system capturing a certain aspect of reality (of phenomena). In this case, reality is more complicated (detailed) than the model, and the correspondence between the model and reality is, metaphorically put, at best 'homomorphic.'[2] For example, in the planetary model of an atom, its nucleus is a mere point mass with charge, and lacks any internal structure as the actual nucleus. Still, to understand the chemical nature of the atom, this model is almost sufficient. That is, this type of model describes the mathematical essence of the system (underlying the phenomenon), and expresses the state of our understanding of it. The Weinberg-Salam model of the electroweak interaction is also this type.

The second usage indicates a (mathematical or materials) model that can substitute the actual system to give realistic representation (of the totality) of a certain aspect of the actual system or phenomenon of interest. The simulation model of a particular power plant, and numerous in silico models in systems biology are examples. If we interpret observation results of a system

[2] ⟨⟨**Isomorphism and homomorphism**⟩⟩ Suppose there are sets A and B with a mathematical structure (say, an algebraic structure such as 'group'). If a map $\phi : A \to B$ is one to one (bijection) and preserves the mathematical structure [e.g., if the structure is a group structure: $\phi(ab) = \phi(a)\phi(b)$], ϕ is said to be an *isomorphism*, and A and B are said to be isomorphic. We have already encountered this in Chapter 2 in the context of dynamical systems. If the map ϕ is surjective (i.e., $\phi(A) = B$) but not one to one, ϕ is called a *homomorphism*. In this case more than one elements in A is bunched together by ϕ, so usually information is lost.

as data sequences or signal sequences, we can say these models aim at information compression. That is, they are systems serving as signal sources that can generate signals mimicking the signals from the actual systems. In contrast to the first usage above, these models are not designed to understand the phenomena as their primary aim; they do not have any intention to pursue mathematical essence of the phenomenon. Since the aspects of the systems outside modeler's direct interest are totally ignored, the relation of these models with the reality is homomorphic (or perhaps we should say projective).

The third usage of the word 'model' indicates something concrete to capture abstract concepts, relations, etc. The usage of the word 'model' in mathematical logic is in this sense. The model of Lobachevskii's[3] hyperbolic geometry in the unit disk due to Poincaré is a typical example of this type.[4] In this case, the existence of a non-Euclidean geometry is demonstrated by the existence of its model itself in a way no one can ever object to.[5] The Turing machine (Section 2.10) is understood as a model of performing 'recursive computations.' This model shows the computability of recursive functions in a way no objection is possible as stressed by Church. These models are the tools of logical demonstration.

In summary, the word 'model' indicates (1) Succinct representations of the level of our understanding of aspects of reality (expression of mathematical essence), (2) Detailed descriptions of aspects of reality (or the results of data compression), or (3) Mathematical systems utilized as tools for logical demonstration. (1) and (2) may be called tools for description, but (1) transcends this role and materializes the essence of phenomena.[6]

For the models in sense (3) the world of ideas is connected to the world of mathematics (as long as we do not make models using actual materials),

---

[3] N. I. Lobachevsky (1792-1856) reached the modern concept of functions slightly before Dirichlet.

[4] See, e.g., 1.3.6.6 of *The Princeton companion to mathematics* (T. Gowers, editor) (Princeton, 2008).

[5] Such a model might be better expressed as 'representation.' Here, this word is used as is used in 'matrix representation of finite groups' or 'coordinate representation of vectors.' Generally, in mathematics 'representation' means to express something abstract in terms of more concrete mathematical objects. I. M. Gel'fand said, "Mathematics is all studies of some sort of representations."

[6] ⟪**Theory, model and artwork**⟫ In this book, what we wish to model are not individual systems but the universal aspect of collections of phenomena. When the object we wish to study is an individual concrete system, it is not a man-made system like a 'particular atomic reactor of a certain power plant' but a god-given system like 'F_1-ATPase.' There is an opinion that theory, model and artwork are distinct (e.g., a theory must be objective, and art is at its antipode), but it is hard to be as simple-minded as this. Boltzmann's formula for entropy or Maxwell's equations are no less aesthetically worth appreciating than artistic masterpieces. Conversely, excellent artistic outcomes would teach us deeper ways to understand the world directly with our natural intelligence.

so there is no problem of connecting the world of words (concepts) and the actual world. In contrast, in (1) and (2) there is a problem of how to connect the actual world and the world of concepts. We will reflect upon this question in the next section. Natural scientists usually do not feel that there is a serious problem here: what we observe usually exists firmly in the world outside us. If the reader thinks this to be without any serious problem, *she can immediately go to Section 4.3.*[7] However, for those who wish to extend the horizon of science it may not be totally useless to read the next section. Through doubting the relation some parts of the humanities have made repeated efforts to undermine the special status of natural science in human culture, and, although the skeptics may not be the majority, they may not be a simple minority, either. The next section is also written for those who wish to claim that natural science is another social construct. It is important as an ABC of science to reflect on whether a given concept or theory is a social construct or not.[8]

Note 4.1. Observation and inevitable disturbance
Although it was written above, 'what we observe usually exists firmly in the world outside us,' those who know quantum mechanics even slightly may not be able to be this 'naive.' We have already discussed the delayed selection experiments (Section 2.4). Any observation involves a physical interaction between an object to be observed and the observer, and it inevitably disturbs the states of both. However, the generated disturbance is also a part of the objective phenomenon. For example, think about the uncertainty relation in quantum mechanics. We could imagine that both the position and the momentum of a particle have definite values but due to inevitable interference the observation cannot avoid uncertainty, but there is no compelling reason that we must adopt the metaphysical (ontological) framework that the position and the momentum are separately meaningful quantities. If we set up an ontologically inappropriate framework, inevitable disturbances occur. Whenever there is an inevitable disturbance, if we confront the reality without prejudice, it only turns out that the interfering quantities or entities cannot be unconditionally taken as separate entities.[9] In the quantum mechanics framework, the position-momentum pair corresponds to reality (ontologically meaningful), but position and mo-

[7] ⟪**Intuition and body**⟫ "The oval and semicircular canals on three mutually perpendicular planes tell us which direction is up and in which way our body is rotating. I think it is a confusing idea that all such organs that were obviously created to preserve species through adaptation to reality and their functions have nothing to do with our *a priori* intuitive cognition of space. They furnish the basis of the intuition of the 3-Euclidean space. Nay, I should declare that, obviously in a certain sense, these organs are intuition itself." (K. Lorenz, *Die Rückseite des Spiegels* (R. Piper & Co. Verlag, München, 1973) a coarse-grained translation).

Incidentally, Mach was one of the chief researchers who clarified the function of the semicircular canals.

[8] A good example is to grasp that the sociology of science is much more influenced by the social background, education, intellectual capability, etc., of the researchers in the field than by the research objective of the field (i.e., science) itself.

[9] Ockham's razor shaves inevitable disturbances off. (It is said that John Duns Scotus (ca 1266-1308) had already emphasized this about 20 years before Ockham (ca 1288-ca 1348)).

mentum individually are meaningless concepts. Candidly speaking, we have not yet understood our external world well; since we think we have completely understood it, surprises and unreasonable things pop up. 'What we can observe exists firmly in the external world,' but what we can really observe is not up to us.

4.2 Correspondence between models and reality

A rule stands there like a sign-post. $\cdots$ Does the sign-post leave no doubt open about the way I have to go? Does it shew which direction I am to take when I have passed it; whether along the road of the footpath or cross-country?[10]

If the reader believes what we observe usually exists firmly in the world outside us, *she can skip this section and can immediately go to Section 4.3.*

In order to model a phenomenon, we must make clear
(A) the domain where the modeling is performed and
(B) the setup of the correspondence relation between this domain and reality. In this book we will not pursue any model made of actual materials. Models discussed will be all mathematical models.[11] Therefore, the question about (A), "what is used to make a model?" is not an important question.[12]

In contrast, (B) is rather subtle. Thus, this section is devoted to the discussion of this point. *If the reader as a serious scientist thinks that this problem actually does not have any subtlety, she may go to the next section from here.* The relation φ between the objects (variables) in actual phenomena and in mathematical models is usually not a problem when we engage in or learn the ordinary established branches of physics or mathematical sciences. A phenomenon is understood to be described in terms of certain quantities that are somehow observable. That is, the correspondence between the world of mathematics and the real world is regarded as well established far before any model or theory is constructed. For example, if we wish to model a system in terms of thermodynamic observables, we need not worry about how these quantities are related to the actual phenomenon (e.g., how to measure entropy). However, if one wishes to establish thermodynamics itself, relating the quantities in the theory and those we measure is a core ingredient of the

---

[10] L. Wittgenstein, *Philosophical Investigations* (Blackwell, 2001; original 1953) 85.

[11] Models formulated in natural languages may be conceivable, but it is hard to formulate them unambiguously, so such models are regarded as heuristic intermediate-step models. However, the author does not have any intention to regard them as unimportant.

[12] There is a problem of what set theoretical axiomatic system to use, but we simply use the conventional framework, the ZFC axiomatic system. It must be the general opinion that natural scientists (empirical scientists) need not worry about such a thing, but we have seen that we might not be able to simply follow this opinion in conjunction to randomness in Chapter 2.

theory. Most scientists never work on such fundamental problems, so it is all too practical for them to assume, just as in many textbooks, that such a problem has been completely solved. One aim of this book is to prepare for expanding the horizon of fundamental science, so it should not be superfluous to reflect on the very basic question of the correspondence between theory and reality.

Suppose we wish to model the *Belousov-Zhabotinsky reaction* famous for its oscillating behavior.[13] In the model appears the concentration x of a certain substance (say, malonic acid). If a chemist is asked to give the value of x, he would measure it with the aid of, e.g., the characteristic absorption spectrum. Before the actual measurement the terms such as 'quantity,' 'concentration,' 'malonic acid,' 'absorption spectrum,' etc., must be clearly understood. In reality, we need not say anything to the chemist. When we make a mathematical model of a chemical reaction, we never worry about how we relate the names or chemical formulas of reactants and the actual chemical substances. We may put these relations in a black box. This is exactly parallel to the situation in which a mathematical ecologist can do meaningful work without being able to identify any organisms.

The black box just mentioned is crucial to make discussions succinct and to allow us to concentrate on the essence of the phenomenon under study. However, we know that pitfalls of mathematical proofs often lie where we write 'obviously.' Thus, analogously, apparently obvious procedures or operations in the black box may still be worth scrutinizing from time to time. Furthermore, we should be conscious of the danger to the spirit of science[14] of a black box the content of which we can never observe. The core of the correspondence between reality and a model is in this black box. For example, 'malonic acid' is a term, so how to interpret it must be shared (agreed upon)

---

[13] ⟪**Belousov-Zhabotinsky reaction**⟫ For the reaction itself, see, for example, `http://online.redwoods.cc.ca.us/instruct/darnold/deproj/Sp98/Gabe/`. This reaction is basically the oxidation of malonic acid by bromic acid $HBrO_3$. If ferroin is used as a catalyst, the oscillation may be observed as a color oscillation of the solution between blue and red. Belousov quit science because he could not publish his fundamental work on this reaction in any established journals. Even after this reaction became famous, he never showed up in any meeting on the reaction. The author is deeply sympathetic to Belousov who quit science for the reason that science (or the science community) was not scientific enough.

S. E. Shnoll was interested in the reaction, so Belousov gave him the prescription. Belousov also promised to publish his report in the annual report of the institute he belonged to. Shnoll told his student Zhabotinsky to study the mechanism (based on `http://people.musc.edu/~alievr/belous.html` (by R. R. Aliev) but the site is no longer accessible).

[14] Here is something symbolizing the schism between contemporary technology and the spirit of science. Now, even if we open up an audio device or a home appliance, we would not understand anything. Parts are just like incantations. Even religious fundamentalists can use laptops. This symbolizes the most serious dilemma of our culture.

by people. However, if being shared is enough (intersubjective agreement is enough), it is not different from mass illusion. Unless the actual substance shows up, this word is only related to another word. Even if we use chemical formulas or 3D molecular models, the situation does not change at all. We can never get out of the world of words (or ideas). We must ask how actual objects and words are related.

To confirm that the given white powder is malonic acid, we must perform various experiments. The description (instruction) of each experiment is given in words, so it cannot bridge the gap between the description of an operation and the operation itself. Note that even the simple operation 'holding a pencil' cannot be specified in words. What is the meaning of 'hold'? However many words you may spend to explain this, the words cannot move a hand. 'Action itself' and 'Description of action' are distinct (they are in different categories). The correspondence between these two must be specified, but the correspondence between the 'real world' and the 'world of words' cannot in principle be described in words alone.[15,16]

The Turing machine (Section 2.10) illustrates the above problem clearly. A Turing machine operates according to a Turing program. For example, suppose '1' in the tape cell being scanned now means 'to move the read/write head to the left by one cell' under the current black box state. Then, actually the Turing machine head moves accordingly. There is no program that interprets the command 'to move the read/write head to the left by one cell' and tells the Turing machine how to obey this instruction. The Turing machine is exempted from this interpretation problem. This is simply because the Turing machine is so constructed that it moves its 'effector' accordingly upon the particular 4-tuple and has no freedom of choice; the implementation has solved the problem (or has removed the possible ambiguity).[17] The point of contact between 'words' and 'actions' must be somewhere in between, so we human beings must have the will-action adaptor, cognition-real world adaptor, etc., corresponding to the black box of the Turing machine.

To objectively specify the correspondence between reality and a model,

---

[15] 'Miseries disappear before the concept of 'miseries.' (P. Nizan, *Les Chiens de garde* (Rieder, Paris, 1932)) is a serious observation.

[16] ⟪**Adaptor**⟫ Many readers must have started to think that a stupid and rather pointless 'philosophical argument' continues on and on. However, as has already been noted, notice that the problem of the relation between amino acids and the genetic codes is a similar problem in that different materials categories are connected. To this end we need substances called *adaptors* that connect these different categories. This is tRNA. Notice that by changing the adaptors, the situation drastically changes. M. Barbieri, *The Organic Code — An introduction to semantic biology* (Cambridge University Press, 2002) is worth reading.

[17] "When I raise my arm I do not usually *try* to raise it." (L. Wittgenstein, *Philosophical Investigations* (Blackwell, 2001; original 1953) 622). When we decide to do something, usually we do not ask whether we should decide or not to decide whether we should do it or not. We have, so to speak, only one layer of consciousness above our unconscious domain.

since the word 'objectively' usually means 'other people also agree,' there must be a considerable amount of agreement between the people beforehand. The core of this agreement is that the adaptors discussed above are universally shared by the people. Then, why are the adaptors universal? Or, why are there so many agreements among us? This must be because there are universal forces that push us to agree. Of course, there are a lot of strong compulsory forces, because our ancestors for the past 4 billion years wished for the welfare of their offspring.[18,19] That is, factual propositions are not independent from value judgments. Notice that no such banal assertions are presented here as "The act of expressing the factual proposition imposes value systems," or as "Factual propositions are accepted with the value judgment based on the former." Here, it is asserted that the value judgment is fundamental. The world forces upon us a certain fundamental value system that makes us recognize facts as such.[20]

The fundamental error of the following assertion by 'science philosopher' Y. Murakami is due to the neglect of our (and our society) being strongly constrained by the external world (e.g., our universal anatomical structure; "The semicircular canals are the form of our intuition"): "Cognition is an interaction between the external world and the subject. The subject does not come into existence without the intersubjective whole. Therefore, the assertion that the obtained 'picture of the external world' is not absolute nor universal is fundamentally different from the denial of the existence of the external world (or even the existence of nature)." Many problems have been solved (decided) by the implementation of our body. One should not make such a crude argument that we are social beings, so our agreements are social, consequently science is socially constructed, and scientific truth is also relative to society. The assertion that there is no universal core in the "picture of the external world" is equivalent to denying the existence of the

[18] Some might say, "This is wrong. The agreed components are due to shared social experiences." However, comparative studies of laws and rules in various cultures demonstrate the existence of common elements that cannot be explained cultural-historically. Cf. K. Lorenz, *Die acht Todsünden der zivilisierten Menschheit* (R. Pier & Co, 1973), Chapter 6.

[19] ⟨⟨**Sapir-Whorf thesis**⟩⟩ There is an assertion called the *Sapir-Whorf thesis* that human cognition is restricted by language, and consequently determined by culture and society. Naturally, the influence of language to some extent must exist, but still we cannot talk black into white. Certainly, there are experiments demonstrating that color cognition is influenced by language, but the non-linguistic element is still large.

[20] Simply put, "the reason why we are forced to recognize something as a fact is that the action according to the recognition that it is not a fact is harmful to us." Then, perhaps one might wish to say that the action need not accord with the recognition. That is, if your action follows a certain 'twisted logic,' you can still avoid harmful actions even if your recognition is wrong (in the usual sense). However, since the world in which we exist is a rather lawful and logical world, to live through 'private logic' does not seem feasible. Simpler logic will allow faster decisions needed for survival.

external world.[21]

A certain model (or a theory) is judged as an effective one for us to understand reality, if our performing an operation dictated by the model, following the specified correspondence between the model and the actual world, gives the result as the model (theory) concludes. In other words, a model (theory) is judged as a failure, if we have a(n eventual) possibility of suffering from a bodily injury (more precisely, if the damage cannot be reduced compared to the case without relying on the model), when we trust the outcome of the model (theory).[22] This is quite similar to the effectiveness of laws. The reason for law to be effective in practice is simply that it is supported by naked violence. Law is meaningful only when it is integrated with a means to inflict bodily pain. This analogy is actually fairly essential. If we err in the judgment of a 'good model,' eventually we will be punished.

What is the conclusion of such a circuitous 'philosophical consideration'? To relate a model (theory) and reality unambiguously, operationally well-defined correspondence is mandatory, and this is realizable thanks to the

---

[21] **⟪Science of *Homo sapiens*, masculine and feminine sciences⟫** However, when we promote science in practice, the emphasis points can naturally depend on the species. For example, human beings are animals of vision, so they are good at classifying birds that are also animals of vision. Or, geometrization of mathematics and physics is likely to have roots in this human special feature. Therefore, we should reflect on whether this is the really correct direction of the progress of fundamental physics (it is possible, however, our being animals of vision might reflect the fundamental feature of Nature that could be geometrical). Similar things can be said about gender. While science is not developed well, masculine and feminine sciences may well exist. Consequently, it is highly likely that our current science is strongly biased by the masculine point of view. However, this should not be taken as an evidence that science is intrinsically restricted by species and gender.

[22] **⟪Evolutionary fundamentalism⟫** As the reader has already sensed and probably felt as being a bit extreme, this book relies heavily on "evolutionary fundamentalism." Thus, one might say the assertions are not at all objective or unbiased. Without (arbitrarily) assuming the existence of the external world natural science does not make sense, and, even worse, evolutionary biology cannot have any basis. A more flexible philosophy must be appropriate for a mature thinker...

The author gives a two-step reply to this criticism. (i) Natural science including evolutionary biology that naively accepts the existence of the external world as objective makes a self-consistent system of *Weltanschauung* (or at least it is tending to this state). However, self-consistency does not necessarily imply 'truth,' so one might not have to obey what is dictated by this world view. (ii) However, there is no freedom of conduct, although one may have any opinion while staying in this world (for example, even if you believe that world mechanics is Aristotelian without inertia, your body and nervous system are wiser, obeying Newtonian and quantum mechanics). The being whose behaviors contradict the concept of 'truth' in the above sense cannot continue to exist in this world, or would be selected out sooner or later. If one wishes to continue to assert the opposite view, she or he must at least behave in conformity with the truth in the above sense, and must be consistently keep the inconsistency between her or his opinion and actual conduct/behavior. We must clearly recognize that the concept of 'truth' does not have any meaning away from the fundamental value system for organisms. Truth does not exist apart from life.

existing sizable amount of *a priori* agreements among human beings.[23] The important portion of these agreements is due not to 'trivial' social agreements but to phylogeny.[24] Henceforth, just as natural scientists have intuitively accepted, we accept the existence of these *a priori* agreements without any reservation. Under this usual assumption, a quantity in science is meaningful only if it is defined operationally.[25,26]

4.3 Models as tools of description

Those who are practical may have only to browse through the last 2 pages of this section (i.e., the second full paragraph on p204).

To make a mathematical model, the quantities appearing in the model and those in reality must be related clearly. Let us express this succinctly as the observability of the quantities (in the model). When a model is constructed to describe an actual phenomenon, quantities not observable must not appear in

---

23 ⟨⟨**Duhem-Quine's thesis**⟩⟩ In conjunction with the discussion around here, some readers might have recalled this thesis (various propositions about the external world are subjected to empirical verification/test not individually but as a whole): In order to determine whether a proposition is true or not other empirical propositions must be taken into account, so individual propositions cannot be verified directly with the sense data. This and the argument of Kuhn about the paradigm shift are rather harmonious (an empirical system holds or not as a whole). It is totally forgotten (even by the proposers of the theses), however, that such arguments are meaningful *only* if the fruits of phylogenetic learning that are built into (implemented in) our natural intelligence are taken into account. When we confront our external world, we mobilize all the experiences of the past 4 billion years, so to speak. Needless to say, we utilize the empirical knowledge in the usual sense, but it is only the tiny tip of a gigantic iceberg, so we have a very stable verification standard. Thus, it is obviously very rare that the empirical data themselves are altered, even if the so-called paradigms change.

24 As the reader has already realized, even social agreements are strongly influenced by biological prerequisites, so there are many nontrivial social agreements as well, precisely speaking.

25 It is dangerous, however, to conclude that quantities that cannot be defined operationally are meaningless, because we cannot unambiguously specify an action in words. Besides, recall that we cannot speak of all we know unambiguously. Thus, we should humbly accept the possibility that the concepts that cannot be specified operationally at present may someday be defined operationally.

26 Up to this point, it has been assumed that the observer and her observing object are distinct. However, isn't there a situation where an observer and her object cannot be separated? This book tries to be maximally conservative, so such a situation is ignored. In such cases objective observational facts are not thinkable. We must first reflect on whether there is an operation deservingly called observation when the observed and the observer are truly inseparable. Of course, we must clarify the concepts of 'observation,' 'object,' 'observer,' 'separation,' etc., so we do not discuss the cases.

the input, and at least one quantity in the result must be observable. However, the requirement that quantities meaningless in reality must not appear in the input is an ideal. In practice, we introduce unobservable quantities and use metaphysical assumptions that no experiment can confirm or refute. In this case, the meaningful outcomes of the model are those invariant under changing metaphysics and unobservable quantities. Notice that the idea of renormalization increases the degrees of freedom in model construction.

A (mathematical) model of a phenomenon is a mathematical structure that has unambiguous[27] correspondence (in other words, operationally clear correspondence) between its various input/output quantities and various quantities in the real world and that the correspondence is a 'homomorphic correspondence.' To make the statement clear, let us use the following diagram:

$$\begin{array}{ccc} A & \xrightarrow{p} & B \\ \downarrow{\scriptstyle \varphi} & & \downarrow{\scriptstyle \varphi'} \\ M & \xrightarrow{\sigma} & N \end{array}$$

Here, A is a set of events (states, observable values, etc.) under study (we do not deal with the world or universe as a whole), p is an (actual) operation (or process) bringing individual events of A to that of B, another event set. The downward-pointing arrow φ is the rule to relate reality to the corresponding set M in the model (the map can be taken as a surjection). φ' is similar. As discussed at length in the preceding section, φ and φ' must be operationally unambiguous for observable quantities. That (M, σ, N) is a model of (A, p, B) is implied by the commutativity of the above diagram.[28] $B = A$, $N = M$, $\varphi = \varphi'$ are allowed. For example, p is a time evolution (e.g., a unit time map); that (M, σ) is a model of (A, p) implies that the model can simulate a certain aspect of the time evolution of the system (A, p). In contrast to the diagram that appeared in the definition of chaos in Chapter 2, φ need not be one-to-one, but can be only surjective. That is, the model may give a coarse-grained description of the actual phenomenon.

In reality, it may not be very easy to make a model that has the structure given in the diagram, but still the diagram must be meaningful as the ideal. For example, that A is a set is not a trifling requirement. This means that the events the model tries to explain are clearly demarcated. When, unfortunately, $\varphi'(p(a)) \neq \sigma(\varphi(a))$, this requirement excludes an apology that a was actually out of consideration for the model. Furthermore, "φ is surjective" is meaningful when we relate the model results to reality; it is useless to discuss anything that does not have any corresponding object in the real world.

In the above commutativity diagram the irreversibility of the correspon-

---

[27] Let us confirm that the concept of 'unambiguous' does evolve. This reminds us of statistical tests of random numbers we encountered in Chapter 2.

[28] Roughly speaking, "A diagram is commutative" implies that any results obtained by any combination of arrows are consistent.

dence φ characterizes models (reality is always richer than models).[29] The reader might say that the commutativity of the diagram implies only that the model does not contradict reality, and the model does not have any predictive power; it is merely explaining what we observe. However, since φ is a map, the actual events are classified into the equivalence classes of the preimages due to φ.[30] That is, the model predicts these equivalence classes. Thus, unless the model is trivial (if M has more than one elements), any model tells us a prediction that is different from total ignorance (however, the results need not be better than total ignorance). Models that cannot satisfy the commutativity diagram above may be judged as 'pseudo models,' which do not contribute to our understanding of the world. If the conditions of the diagram are met, then can we say the model is a good one? Even a mere curve-fitting result of experimental data satisfies the conditions for the diagram.

Usually, scientists scorn curve fitting. However, is a model that gives us deep insight really distinct from 'mere' curve fitting? Let us look at the moment of the birth of the formula by Planck (1858-1947) for blackbody radiation; the formula for the energy distribution u_ν (the energy density carried by the electromagnetic wave with frequency between ν and $\nu+d\nu$) of the electromagnetic wave in equilibrium with a blackbody maintained at temperature T. Wien (1864-1928) had proposed an experimental formula for $\nu \to \infty$:

$$u_\nu \sim A\nu^3 e^{-B\nu/T}, \tag{4.1}$$

where T is the absolute temperature, and A, B are positive constants. Planck also knew the experimental result in the $\nu \to 0$ limit:

$$u_\nu \propto T. \tag{4.2}$$

Planck proposed the interpolation formula between these two:

$$u_\nu = \frac{A\nu^3}{e^{B\nu/T} - 1} \tag{4.3}$$

during a conference (on October 19, 1900) in the discussion after a presentation by Rubens (1865-1922). Actually, he knew (4.2) in the morning of the day when Rubens gave the talk.[31] The next morning, Rubens came to tell Planck that (4.3) was satisfied well for all the frequencies, and Planck's the-

---

[29] However, this need not be ascribed to microscopic details.

[30] Since φ is homologous in the metaphorical sense, there is a natural equivalence relation introduced by the map. That is, for any element m of M there is its preimage $\varphi^{-1}(m) \subset A$ and this defines a partition of A.

[31] The story here is copied from H. Ezawa, "Quantum mechanics, from its birth to re-examination," in *Physics of 20th Century* edited by H. Ezawa (Saiensu, 1998) p62. Notice that (4.2) was theoretically derived only later by Rayleigh and Jeans and by Einstein from classical electrodynamics.

oretical efforts began. That is, an extremely successful experimental formula is much more important than many theoretical results.

The standard theory of elementary particles may appear to be an antipode of mere curve fitting, and deep physics insights such as gauge symmetry and renormalizability are built into it, but these insights might be understood only as means to narrow the space in which curves are fit. We tend to think that curve fitting and modeling are fundamentally different. The former has no 'philosophy,' but a model has an idea to construct it, so the success of a model lends some support to the idea on which the model is based. This sounds sensible, but we should not forget that even in the curve-fitting procedure certain ideas about the world are already incorporated. For example, we usually use smooth curves. Curve fitting and models make a continuous spectrum. If the phenomenon becomes complicated, the model obtained by mere data compression could be non-trivial. Therefore, to understand complicated or complex phenomena even a simple curve fitting could be meaningful as a tool of nontrivial description.

Now, what do we mean by the statement that a model of a phenomenon is an expression of understanding of the phenomenon? Take Planck's example above. Even if we know the model (formula) (4.3), we do not have any insight into blackbody radiation. With this formula alone, the only reason to justify it is that it agrees with experiments. Although the commutativity of the diagram above is required for objective confirmation of the correctness of the insights, whether a model gives us insight is not in the commutativity but in the way the model is constructed. The goodness of a model is strongly influenced by how logically directly it is connected to clear ideas (the requirement of logical leanness).

The scarcity of adjustable parameters can be a feature of logical cleanliness of the model. In natural science, a model is usually for a class of systems and not for a particular system; consider, for example, the Navier-Stokes equation (see Example 3.1) with only two parameters. The extent of nontriviality and generality distinguishes models from 'mere' curve fitting. These two requirements tend to contradict each other; a good model is a balance between these two.

To reduce the number of adjustable parameters, there are two ways: (1) to reduce the accuracy; (2) to restrict the scope. Both are common practices, so they may not require any discussion. For (1) information-theoretical criteria for rational choice of the number of parameters have been discussed.[32] However, the choice dictated by such criteria pays no attention to the phenomenon itself. Therefore, although practically useful, these criteria do not directly contribute to our obtaining insight. Statistical hypothesis testing is often exploited to make desirable assertions accepted by the audience, instead

[32] The so-called XIC (X = A, B, $\cdots$) are typical examples.

of gaining insight.[33]

(2) is also often done but if it is done by arbitrarily giving up various aspects of the phenomenon to explain, then it is not very different from giving up accuracy to some convenient degree. As the reader should have already realized, the 'royal road' to this reduction is to make a phenomenology of the phenomenon or system under study. If we may expect a phenomenological description, then try to make a model that captures its universal aspect. Thus, the minimal model explained in the last chapter is the good model. However, we should not forget that whether there is such a model is an intrinsic property of the phenomenon we study. If the phenomenon of interest becomes complex, to attempt to extract a phenomenological description can be a shortcut to insight. We could even say that a clear phenomenological description could be the insight itself. Thermodynamics is a good example.

The modeling as a means to express our understanding explained above is quite restricted from the viewpoint of engineers. In engineering, modeling is used to simulate a particular system, e.g., to model the operational characteristics of a plant instead of understanding phenomena. Therefore, the success of modeling does not necessarily imply the understanding of the system being modeled. There are many cases for whose practical studies no understanding is needed. Our industrial civilization has developed to the current level, because natural phenomena can be exploited without understanding them. However, as its natural fallout, tragedies engendered by simplified models are numerous from pollution to political systems; 'a little learning is a dangerous thing.' If we reflect on such experiences, we understand the fundamental (and even ethical) significance of constructing models that allow us to obtain insight. Furthermore, this naturally dictates how we should deal with models. This is the very reason why this book concentrates on understanding the world and wishes to avoid the arrogance of attempting to change the world.[34]

These days, there seems to be an interesting tendency in descriptive models. Thanks to the popularization of high-performance computers, the attempts are becoming meaningful to construct truly realistic models describing some aspects of complex phenomena. Detailed modeling of biochemical phenomena in a single cell and modeling of early stages of developmental processes are typical examples. These models may currently aim at detailed realization of the target phenomena more than getting insights about them, but they have an aspect fundamentally different from analogous modeling ((2) in Section 4.1) in conventional engineering. These new approaches analyze phenomena to their elementary processes and then model all of the

---

[33] The 'p' for P-values is often the 'p' for politics.

[34] The originator of this philosophical arrogance seems to be, "The philosophers have only interpreted the world, in various ways; the point is to change it." (K. Marx, *Theses on Feuerbach* IX (1845); `http://www.marxists.org/archive/marx/works/1845/theses/theses.htm`)

elementary processes (each chemical reaction, each detailed regulatory relation among genes, etc.). If this approach were successful, for example, by the complete modeling of biochemical network of the cell, any pharmacological tests could be completely simulated. If such a model could actually be realized,[35] to understand its mathematical essence would make an important contribution to the understanding of the essence of phenomena. However, the models become huge, and the outputs are such as the concentrations of all chemicals in the system at every second, even if all the parameters could be determined. Thus how to cope with the deluge of numbers becomes the key issue. Again, eventually, phenomenological summary must be the right approach.

---

[35] Is this really possible? Some people would say certain examples already have been realized. However, it is highly likely that such models still have ignored crucial elements, because many complicated systems have not yet been described properly as the following examples clearly demonstrate. Ecological system models often forget parasitic organisms (for their importance see, e.g., A. M. Kuris, et al., "Ecosystem energetic implications of parasite and free-living biomass in three estuaries," Nature **454**, 515 (2008)), models of vegetation transitions often ignore fungi, most genetic models of development still take ncRNAs (non-coding RNAs; RNAs that do not code proteins) into account incompletely. Furthermore, it is quite likely that parameters in the models cannot be determined. Models purportedly verified are still likely to be those that have meaning 'only in paradise.'

⟨⟨Are large-scale models of natural phenomena harmful?⟩⟩ It is a lesson we should not forget that modeling of complicated natural phenomena has not been successful. Quantitative mathematical models are often the escape from reality. See O. H. Pilkey and L. Pilkey-Jarvis, *Useless Arithmetic—Why environmental scientists can't predict the future* (Columbia University Press, 2007). See the following review of this book: D. Simberloff, "An angry indictment of mathematical modeling," Bioscience **57**, 884 (2007).

We must humbly listen to the criticism that apparently detailed mathematical models are easily exploited by politics and have caused more harm than good.

It is convenient to classify quantitative models of complicated phenomena into two categories. One category consists of models of phenomena with numerous elementary processes that are not understood well (most biological and environmental models) and the other of phenomena whose basic physics is well understood but systems are huge (a typical example is weather forecasting). That the former category is not successful is all too natural as discussed at the beginning of this footnote. The latter category fails when not enough data are available. For example, in the case of weather forecasting, if the computer power is sufficient (it is still not sufficient in 2011), the weather tomorrow must be perfectly forecasted if detailed data today are available. Another review of *Useless Arithmetic*: C. K. R. T. Jones, "Is mathematics misapplied to the environment?" Notices Am. Math. Soc. **55**, 481 (2008) appreciates the latter category of models. The review expresses apprehensions about abuse of environmental models, but concludes that there may be a reason to discard abuses but not the models. However, this is a rather naive opinion, because the existence of purportedly reliable models has been harmful. The critical spirit mandatory to science is often censored by money; development of such models is mired down in big money, so the researchers involved in such projects usually do not commit any suicidal criticism of the defects of such modeling.

4.4 Models as tools of deduction

Another important usage of models is as a means to demonstrate assertions or to check the consistency of concepts and ideas. A simple example appeared at the end of Chapter 2 to consider the distinction between making and understanding (a system). The modeling there may not be a very good one, but still we can say that a RENR set (see Appendix 2.14A) captures at least some aspects of complex things and can give us some insight into the difficulty in making a device to produce complex systems. Therefore, although this is a kind of a parable, it still is slightly better than a parable in ordinary natural language.

In modeling, the crucial element is the correspondence between the mathematical quantity/concept and the quantity/concept external to mathematics. Even inside mathematics, between what systems or how to make homomorphisms is often crucial as can be seen in Poincaré's model of non-Euclidean geometry, but such relations are not revealed by formal logical manipulations. We could interpret making correspondence as a sort of conceptual analysis. If the reader recalls the definition of chaos, the definition was to give a mathematical model capturing the essence of 'chaos.' A definition is a window a formal system opens to the external world. Consequently, giving a clear definition of a concept can be an important part of the efforts to understand reality. That is why conceptual analysis was discussed at length in Chapter 2.

Ito's model of great earthquakes that appeared at the beginning of Chapter 2 was originally designed to study the interference effect between blocks. However, the correspondence between the parameters in the model and quantities in reality is unclear. Therefore, irrespective of its motivation, the model should be understood as a tool to demonstrate a certain proposition or to answer some conceptual questions. The demonstrated proposition is that if (i) there is a constant input of strain energy to blocks from the plate and earthquakes occur when the strain energy reaches a fixed threshold, and if (ii) there are inhibitory interactions among blocks, then the earthquakes occur chaotically. This conclusion is structurally stable in the sense that the input strain rate may not be strictly constant or the interference need be only generally inhibitory. Furthermore, the model allows exact geometrical analysis of the reason why the system is chaotic (deterministic but apparently stochastic). Therefore, this model is a respectable model as a tool for logical demonstration.

A lesson we should learn from the model is that even though a model may not explain anything that has a counterpart in the actual world, a lean model that can exhibit nontrivial behaviors can be as useful as or perhaps more useful than the models that explain particular reality. We could cite the Lorenz model as an example. It may not be surprising that these two models are closely related, as we have seen in Chapter 2.

4.5 Examples of modeling—examples of abduction—

We have discussed basic questions relevant to modeling one by one, so now we return to the original aim of this chapter: to illustrate modeling in terms of successful examples of abduction. Let us construct a model without assuming almost anything. The example is spinodal decomposition (Example 3.4, but for the solid alloy case). Suppose a disordered phase at a high temperature is brought to a point in the two-phase coexisting region of the phase diagram. How is the phase ordered?[36] In this section, we wish to construct a minimal model using this phenomenon as an example.

Let us assume that the system consists of two equal amounts (number of molecules) of components A and B; A and B avoid each other, but except for this property they are idealized to be indistinguishable. That is, even if we completely exchange A and B, the property of the system does not change. At high temperatures due to vigorous thermal motion, A and B mix well and the correlation length is small, so if we define the order parameter field through coarse-graining the system on a rather small scale (covering a few molecules), it fluctuates slightly around zero.

Now, let us quench the system to a low temperature far below the critical point (See state B in Fig. 3.1). At the moment nothing changes, so the order parameter only slightly fluctuates around zero all over the system. However, thermal fluctuation has been reduced, and the molecules of the same species start to gather. Thus, tiny A-rich or B-rich domains start to form. Next, each domain tries to grow larger, and the size ℓ of the segregated domains gradually increases. If the system size is sufficiently large, then doesn't the growth mode of ℓ settle down to a certain function of time as an intermediate asymptotic state (Appendix 3.5B)? Then, isn't there a general law for the correlation function (or its Fourier transform, the form factor $S(k)$)? (This was expected in Example 3.4.)

Such space-time pattern problems are usually modeled in terms of partial differential equations, but the resultant equations are often nonlinear (they will appear later) and defy any analytical research. If we cannot avoid the use of computers, shouldn't we make a model that is adapted to computers and can be computed efficiently? The parallelization of computers is in rapid progress, so a model that can be efficiently computed with a processor lattice of, say, $1,000 \times 1,000 \times 1,000$ is promising. The dynamics in each 'cell' at the lattice point can be indefinitely complicated. Thus, a model consisting of a lattice of cells must be the best model of extended systems.[37] In short, as a modeling tool for spatially extended systems, dynamical systems that can be

---

[36] An excellent and authoritative textbook is A. Onuki, *Phase Transition Dynamics* (Cambridge University Press, 2002).

[37] Y. Oono and S. Puri, "Study of phase separation dynamics by use of cell dynamical systems I. Modeling," Phys. Rev. A **38**, 434(1988).

described by boxes (cells) piled up to make a lattice (*cell dynamical systems*, CDS) must be convenient (the example here requires, however, very simple dynamics in each cell).

A remark must be added here to avoid any misunderstanding. Here, it is never asserted that discrete models are fundamental and that continuous models are only artificial. For macroscopic phenomena, whether Nature is actually discrete or continuous can often be irrelevant.[38] Thus, often the mathematical essence of the phenomenon transcends the distinction. However, regretfully, we do not at present have any means to express a formula transcending the distinction of whether the variables are continuous or not.

Now, let us try to model the phase separation process. The cell size is roughly the scale of coarse-graining (if one wishes to start with a microscopic model), so in each cell at location $\boldsymbol{n}$ lives an order parameter $\varphi_t(\boldsymbol{n})$,[39] where t is (discrete) time.

To understand a thing is an action to select a hypothesis that matches reality from the set of hypotheses generated ("this or that?") by the person who wishes to understand it.[40,41] Modeling is no exception. It is not a process of simple induction from detailed observation of the relevant phenomenon.[42] There is a considerable freedom in what factors to regard as essential when

---

[38] ⟪**Cases for which discreteness is crucial**⟫ However, discreteness can be crucial in such examples as something spreading while wetting a surface of a material, burning of a fuse or multiplication of germs, a nucleus of certain unstable phenomenon governing the phenomenon, etc. M. Shnerb, Y. Loouzoun, E. Bettelheim and S. Solomon, "The importance of being discrete: life always wins on the surface," Proc. Nat. Acad. Sci. **97**, 10322 (2000) is a simple but instructive example. In the case of the chemical reaction of burning, the continuous model has only one length scale of the thickness of the burning front, but if molecular discreteness is explicitly taken into account, two distinct scales, the length related to the reaction cross section and the mean distance between molecules, can appear. Therefore, the usual continuous models cannot describe phenomena depending on the ratio of these two length scales. See Y. Togashi and K. Kaneko, "Discreteness-induced stochastic steady state in reaction diffusion systems: self-consistent analysis and stochastic simulations," Physica D **205**, 87 (2005).

[39] As was explained in Example 3.2 or in the above we can assume that the order parameter is a linear function of the local concentration deviation of A from the mean such that it is zero if A and B are equally contained in the cell, 1 if A only, and -1 if B only.

[40] K. Popper, "Evolutionary epistemology," in *Evolutionary Theory: paths into the future* (edited by J. Pollard, Wiley, 1984) Chapter 10. He also quotes Lorenz's *Behind the mirror.*

[41] Our understanding of language also seems active: M. J. Pickering and S. Garrod, "Do people use language production to make predictions during comprehension?" Trends Cognitive Sci. **11**, 105 (2007). More generally, the categorization of sensory data is also accelerated with the aid of prediction: T. Serre, A. Oliva and T. Poggio, "A feedforward architecture accounts for rapid categorization," Proc. Nat. Acad. Sci. **104**, 6424 (2007).

[42] It cannot be a simple deduction, either. The Master said, "He who learns but does

we make a model, and the choice is often a matter of taste. However, there are good and bad tastes, and the matter of taste could be a matter of the utmost importance in research.

What we practice below is an example of the general strategy:
(1) Respect general principles (e.g., conservation laws, symmetry, etc).[43]
(2) Think as if you are the element governed by the model (e.g., what will you do if you are placed in this potential well?)
(3) Pay due attention to the aesthetics of the model.[44]

In practice, it is convenient to take a single cell in an ideal environment, and then to take its interaction with its surroundings subsequently. That is, the neighboring cells can be ignored or assumed to be conveniently cooperative to the time evolution of the cell under consideration. Then, the time evolution of the order parameter in the cell may be described by a map f from the order parameter at time t to that at time $t+1$:

$$\varphi_{t+1}(\boldsymbol{n}) = f(\varphi_t(\boldsymbol{n})). \tag{4.4}$$

We can impose some constraints on f from general considerations. First of all, after quench the order parameter has a tendency to evolve away from zero toward ± 1 (toward phase separation). This change is monotonic, and once separation is complete, there will be no further change. This is easy to understand if the reader pretends that she is a molecule (anthropomorphic considerations are not bad; we are all materials). Therefore,
(1) $f : \boldsymbol{R} \to \boldsymbol{R}$ is a homeomorphism (= one to one and continuous in both directions) on an appropriate region containing the interval $(-1, 1)$ with ± 1 as attractors and 0 as a repeller.
We assume that the stability of the phases are understandable by linearization, because the stability of each phase is definite. Therefore,
(2) f is hyperbolic.[45]
Our system is idealized so that exchanging A and B does not change the dynamics, so
(3) f is an odd function.
Thus, for example, we can choose $f(x) = 1.3 \tanh x$: in each cell

---

not think, is lost. He who thinks but does not learn is in great danger." *Analects* Book 2, 15 [A. Waley, *The Analects of Confucius* (Vintage, 1989)].

43 It is not said that they must be incorporated into the model, because if we could satisfy them with high precision without explicitly demanding them, this is often much better as we will see.

44 "Invention is choice; this choice is imperatively governed by the sense of scientific beauty." (J. Hadamard, *The Psychology of Invention in the Mathematical Field* (Dover, 1954), III The Unconscious Discovery, p31. This book is highly recommended.

45 That a map is *hyperbolic* means that around each critical point (= zero or fixed point) its local behavior and the result of linearization are homeomorphic. Roughly speaking, the original behavior and the behavior of the linearized system can be superposed with a continuous deformation of the linearized system.

$$\varphi_{t+1} = 1.3 \tanh \varphi_t. \tag{4.5}$$

Here, the coefficient 1.3 is chosen to locate the stable fixed points close to ± 1.

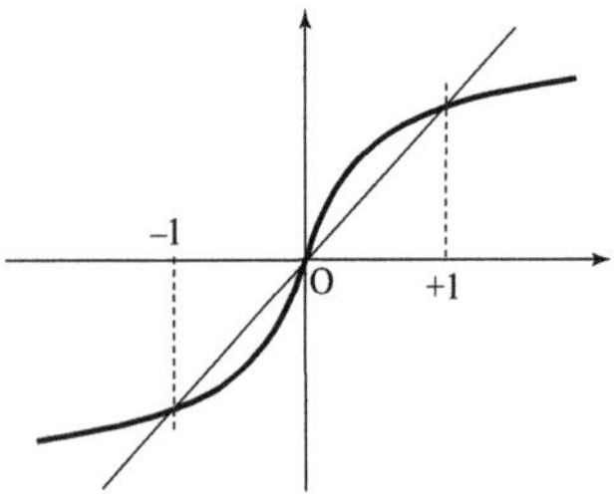

Fig. 4.1 The form of f satisfying (1)-(3). The resultant time evolution may be seen by a method to chase the chaotic history graphically as explained in Fig. 2.14. O is an unstable fixed point and ± 1 are stable fixed points. Both are hyperbolic. The former represents a well-mixed (disordered) phase, and $+1$ (respectively, -1) the A-rich (respectively, B-rich) phase after phase separation.

Needless to say, the interactions with neighboring cells is important. Once phase separation occurs, the order parameters in the cells within a bulk phase never spontaneously assume different values. Therefore, there must be an interaction that spatially homogenizes the order parameter field (that increases entropy). The simplest way to model this interaction is to add a term that drives the value of the order parameter in a cell toward its local average value. To this end let us introduce a suitable local average $\langle\langle \varphi_t(\boldsymbol{n}) \rangle\rangle$ of the order parameter values in the neighborhood cells of the cell at $\boldsymbol{n}$, and then modify (4.5) as

$$\varphi_{t+1}(\boldsymbol{n}) = f(\varphi_t(\boldsymbol{n})) + D[\langle\langle \varphi_t(\boldsymbol{n}) \rangle\rangle - \varphi_t(\boldsymbol{n})], \tag{4.6}$$

where D is a positive constant analogous to the diffusion constant. Indeed, if the surrounding average is larger than itself, φ changes in the increasing direction. There are many possibilities to calculate or define the average of the surrounding values, but isotropy is crucial. In 2D (on the square lattice) the following form is known to be optimal:[46]

$$\langle\langle \varphi \rangle\rangle \equiv \frac{1}{6} \sum \text{nearest neighbor } \varphi + \frac{1}{12} \sum \text{second nearest neighbor } \varphi. \tag{4.7}$$

Let us observe what happens to a thus-constructed CDS model (4.6) with a small random initial condition. Notice that the model is deterministic, so if we assume $\varphi_0(\boldsymbol{n}) \equiv 0$, nothing would happen. Thus, we add small random

[46] H. Tomita, "Preservation of isotropy at the mesoscopic stage of phase separation processes," Prog. Theor. Phys. **85**, 47 (1991). Especially for long calculations, accurate isotropy is required; if not, the patterns become increasingly squarish. In 2D, if a triangular lattice is adopted, isotropy is easier to realize than for a square lattice. However, in general dimensional spaces, (hyper)cubic lattices are the simplest, so let us stick to a scheme that is good even on the square lattice.

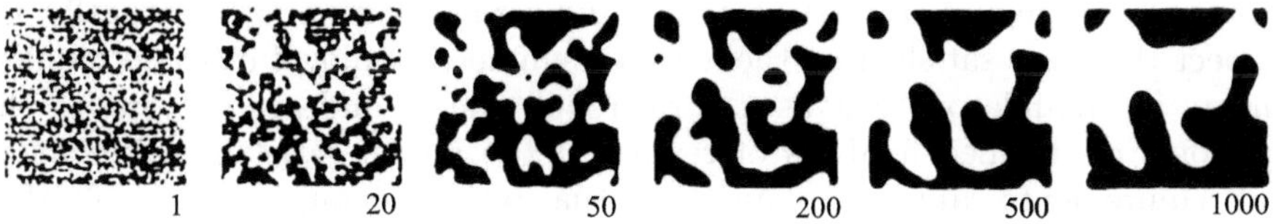

Fig. 4.2 An example of time evolution of the model (4.6) that forgets conservation of matters with the initial random pattern with small fluctuations around zero. The numbers below the figures are the numbers of steps from the initial time. In this case 'white' eventually wins. The model (4.6) may be suitable as a model of a magnet with the interpretation of black and white as the up and down directions of spins, but it is not good for a model of binary alloys.

noise everywhere to the zero initial condition. An example of the time evolution is in Fig. 4.2. This is, however, the result that 'allows alchemy,' and is quite nonsensical as a model of spinodal decomposition.

The reason for this nonsense is trivial. For example, the map f tells us that a cell with more A than B will have more A. However, A cannot be created, so A must decrease somewhere else. This important law of conservation of materials is ignored in this model.

Let us write (4.6) as

$$\varphi_{t+1}(\boldsymbol{n}) = \varphi_t(\boldsymbol{n}) + \mathcal{I}_t(\boldsymbol{n}), \tag{4.8}$$

where $\mathcal{I}_t(\boldsymbol{n})$ is the increment of the order parameter in cell $\boldsymbol{n}$ between times t and $t+1$:

$$\mathcal{I}_t(\boldsymbol{n}) = f(\varphi_t(\boldsymbol{n})) - \varphi_t(\boldsymbol{n}) + D(\langle\langle \varphi_t(\boldsymbol{n})\rangle\rangle - \varphi_t(\boldsymbol{n})). \tag{4.9}$$

The molecules of A and B do not jump a long distance within a single time step, so $\mathcal{I}_t(\boldsymbol{n})$ must be supplied locally (the requirement of local conservation). However, the neighboring cells are just the same kind of cells as cell $\boldsymbol{n}$, so they also have their own similar 'desire.' Thus, cells compete for molecules (for increments of the order parameter). We have already coarse-grained the system, so we may describe this competition on the average. That is, cell $\boldsymbol{n}$ can obtain only $\mathcal{I} - \langle\langle \mathcal{I}\rangle\rangle$ instead of $\mathcal{I} = \mathcal{I}_t(\boldsymbol{n})$. That is, it can obtain the difference between its own 'desire' and the average 'desire' of the neighboring cells. The simplest model respecting the conservation of matter reads[47]

$$\varphi_{t+1}(\boldsymbol{n}) = \varphi_t(\boldsymbol{n}) + \mathcal{I}_t(\boldsymbol{n}) - \langle\langle \mathcal{I}_t(\boldsymbol{n})\rangle\rangle. \tag{4.10}$$

The model is constructed based on a very general consideration. Roughly speaking, the model uses only the local atomic preference (local time evolution rule), symmetry (spatial isotropy, translational symmetry, and symmetry

[47] This model may be understood as a coupled map lattice, but the interpretation is of no use.

in the exchange of A and B), stability, and a conservation law. Naturally, we expect these are satisfied by the actual (symmetric) alloy, but these requirements are mandatory just as natural requirements and are not due to the outcome of detailed analysis of reality. If the model satisfying such natural requirements cannot explain the actual phenomenon, it is possible that considerably important factors have been ignored. We will consider in what sense the model is good (or bad) in the next section. Although it may be premature to discuss the significance of the present model before discussing how good it is, we can say that the model is computationally much more efficient than the partial differential equation (4.14) that supposedly describes the same phenomenon, so the model can be conveniently used to approach the asymptopia relatively easily. The above discussion also illustrates, as we will come back to again in Section 4.7, the general idea that considering the phenomenon itself is perhaps more effective in numerical analysis than the usual numerical solution methods for partial differential equations.

Discussion 4.1. The operation $\langle\langle\phi\rangle\rangle - \phi$ on the 'discrete field' ϕ in (4.10) is proportional to the difference approximation of the Laplacian. Reviewing salient features of *harmonic functions*, understand its important property, the (spherical) mean-value property, intuitively.[48] □

Discussion 4.2. What happens in the above CDS, if the map f is replaced by the following? What is the physical meaning of this replacement? Is there any system that may be modeled with the aid of the following f?

$$f(x) = \begin{cases} 1.3\tanh x & \text{for } x \geq 0, \\ x & \text{for } x < 0. \end{cases} \tag{4.11}$$

□

A traditional way to model such systems undergoing phase transitions is to use the *Ginzburg-Landau free energy* (explained below) $H[\varphi]$. The resultant partial differential equation model reads

$$\frac{\partial}{\partial t}\varphi(\boldsymbol{r},t) = D\Delta\frac{\delta H[\varphi]}{\delta\varphi(\boldsymbol{r},t)}, \tag{4.12}$$

where $\varphi(\boldsymbol{r},t)$ is the order parameter at position $\boldsymbol{r}$ and time t, Δ is the Laplacian to impose the conservation law,[49] D (> 0) is a parameter describing the rate of the change, and we can absorb it in an appropriate choice of the unit of time (so henceforth $D = 1$). $\delta/\delta\varphi$ is the functional derivative (see Note 4.2). If we adopt as H the standard φ^4 model (Ginzburg-Landau free

---

[48] In fact, the (spherical) *mean value theorem* motivated our discretization scheme. It is an example of the discretization motivated by crucial theorems relevant to the system.

[49] Without this $\int \varphi\, d^3\boldsymbol{r}$ is not kept constant. This corresponds to the replacement $\mathcal{I} \to \mathcal{I} - \langle\langle\mathcal{I}\rangle\rangle$ in (4.10).

energy)[50]

$$H[\varphi] = \int d^3\boldsymbol{r} \left[\frac{1}{2}(\nabla\varphi)^2 + \frac{\tau}{2}\varphi^2 + \frac{1}{4}\varphi^4\right], \tag{4.13}$$

where τ is a numerical parameter, we obtain the *Cahn-Hilliard equation*:

$$\frac{\partial}{\partial t}\varphi(\boldsymbol{r},t) = \Delta\left[-\Delta\varphi(\boldsymbol{r},t) + \tau\varphi(\boldsymbol{r},t) + \varphi(\boldsymbol{r},t)^3\right]. \tag{4.14}$$

If $\tau = -1$, notice that $\varphi \equiv \pm 1$ satisfies the equation.[51] It is not easy to obtain the asymptotic behavior of this system even numerically.

The model corresponding to (4.6) that does not satisfy the conservation law reads

$$\frac{\partial}{\partial t}\varphi(\boldsymbol{r},t) = -\frac{\delta H[\varphi]}{\delta\varphi(\boldsymbol{r},t)} = \Delta\varphi(\boldsymbol{r},t) - \tau\varphi(\boldsymbol{r},t) - \varphi(\boldsymbol{r},t)^3. \tag{4.15}$$

Note 4.2. Functional derivative
Let $H[\varphi]$ be a functional of a function φ of position. We can see how much H changes when φ is modified by $\delta\varphi(\boldsymbol{r})$ in the very small neighborhood of $\boldsymbol{r}$ with a variational calculation. δH depends on $\delta\varphi$ linearly, so, with the aid of a linear operator L, we can write it as

$$\delta H = L\delta\varphi. \tag{4.16}$$

Mathematicians call this linear operator itself the functional derivative,[52] but physicists use the fact that at least formally a linear operator can always be written in terms of an integral kernel K as

$$\delta H[\varphi] = \int d^d\boldsymbol{r} K(\boldsymbol{r})\delta\varphi(\boldsymbol{r}), \tag{4.17}$$

where $d^d\boldsymbol{r}$ is the d-dimensional volume element, and call K the functional derivative, which is symbolically expressed as

$$K(\boldsymbol{r}) = \frac{\delta H[\varphi]}{\delta\varphi(\boldsymbol{r})}. \tag{4.18}$$

We saw equilibrium critical phenomena as an example in Chapter 3 (Example 3.2). A universality class of the critical phenomenon is uniquely determined by the general smoothness condition for continuous models and the following

---

[50] The form of the free energy is proposed in the present context in J. W. Cahn and J. E. Hilliard, "Free energy of a nonuniform system I. Interfacial free energy," J. Chem. Phys. **28**, 258 (1958). The equation itself appeared in J. W. Cahn, "Phase separation by spinodal decomposition in isotopic systems," J. Chem. Phys. **42**, 93 (1965).

[51] R. V. Kohn and F. Otto, "Upper bounds on coarsening rates," Commun. Math. Phys. **229**, 375 (2002) is perhaps the most interesting mathematical paper to physicists on this equation so far published; the exponent generally believed is given as the possible upper bound value.

[52] The properly formulated version is the *Frechét derivative.*

four conditions: (1) the order parameter is a scalar quantity, (2) the system is invariant under the sign change of the order parameter, (3) the interaction between order parameters at different spatial positions are local (short-ranged) and they locally tend to uniformize the order parameter, (4) if the order parameter is spatially uniform, then the model has a parameter controlling a *pitchfork bifurcation*.[53,54] Spatial translational and rotational symmetries are always required.

Ginzburg-Landau free energy was constructed to describe the equilibrium state consistent with (1)-(4) (and fluctuations around it). To justify the model from a really microscopic atomistic model is at present impossible. Rather, apart from such justification, it is better to understand that it is a result of abduction just as was the model of phase separation we derived. First, according to (1) let the order parameter be a scalar field φ, and assume that $\varphi = 0$ corresponds to the disordered phase. Let τ be the bifurcation parameter in (4). Then, for $\tau > 0$ $\varphi = 0$ is the stable fixed point, and for $\tau < 0$ this state (disordered phase) loses its stability. We first consider the spatially uniform case, and do not write the $\boldsymbol{r}$ dependence of φ (recall that space is assumed to be translationally and rotationally symmetric). Then,

$$H \propto \frac{1}{2}\tau\varphi^2 \tag{4.19}$$

(partially) realizes the requirement stated above. For $\tau < 0$ $\varphi = 0$ is not only unstable, but the energy ceases to be bounded from below (the system becomes thermodynamically unstable). Therefore, to prevent φ from becoming too large, large magnitudes of φ should increase H (i.e., we impose an (free-)energetic penalty). The simplest way is to add a positive definite higher-order term as:

$$H \propto \frac{1}{2}\tau\varphi^2 + \frac{1}{4}g\varphi^4, \tag{4.20}$$

where g is a positive constant. Here, it is implicitly required that when φ is spatially uniform, H is a smooth function of φ. The remaining condition is (3). Being ferromagnetic (or spins wanting to align, or the same species wanting to gather) implies that the order parameter tends to be spatially uniform. This implies (since the interactions are short-ranged) the local spatial nonuniformity of the order parameter (the gradient $\nabla\varphi(\boldsymbol{r})$ of the field $\varphi(\boldsymbol{r})$ being nonzero) increases energy. Therefore, H should depend on $|\nabla\varphi|^2$ (recall that space is assumed to be isotropic). Since H should be an extensive

[53] This means that, if the control parameter is written as τ, for $\tau > 0$ the system has a single stable fixed point, and for $\tau < 0$ the fixed point becomes an unstable fixed point and two stable fixed points emerge on both its sides.

[54] This may not have been stated clearly as a mathematical assertion. Long ago Professor Kadanoff told the author that initially it was assumed that there was a single universality class for all critical phenomena. As the study progressed, it was recognized that, e.g., if the order parameter is a vector, the exponents are different.

quantity (i.e., proportional to the volume of the system) and $\varphi(\boldsymbol{r})$ is locally defined, a natural energy functional should read

$$H = \int d^d\boldsymbol{r} \left\{ \frac{1}{2}|\nabla\varphi|^2 + \frac{1}{2}\tau\varphi^2 + \frac{1}{4}g\varphi^4 \right\}. \tag{4.21}$$

Here, the numerical coefficient for the newly added term dependent on $\nabla\varphi$ can be normalized to be $1/2$ by choosing the spatial unit appropriately. Thus, the *Ginzburg-Landau model* has been constructed.

Such models are by themselves conclusive (self-sufficient) models, and need not be regarded as the result of coarse-graining of some more microscopic models. To understand critical phenomena (or more generally, phenomena whose prerequisites are (1)-(4) above) in general terms, a 'simple' model such as this is enough (it is indeed a minimal model). That is, in Nature a mathematical structure exists that can be captured by such a model.

Can we call the above procedure 'modeling'? What we have performed is a typical abduction. The lesson is that such abductive *a priori* modeling is often very effective. Lanczos remarked about Einstein's equation of gravity: "In fact, no amount of experimentation could ever lead to such a complex and highly nonlinear set of equations. $\cdots$ It demonstrated that possibly the experiments give us only the *tip of the iceberg*."[55] That is, even if we scrutinize experimental results, we are not guaranteed to obtain a nice model.

If one wishes to think microscopically, one can make a model of the above critical phenomenon with microscopic entities attached to lattice points. These variables point up or down (called *spins*) and have a tendency to point in the same direction as their own neighbors (i.e., the parallel states are more stable energetically than antiparallel states). If the temperature is high, they cannot align, but as the temperature is lowered, alignment becomes possible. Lenz[56] saw this and gave a model to his student, Ising,[57] that has the

---

[55] C. Lanczos, *The Einstein Decade (1905-1915)* (Paul Elek Ltd. London, 1974) Chapter 2, p14 (the italicized part is in the original). Read the book. Lanczos was an assistant to Einstein (1928-9). He published this book when he was 80 (one year before he died). Cf. `http://www-groups.dcs.st-and.ac.uk/~history/Biographies/Lanczos.html`.

[56] ⟪**W. Lenz 1888-1957**⟫ Wilhelm Lenz, different from Lenz of Lenz's law, but the Lenz of the Laplace-Runge-Lenz vector closely connected to the O_4 symmetry of the Kepler problem. He was a student of Sommerfeld, and was his assistant for a while. Pauli and Jordan were his assistants. He played an important role in the development of atomic physics in Germany.

[57] About E. Ising 1900-1998 (`http://www.bradley.edu/academic/departments/physics/why/ising.dot`) The author did not know until the mid-90s that he lived in Peoria (it is a 'next nearest neighbor' town of the town where the University of Illinois is). The name 'Ising model' was coined by R. Peierls (1907-1995). Lenz gave the model as a model of phase transition (although not restricted to the nearest neighbor interactions) to Ising, who solved the 1D nearest neighbor version to show

following energy function (the Ising model):

$$H = -J \sum_{\langle i,j \rangle} \sigma_i \sigma_j, \tag{4.22}$$

where the sum over $\langle i, j \rangle$ denotes the sum over all the nearest neighbor pairs of lattice points i and j, and J is a positive constant (called the *coupling constant*), and σ_i is the spin at lattice point i; the up state is $+1$, and down -1.[58] If the temperature is sufficiently low, the spins tend to order in the up or down state. These ordered phases are called the *ferromagnetic phase*. However, if the temperature is high enough, due to thermal fluctuations, the directions of the spins are randomized (the disorder state, the paramagnetic phase in this case).

From the just-stated microscopic point of view, the Ginzburg-Landau model ought to be obtained by coarse-graining the Ising model.[59] From the phenomenology point of view, however, the Ising model is expected to be merely one way to realize the Ginzburg-Landau model. Here, we need not decide which is a more fundamental viewpoint. However, there must be consistency between them. For Nature to realize (1)-(4), She must use limited materials, so as 'technical details' various 'unnecessary' complications are introduced.[60] For materials science, these 'unnecessary complications' are its staple, but the mission of fundamental science is to study first the universality independent of such details, and then to study the universal features (laws) found in the diversity. It is the mission of fundamental science to study larger universality classes first and then to study smaller universality classes in each of them, descending the hierarchically nested structure. The study of diverse individual phenomena comes only after such studies.

The phase-transition temperatures T_c cannot be quantitatively explained by such a crude model as minimal models. Such a quantity is sensitive to microscopic details. Then, are minimal models only restricted and incomplete models? Let us recall our consideration as to what we must first explain by models. We must first study the phenomenological universal features. Critical indices we have already encountered are universal constants. Just as π they are dictated by the structure of the (logical) world. In contrast, T_c is an inessential number under the influence of numerous contingencies. Therefore, to understand universal features such as the critical indices is a supreme command, but to be particular about such as T_c is off the mark. Michael Fisher

---

there is no phase transition and concluded that 3D was the same. We of course know now that Lenz's insight was correct in 2-space and beyond.

[58] There are magnets that can be well approximated by the model.

[59] No one has done this in a mathematically respectable fashion.

[60] These are 'materialistic obstacles' to realize ideal natural laws, as Galileo said.

calls such a research attitude that is particular about T_c *fetishism*.[61]

4.6 What characterizes good models?

We have been discussing models of real phenomena, so the models that do not agree with reality are useless. Therefore, we must require, to begin with, (0) The consequences of the model agree with the target (aspects of the) phenomenon within the expected error bounds.

However, if a good fit is just because of many adjustable parameters, agreement with reality is rather trivial. Therefore, it is sensible to demand: (1) The number of adjustable parameters is small, which may be determined experimentally, or may be eliminated when the results are compared with experimental results.

The standard observable of the spinodally decomposing systems considered in the preceding section is the form factor $S(\boldsymbol{k}, t)$, which is the spatial Fourier transform of the order parameter field autocorrelation function $g(\boldsymbol{r}, t)$ at time t. If we write the order parameter field at time t as $\varphi(\boldsymbol{r}, t)$, thanks to the translational symmetry (the expectation values do not depend on the absolute position),

$$g(\boldsymbol{r}, t) = \langle \varphi(\boldsymbol{r}, t)\varphi(\boldsymbol{0}, t)\rangle, \tag{4.23}$$

where $\langle \ \rangle$ is the ensemble average. The form factor has a shape of a single mound as in Fig. 4.3. Its peak location $\langle k\rangle_t$ at time t is, since the dimension of k is the reciprocal of length, proportional to the reciprocal of the representative scale of the pattern at the time. According to dimensional analysis,

$$S(\boldsymbol{k}, t) = \langle k\rangle_t^{-d} F(k/\langle k\rangle_t, t) \tag{4.24}$$

[61] We read on p4 of M. E. Fisher, *Scaling, Universality and Renormalization Group Theory*, Lecture Notes in Physics **186** (1983) as: "What is the task of theory? It is worthwhile, when embarking on theory to have some viewpoint as to what theory is. There are different opinions on this subject. Some people feel the task of theory is to be able to calculate the results of any experiment one can do: they judge a theory successful if it agrees with experiment. That is not the way I look at a theory at all. Rather, I believe the task of theory is to try and understand the universal aspect of the natural world; first of all to identify the universal; then to clarify what they are about, and to unify and inter-relate them finally, to provide some insights into their origin and nature. Often a major step consists in finding a way of looking at things, a language for thinking about things — which need not necessarily be a calculational scheme." (underlines in the original)

must be the general form.[62] It is numerically inferred that for a given system the function F ceases to be explicitly dependent on t after a while,[63] and this is also experimentally supported. That is, for a sufficiently large t_0, if $t \geq t_0$,

$$F(k/\langle k \rangle_t, t) \simeq F(k/\langle k \rangle_t, t_0) = \hat{F}(k/\langle k \rangle_t). \tag{4.25}$$

We can obtain a single master curve, if we scale the horizontal and vertical axes to superpose the positions of the peak (Fig. 4.3). This curve must be $\hat{F}$. If we solve (4.10) with an initial condition that is a small random fluctuation around zero, the master curve of the form factor can be obtained satisfactorily. There is an experimental method to reconstruct the 3D structure of the specimen by studying the cross-sectional spatial composition distribution by

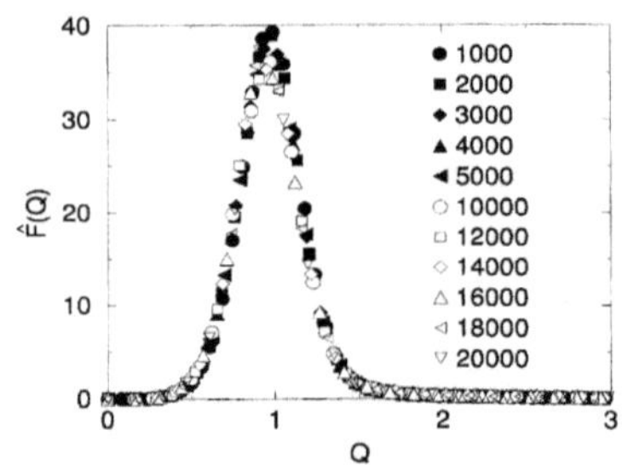

Fig. 4.3 The master curve $\hat{F}$. [A. Shinozaki and Y. Oono, Phys. Rev. E **48**, 2622 (1993), Fig. 22]

stripping atomic layers one by one by a kind of an improved ion microscope. The experimental structure thus obtained and the numerical result (Fig. 4.4) are not easy to distinguish. Therefore, the model seems to be semiquantitatively successful. If an abstract model is really successful, the outcome must not depend on the details required to implement the model (e.g., in the present example the form of the map f). To expect that $\hat{F}$ can be computed with our model is to expect a fairly strong universality (materials independence). Then, is the outcome indifferent to any map f satisfying (1)-(3)? This is a true universality assertion independent of individual materials systems. The assertion is stronger than the existence of the master curve seen above for each system, since here the identity of all the master curves is asserted.

Ideally, the universality should be mathematically (analytically) demonstrated. However, at present, this is impossible. If we take various forms of f and perform numerical simulations, within (numerical) experimental errors,

[62] Fourier transformation is performed by integration with respect to the volume element after multiplication of a dimensionless factor $e^{i\boldsymbol{k}\cdot\boldsymbol{r}}$ to a target function, so if the target function is dimensionless, its Fourier transform has a dimension of length (k^{-1}) to the power of d. Since we assume that the system is homogeneous and isotropic, the form factor turns out to be a function of the magnitude k of the wave vector $\boldsymbol{k}$.

[63] That is, there is an intermediate asymptotic regime. J. L. Lebowitz, J. Marro, and K. H. Kalos, "Dynamical scaling of structure function in quenched binary alloys," Acta Metall. **30**, 297 (1982) asserted this explicitly first.

the asserted universality seems to hold. There is a physics reason for the universality to be expected. It is guessed that the motion of the interfaces governs the evolution of the patterns, and the mean curvature governs the interface dynamics. The detail of f controls the interface structure perpendicular to the interface (the spatial compositional change from one phase to the other), but such details should be irrelevant to the interface dynamics governed by the mean curvature, if the curvature radius is much larger than the interface thickness.[64]

Fig. 4.4 An example of the time evolution of the system (4.10). The actual structure and these pictures are hardly distinguishable. 1000, 2000, 4000, 7000, 12000, 20000 time steps are shown. In these figures one component is illustrated after the other component has been etched out. [A. Shinozaki and Y. Oono, *ibid.*, Fig. 16.]

(1) mentioned above may be restated as

(1') When the model is constructed, there are not many arbitrary elements.

Our model of spinodal decomposition seems to satisfy this. However, the models satisfying (0) and (1) or (1') do not really go beyond the level that the models are useful in organizing experimental data; they do not give insight into the phenomenon being modeled. The models satisfying the above conditions can give insight only if they are constructed logically based on clear ideas or a clear picture:

(2) The logic to construct the model is clear.

If a model is constructed logically clearly, even when it is not successful, the reason for the discrepancy could often be understood clearly, or we can expect that there is another phenomenon in Nature that can be explained by the model. Therefore, a good lesson is that to make clean but unsuccessful models is a much more important contribution to science than to make dirty but 'successful' models.

From the point of view of (1) and (2) (4.10) is undoubtedly a good model. How about (0)? No comparison with real data has been given yet. Most researchers were contented with the existence of the master curve and approx-

[64] For example, see Y. Nishiura, *Far-from-equilibrium dynamics* (Translations of Mathematical Monographs vol 209, Am. Math. Soc., 2002; original Japanese edition 1999) Chapter 5.

imate reproduction of the $t^{1/3}$-law. However, if we try to superpose master curves available from various experiments for solid systems, we see in Fig. 4.5 Left (quoted experiments are in the footnote[65]) that there is no universality at all.

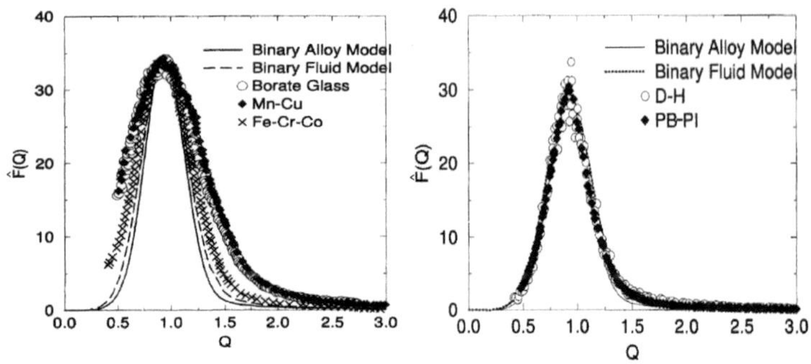

Fig. 4.5 The master curve $\hat{F}$: Left: There is no universality in solid systems. However, as shown at Right, fluid systems agree with each other much better. Furthermore, even the solid systems, if solid specific peculiarities are ignored, agree with fluid systems better than with actual solid systems. [A. Shinozaki and Y. Oono ibid., Figs. 50, 52]

That is, our model does not agree with reality. Solid systems have their peculiar complications such as mismatch of crystal structures between different phases, elastic interactions, anisotropy, etc. Therefore, it is not so surprising that they lack universality. Inessential factors have too large effects. Then, we should use fluid systems. Indeed, the master curves obtained from fluid systems superpose much better even between polymer systems and low-molecular-weight compound systems. It is possible to construct a CDS model which takes into account hydrodynamic interactions, and its results agree well with experiments, although computationally it is one order more costly.[66] Thus, it seems that we have grasped the (quantitative) mathematical essence of spinodal decomposition (of course, Cahn and Landau saw this long ago). Numerical simulations cannot, however, assert anything general,

---

[65] Fe-Cr-Co system: S. Spooner, in *Kinetics of Aggregation and Gelation*, edited by F. Family and D. P. Landau (North-Holland, Amsterdam, 1984); Mn-Cu system: B. D. Gaulin, S. Spooner, and Y. Morii, Phys. Rev. Lett. **59**, 668 (1987); boric glass system: A. F. Craievich, J. M. Sanchez, and C. E. Williams, Phys. Rev. B **34**, 2762 (1986); D-polystyrene-H-polystyrene system: F. S. Bates and P. Wiltzius, J. Chem. Phys. **91**, 3258 (1989); Polyisoprene-polybutadiene system: M. Takenaka and T. Hashimoto, J. Chem. Phys. **96**, 6177 (1992). It is not shown in the figure, but the agreement of the polymer system with the following low-molecular-weight compound systems is good: Isobutyric acid-water system: N.-C. Wong and C. M. Knobler, J. Chem. Phys. **69**, 725 (1978); Lutidine-water system: Y. C. Chou and W. I. Goldburg, Phys. Rev. A **23**, 858 (1981).

[66] For CDS spinodal models see A. Shinozaki and Y. Oono, "Spinodal decomposition in 3-space," Phys. Rev., **48**, 2622 (1993) for both solid and fluid systems.

so the problem is far from conclusive.[67]

When we speak of the agreement of models with experiments, 'agreement' means various things: from quantitative agreement such as we have aimed at above (the quantitative agreement of a certain function (say, the form factor) with experimental results) to rough qualitative 'agreement.' Why can our model agree quantitatively with reality to a good extent? For quantitative agreement without fine tuning to be possible the target phenomenon must have considerable universal features. It is in the first place reckless to try to understand phenomena without universality quantitatively with a model without tuning parameters (so we failed in the solid case). In short, modeling is most successful when we can describe the phenomenon 'roughly accurately' (or 'qualitatively accurately'). Whether modeling is successful or not cannot be changed by the capability of the modeler, but is dictated by the target phenomenon itself.[68] Understanding a set of phenomena and making a nice model for the set are virtually the same thing. If there is no phenomenological structure, modeling is impossible. Here, modeling means the construction of minimal models, so the reader may object to this general assertion, but if one does not wish to contrive models by piling up adjustments, there is no choice.

If the agreement is not perfect, it is a difficult question what agreement is regarded as good. Also, even if we have quantitative agreement to some aspects of the phenomena, how can we check that the idea on which the model is based is not a mere fairy tale but a real insight?[69] We must exclude the possibility that the agreement is due to some fortuity or due to some other reason than intended. To this end, we have no other strategy but to make requirements for the model more stringent. Typical additional requirements are:

(3) The response of the model to (structural and ordinary) perturbations agrees with that of the actual system being modeled.
(4) The fluctuations (noise structure) given by the model is realistic.

The requirement of universality and structural stability are special (but very important) cases of (3). When we make a model, it consists of some elements (say, cells or spins). Therefore, a certain idea or interpretation has been incorporated about smaller scales[70] than those at which we observe the system. Consequently, there is often a noise due to the nonuniformity of these un-

---

[67] These days we have powerful computers and efficient algorithms, so we wish to expect that many stronger assertions can be made, but many papers do not even check whether they really reach asymptopia.

[68] What to model or what aspect to model is an important question, and the crucial factor called modelers' taste decides this.

[69] This paragraph is motivated by this question of Professor N. Wakabayashi of Keio University during the author's lectures in 1996.

[70] Actually, this need not be literally smaller scales. The point is that we introduce some elements or factors we cannot directly observe into the model.

observable elements in the model. This noise should not be unrealistic; this is what (4) demands. For example, the Ising model correctly captures the critical fluctuations, so it is trusted.

As has already been mentioned, the CDS spinodal model gives natural results when anisotropy or strain effects are incorporated.[71] The model is deterministic, but noise that does not contradict the conservation laws can be introduced and does not alter the universal features obtained by the original deterministic model. To study higher-order correlation functions is also a possible approach to (4) and there are some publications, but actual experiments are nontrivial.

Can one ever say that a model is the 'true model' of a given phenomenon? In mathematical logic, all the cases to compare can be checked, so the statement 'it is a model' has a mathematical meaning. However, if we must compare the result with real observations, it is impossible to study all cases. Even if we say all the representative cases, we cannot avoid the question what 'representative' really means. Thus, when we discuss actual phenomena, the term 'true model' is in most cases meaningless, strictly speaking, and the goodness of a model or the extent of success of a model is a matter of degree. Still the difference between good and bad models is enormous. There are even models that give reasonable results only with wrong approximations. Such models are much worse than mere curve fitting, because they pretend to furnish some insight. However, there are cases in which we can say the model is almost true. Such models are possible when the phenomenology is clear and captures reality well and when models express its mathematical essence (when we can say the models are minimal models).

As an example of approximate phenomenology how good is the van der Waals equation of state (Example 3.5) as a model of imperfect gases? There is no real gas that can be described by this equation. However, the equation with Maxwell's rule can (qualitatively) describe even the gas-liquid phase transition. The idea of the model is clear, the number of parameters is small, and they are unambiguously determined if we know the critical point. The evaluation of this model may pragmatically rely on the goodness of fit to the real data. However, we must not forget that there is a very important (or perhaps the only meaningful) criterion for approximate models. That is:

(5) The model is related to a certain idealization limit.

For example, the van der Waals equation of state (or more precisely the augmented van der Waals equation of state) + Maxwell's rule can be obtained as an exact result of the imperfect gas with the *Kac interaction potential* = hard core + infinite range infinitesimal attractive interaction.[72] In this sense, the

[71] The book by A. Onuki quoted above contains many succinct mathematical expressions to describe essential features of various factors associated with phase transitions.

[72] This is the theory of the Kac model: P. C. Hemmer, M. Kac and G. E. Uhlenbeck, "On the van der Waals theory of the liquid-vapor equilibrium, I-III," J. Math. Phys. **4**, 216, 229 (1963), **5**, 60 (1964). This is for 1D. The 3D case is also treated later.

van der Waals equation of state captures some sort of mathematical essence of imperfect gases.

4.7 By-products of modeling

Good models have many useful spinoffs other than contributing insight into the target phenomena. Most simply, the model (with small modifications) could explain related phenomena. The approach successfully used to construct the model could be translated or mimicked to model other systems.

In the CDS spinodal model the size of the pattern increases without bound as time goes on. However, there are many cases where the pattern size cannot grow indefinitely. Physically the simplest example may be the *block copolymer* (Fig. 4.6). Chain A and chain B are connected with a covalent bond into a single polymer chain. Such polymers are increasingly important simply because they can spontaneously form mesoscale patterns. That is, the block copolymer system is a self-organizing system.

When A and B mix well at high temperatures, the existence of the covalent bond between the A block and the B block does not cause much problem. However, if quenched, after A and B have segregated to some extent, due to the connection further segregation is hindered. Still, before long (before the molecules realize the serious consequence of the covalent bond), the CDS spinodal model (4.10) should be roughly all right, so the system is expected to be modeled with a small modification of this model.

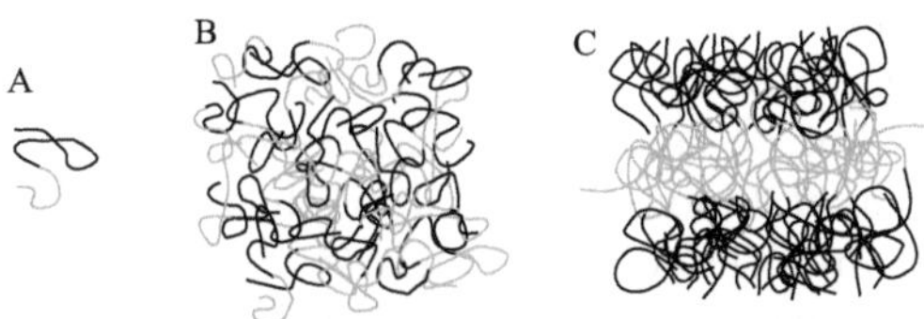

Fig. 4.6 Block copolymer. **A** illustrates a single molecule of block copolymer. The black and the gray 'partial' polymers (blocks) are connected by a covalent bond. They do not wish to be together, but cannot be apart due to their 'fatal bond.' They tolerate this when the temperature is high and mix together as in **B**. However, as the temperature decreases getting along becomes increasingly unpleasant, and they wish to divorce each other, but unfortunately they cannot due to the 'fatal bond,' so they must make a layered structure as in **C**.

What is the most important difference between the systems with indefinitely large domains and those without them? It is that the interfaces between A-rich and B-rich phases do not disappear indefinitely or not. Then, to model a system in which the pattern growth is eventually arrested, we have only to stabilize the interfaces. Since the interface is (for the A-B symmetric case) $\varphi = 0$, we should stabilize $\varphi = 0$. The simplest way to accomplish this is to modify (4.10) with a small positive number C as[73]

$$\varphi_{t+1}(\boldsymbol{n}) = (1 - C)\varphi_t(\boldsymbol{n}) + \mathcal{I}_t(\boldsymbol{n}) - \langle\langle \mathcal{I}_t(\boldsymbol{n}) \rangle\rangle. \tag{4.26}$$

If a bulk phase were formed, where A or B dominates, $\mathcal{I}$ should be close to zero, so a map $\varphi_{t+1}(\boldsymbol{n}) = (1-C)\varphi_t(\boldsymbol{n})$ making $\varphi = 0$ as its stable fixed point would dominate the system evolution for a while at least. A 2D simulation result is in Fig. 4.7. The experimentalists of block copolymer films cannot

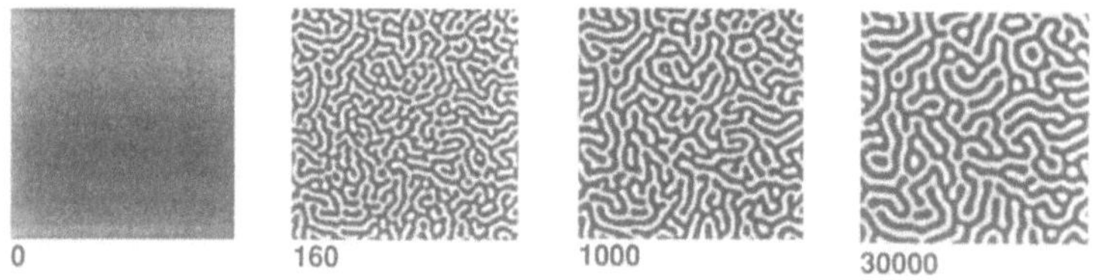

Fig. 4.7 Phase separation in the block copolymer system. The numbers denote the number of time steps from the initial state, which is a random small fluctuation around zero. In this case patterns would not change any more beyond the last one. The situation is the same as in the actual film experiment. Really equilibrium patterns must be a regular black and white striped structure as suggested by C in Fig. 4.6. [Fig. 2, M. Bahiana and Y. Oono, Phys. Rev. A **41**, 6763 (1990)]

distinguish these figures and real film results. Thus, at least qualitatively, the outcome of the model is indistinguishable from the reality. However, good looks are not enough. To be quantitative, we must understand the meaning of C. Here, the modeling strategy is abstract. We considered the problem, "What is the simplest modification of (4.10) to stabilize $\varphi = 0$?" The physical meaning of C has never been considered in the modeling process. If there is no way to determine C, (4.26) may not be said to be a model of an actual phenomenon. We can infer with the aid of dimensional analysis that C and the molecular weight M of the polymer are related as $C \propto 1/M^2$.[74] The model can even be applied to the phase-transitions with chemical reactions,[75] and

[73] This idea is used in Y. Oono and Y. Shiwa, "Computationally efficient modeling of block copolymers and Benard pattern formation," Mod. Phys. Lett. B **1**, 49 (1987). A much more materials-scientifically respectable approach is in T. Ohta and K. Kawasaki, "Equilibrium morphology of binary copolymer melts," Macromolecules, **19**, 2621 (1986).

[74] M. Bahiana and Y. Oono, "Cell dynamical system approach to block copolymers," Phys. Rev. A **41**, 6763 (1990). Again, physically respectable studies are in T. Ohta and K. Kawasaki, Macromolecules, **19**, 2621 (1986) and in T. Ohta and K. Kawasaki, "Comment on the free energy functional of block copolymer melts in the strong segregation limit," Macromolecules, **23**, 2413 (1990).

[75] A review may be found in Qui Tran-Cong, "Self-organization of reactive polymer systems," Butsuri **56**, 590 (1996).

the surface melting of semiconductor with laser.[76]

Another development of such coarse-grained phase-transition dynamics models is to explain much more complicated systems. For example, the model of the iron-aluminum alloy system beautifully reproduces various patterns.[77] Although in this case only patterns are studied, already it is beyond the level of simple good looks. The phase separation dynamics of the lithium-aluminum alloy that is important in engineering as a high-strength light alloy can be modeled almost quantitatively.[78] These topics may be useful for metal engineering but are irrelevant to this book, since they do not offer anything new in the mode of thinking.

Although there is nothing new in the mode of thinking as above, to apply the coarse-grained pattern dynamics models to *biological morphogenesis* is not without interest. The partial differential equation models due to Meinhardt of the shell patterns are instructive.[79] In this case, the outcomes are not only good looking, but there is good reason to believe that they capture some sort of essence of these morphogenesis samples. As Meinhaldt himself emphasizes, when some injury is inflicted on the shell, perturbed patterns allow some insight into the mathematical mechanism of the pattern formation. That is, his models give realistic perturbation results.

However, there is also an instructive 'counter example.' Our upper arm contains one bone (humerus), and the lower part contains two (radius and ulna). The basic problem is to explain why this is so. Our limb develops from the limb bud which is a small protrusion on the body side (it is 0.2mm for

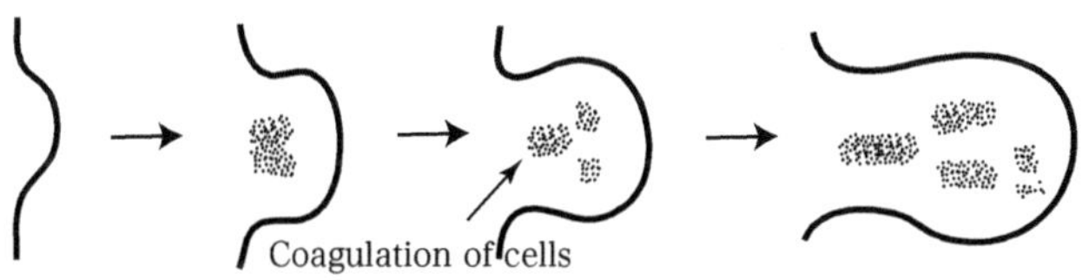

Fig. 4.8 Formation of cartilages in the limb bud. The cells forming cartilages are supplied from the growing edge of the limb bud. Then, inside the limb bud their surface property changes and they become sticky with each other. Thus, something like a phase separation with the conservation of particles happens.

[76] C. Yeung, and R. C. Desai, "Pattern formation in laser-induced melting," Phys. Rev. E **49**, 2096 (1994).

[77] T. Ohta, K. Kawasaki, A. Sato and Y. Enomoto, "Domain growth in a quenched tricritical system," Phys. Lett. A **126**, 94 (1987).

[78] R. Goldstein, M. Zimmer, and Y. Oono, "Modeling mesoscale dynamics of formation of δ'-phase in Al_3Li alloys," p255 in *Kinetics of Ordering Transformations in Metals, a symposium held in honor of Professor Jerome B. Cohen*, edited by H. Chen and V. K. Vasudevan (TMS, 1992).

[79] H. Meinhardt, *The Algorithmic Beauty of Sea Shells*, Third Edition (Springer, 2003).

chicks, Fig. 4.8). A limb bud is covered with the exoderm and its inside is filled with cells from the mesoderm. It is known that for the development of bones, the pattern-formation processes of muscles and vessels are not essentially related. The bone pattern is formed first as the cartilage layout. The cartilage formation proceeds from the proximal to the distal part of the limb bud. The cells that form cartilages (chondrogenic cells) are supplied by the progress zone at the tip of the limb bud, and we may roughly assume that the number of these cells does not increase in the region where the cartilage pattern is formed. Therefore, the cartilage pattern formation may be understood as a coagulation process of chondrogenic cells segregating out from the cells to form connective tissue. Then, this is a segregation of cells of different kinds and the CDS spinodal model may be used. An example of a simulation based on this idea is in Fig. 4.9. In the illustration the model may be designed to describe bones close to the wrist with tuning parameters, but to explain the original problem of the bone numbers (why one or two) is quite easy without any fine tuning of the parameters.[80]

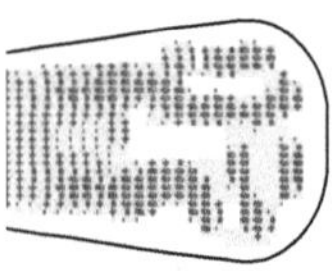

Fig. 4.9 An illustration of the chondrogenesis model. Something suggestive can be done easily. In this illustration no tuning of several parameters in the model has been done, and so the pattern is not realistic and is uninteresting. The illustration is only for the sake of conceptual explanation.

But, is it really a model that gives us insight into the cartilage pattern formation?

In reality, every small bone is meticulously specified by a set of homeobox gene expressions as shown for chickens. The proximo-distal coordinate is specified by *Chox-1*, and the branching pattern is closely related to the *Chox-4* gene expressions.[81] Even if a mathematical principle works common to the phase-separation process, the pattern is not determined by such a mechanism.[82] Even if we could reproduce small bone pieces by the model

---

[80] work with J. Mittenthal quoted in Y. Oono and A. Shinozaki, Cell dynamical systems, Forma **4**, 75-102 (1989) [a CDS review].

[81] Y. Yokouchi, H. Sasaki, and A. Kuroiwa, Homeobox gene expression correlated with the bifurcation process of limb cartilage development, Nature **353**, 443 (1991); A. Graham, "The *Hox* code out on a limb," Current Biol. **4**, 1135 (1994).

[82] Notice that even the pattern of our original concern, that is, whether the number of bones is one or two, is also in detail governed by the *Hox* genes: D. L. Wellik and M. R. Capecchi, "*Hox10* and *Hox11* genes are required to globally pattern the mammalian skeleton," Science **301**, 363 (2003). See also M. Kmita, B. Tarchini, J. Zàkàny, M. Logan, C. J. Tabin and D. Duboule, "Early developmental arrest of mammalian limbs lacking *HoxA/HoxD* gene function," Nature **435**, 1113 (2005).

mentioned above, that has little significance.

Even in biological pattern-formation processes universal mathematical principles should be working.[83] Resourceful biological systems always use whatever they can exploit. It is meaningful to understand the universal features behind various phenomena, but most of them should be recognized before doing biology. Really biological questions are about what happens beyond these elementary physics mechanisms in biological systems. Therefore, the modeling approach adopted here is hardly meaningful biologically. We can say a similar thing about *dissipative structures*. It has no essential relation to biology. Needless to say, biological systems must be maintained away from equilibrium, but it is quite a general prerequisite and has only the information just as a banal statement "scientists eat" has.

There have been numerous models with the aid of the *reaction-diffusion systems*[84] that try to describe biological pattern-formation processes since Turing. Meinhardt's shell pattern dynamics, already mentioned, is perhaps the most successful one. Recent work on fish skin patterns is also of interest.[85,86] However, as can be seen in the developmental processes of *Drosophila* (Fig. 4.10), reaction-diffusion mechanisms are never used in vital developmental processes. All the important structures are rigorously and meticulously dictated by genes, and reaction-diffusion systems are used only where no precision is needed but varieties of outcomes are desirable such as the skin pattern.[87] The important message is that biological systems never trust physics nor chemistry, in short. Since strict regulation of self-organizing capabilities is the crucial element of biological pattern and structure formation, biological processes cannot simply be understood by physico-chemical considerations. Self-organizing materials are effectively utilized by biological systems, but the

---

[83] However, such an obvious truism does not seem to have been recognized clearly for a long time in biology, so it was quite meaningful to point out its significance. For example, S. A. Newman and W. D. Comper, "'Generic' physical mechanism of morphogenesis and pattern formation," Development **110**, 1 (1990). There is an article supporting the above model in S. A. Newman, Sticky fingers: *Hox* genes and cell adhesion invertebrate limb development, BioEssays **18**, 171 (1996).

[84] reaction-diffusion system

[85] A recent report is M. Yamaguchi, E. Yoshimoto and S. Kondo, "Pattern regulation in the stripe of zebrafish suggests an underlying dynamic and autonomous mechanism," Proc. Nat. Acad. Sci., **104**, 4790 (2007). Biological systems use anything useful: P. K. Maini, R. E. Baker, C.-M. Chuong, "The Turing model comes of molecular age," Science **314**, 1397 (2006).

[86] However, even if the system is described by a reaction-diffusion equation, chemical reaction and diffusion need not actually exist in the system. See, for example, A. Kicheva, P. Pantazis, T. Bollenbach, Y. Kalaidzidis, T. Bittig, F. Jülicher, M. Gonzálex-Gaitán, "Kinetics of morphogen gradient formation," Science **315**, 521 (2007).

[87] W. L. Allen, I. C. Cuthill, N. E. Scott-Samuel, and R. Baddeley, "Why the leopard got its spots: relating pattern development to ecology in felids," Proc. Roy. Soc., **278**, 1373 (2011) is another excellent example, clearly showing that the patterns, which are explained by a reaction-diffusion system, change very rapidly phylogenetically.

characteristic feature of biological systems does not reside there.[88] We will see in the last chapter how off the mark the physicists' study of the so-called complex systems have been.

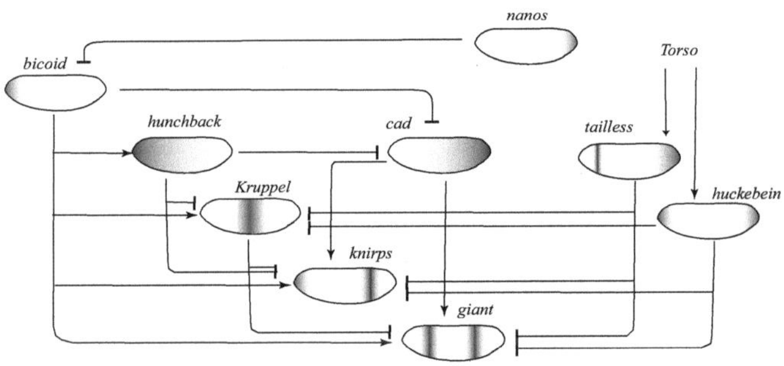

Fig. 4.10 The early developmental process of *Drosophila*, especially the action of gap genes. Notice that only simple switches are used. Based on the concentration gradients of maternal genes (*bicoid*, *nanos*), the pattern formation proceeds. The arrows denote transcription enhancement and bars inhibition. There is no dissipative structure. The expression patterns of the genes are based on `http://bdtnp.lbl.gov/Fly-Net/bidatlas.jsp` (Berkeley Drosophila Transcription Network Project). The interaction among genes roughly follows S. B. Carroll, J. K. Grenier and S. D. Weatherbee, *From DNA to Diversity, Molecular genetics and the evolution of animal design* (Blackwell Science, 2001).

A more meaningful byproduct is obtained by reflecting on the ideas used in the modeling. The initial motivation of the CDS spinodal decomposition was that if a numerical solution cannot be avoided to study the partial differential equation obtained by continuum modeling, we should make *numerical-computation-adapted* discrete models from the beginning. In the case of spinodal decomposition the results of the Cahn-Hilliard equation (4.14) agrees with those of the CDS model in the time range so far as the Cahn-Hilliard equation is computed. However, it is better to know the relation between these two models more directly than through numerical results, albeit heuristically. Here, let us consider the simpler non-conserved case, the relation between (4.15) and (4.6) (for spinodal decomposition, we have only to impose the conservation law). Let us solve (4.15) for a short time. To this end, the equation is separated into two parts, the local nonlinear dynamics:

$$\frac{\partial \varphi}{\partial t} = \varphi - \varphi^3 \tag{4.27}$$

[88] Proc. Nat. Acad. Sci. **99**, suppl. 1 is a special issue dedicated to 'self-organized complexity.' Its introduction is: D. L. Turcotte and J. B. Rundle, "Self-organized complexity in the physical, biological, and social sciences," *ibid.*, 2463 (2002). It is said a classic example is the power law in the relation between the magnitude and the frequency of earthquakes. The so-called $1/f$-noise and related topics and chaos are also discussed in this issue. The topics discussed in this issue are a mixture of true and *pseudo-complex phenomena*. Many are only about scaling (cf. Note 5.1).

and the diffusion process:

$$\frac{\partial \varphi}{\partial t} = \Delta \varphi. \tag{4.28}$$

Note that we are not making a system of simultaneous equations. First, we solve (4.27) at each spatial point from a given initial condition for a short time δt (during this procedure the spatial coordinates are mere parameters). Then, using the obtained solutions as the initial condition, we solve (4.28) for a short time δt. Next, we solve (4.27) at each point with the obtained solution as the initial condition, $\cdots$. We repeat this procedure and can solve (4.15).[89]

If we solve (4.27) from t to $t + \delta t$, we obtain

$$\varphi_{t+\delta t} = \frac{\varphi_t e^{\delta t}}{\sqrt{1 + \varphi_t^2 (e^{2\delta t} - 1)}}. \tag{4.29}$$

Its graph is very similar to the graph of (4.5) in Fig. 4.1. That is, (4.6) solves (4.15) virtually with the *nonlinear Lie-Trotter method.* If we utilize the fact discussed before that the form of the potential does not affect the form factor, (4.29) may be appropriately replaced by a map of the type (4.5). Thus, the CDS model may be understood as a special numerical method for (4.15). This is a method to compute accurately the form factor and related quantities only, and it is not a numerical scheme to compute $\varphi(t, \boldsymbol{r})$ accurately. As a solution method for partial differential equations CDS is a new type of numerical scheme. It is inaccurate in the usual sense, but certain aspects are very accurate. This implies that the idea of a *qualitatively accurate solution* is meaningful. A take-home lesson is that if a partial differential equation captures physics correctly, we can invent such a numerical scheme. It is also clear that modeling (thus, phenomenology) and numerical analysis are closely related.

In the extension of this lesson is a 'numerical scheme motivated by the phenomenon.' It is to make a discrete representation that captures in the computer the key elements of physics of the phenomenon, which is usually modeled in terms of a partial differential equation, without paying any attention to the partial differential equation. For example, if we have to study propagation of waves or not slow flows (i.e., hyperbolic equations), inertial effect must be correctly taken into account. The translational momentum is conserved if the space is translationally symmetric. Thus, the key physics must be the translational symmetry of the space. However, numerical lattices are discrete, so translational symmetry of continuous space is not naturally expressed with the aid of discrete lattices. Refining the numerical lattice (e.g., adaptive mesh) is not an elegant approach. The most natural approach is the

---

[89] This idea is common in numerical analysis, and is the same as *Trotter's formula* (the basic idea is due to Sophus Lie). Therefore, it is appropriate to call the method the *nonlinear Lie-Trotter method.*

interpolation-resampling method illustrated in Fig. 4.11. The values on the numerical lattice points are used to reconstruct the continuous curve. Then, it is translated and the values at the numerical lattice points are resampled.

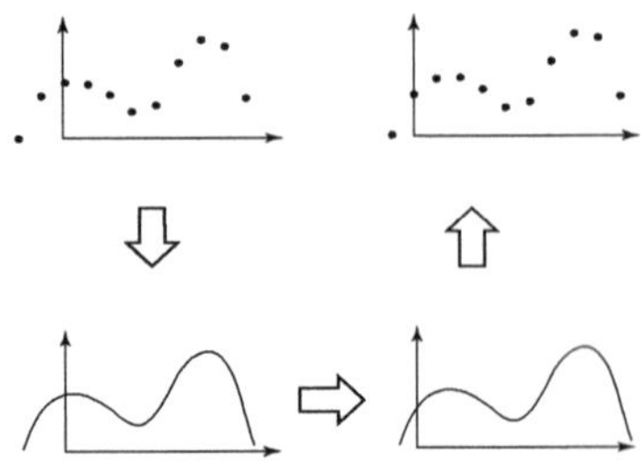

Fig. 4.11 Interpolation-resampling scheme to capture spatial translation symmetry [Fig. 1 of L. San Martin, and Y. Oono, "Physics-motivated numerical solvers for partial differential equations," Phys. Rev. E **57** 4795 (1998) (submitted in 1995)].

If the reconstruction is successful, then this method solves, for example, the following elementary differential equation

$$\frac{\partial f}{\partial t} + c\frac{\partial f}{\partial x} = 0 \tag{4.30}$$

accurately. This equation is analytically trivial, but numerically highly nontrivial, because usual numerical methods suffer from the so-called *numerical diffusion* that blurs the function profile. We can use efficient numerical schemes based on the above idea for hydrodynamic equations and wave equations.[90]

In the case of *physics-motivated* numerical schemes, if the numerical result and the result of the partial differential equation disagree, it must be because the partial differential equation fails to grasp physics.[91]

The mainstream numerical approaches also try to capture the core of physics of the phenomenon accurately. For example, if there is a conservation law, usually they make a discretization scheme that exactly satisfies the conservation law. However, even if the conservation law is rigorously satisfied,

---

[90] The method to solve the Navier-Stokes equation in the following paper is best understood from the interpolation-resampling view: T. Yabe and T. Aoki, "Shoemaker-Levi comet and Jupiter collision simulation," Butsuri **49**, 909 (1994). See the review: T. Yabe, F. Xiao, and T. Utsumi, "The Constrained Interpolation Profile Method for Multiphase Analysis," J. Comp. Phys. **169**, 556 (2001). Attempts to study wave equations are in L. San Martin, and Y. Oono, "Physics-motivated numerical solvers for partial differential equations," Phys. Rev. E **57** 4795 (1998).

[91] "A recurring theme throughout this sermon has been the need to be perpetually cognizant of the physical basis for any numerical computation. In fact, this might profitably be adopted as the First Commandment of Numerical Computation. Of course, this requires an in-depth study of the physical foundations of a mathematical problem — a sojourn upon which many Mathematicians seem reluctant to embark." (B. J. McCartin, "Seven deadly sins of numerical computation" Am. Math. Month. December 1998, p929).

this does not guarantee the accuracy of the algorithm. Within the philosophy of 'numerical scheme motivated by the phenomenon,' we do not develop algorithms through considering the satisfaction of conservation laws as one of the key constraints. For example, when we devise algorithms to capture the inertial effect in wave equations and due to the term $(\boldsymbol{v}\cdot\nabla)\boldsymbol{v}$ in the Navier-Stokes equation, as we have seen above, we never directly demand the conservation law of momenta, but we make utmost efforts to realize the spatial translational symmetry.[92] Thus, in consequence, we can substantially satisfy the conservation law with high precision. We can even sense the reliability of our numerical calculation to some extent by checking the conservation law.

References

Allen WL, Cuthill JC, Scott-Samuel NE, Baddeley R (2011) Why the leopard got its spots: relating pattern development to ecology in felids. Proc Roy Soc Lond, Ser B: Biol Sci 278:1373-1380

Bahiana M, Oono Y (1990) Cell dynamical system approach to block copolymers. Phys Rev A 41:6763-6771

Barbieri M (2002) The organic code—an introduction to semantic biology. Cambridge University Press

Bates FS, Wiltzius P (1989) Spinodal decomposition of a symmetric critical mixture of deuterated and protonated polymer. J Chem Phys 91:3258-3274

Cahn JW (1965) Phase separation by spinodal decomposition in isotopic systems. J Chem Phys 42:93-99

Cahn JW, Hilliard JE (1958) Free energy of a nonuniform system I. Interfacial free energy. J Chem Phys 28:258-267

Chou YC, Goldburg WI (1981) Angular distribution of light scattered from critically quenched liquid mixtures. Phys Rev A 23:858-864

Craievich AF, Sanchez JM, Williams CE (1986) Phase separation and dynamical scaling in borate glasses. Phys Rev B 34:2762-2769

Ezawa H (1998) Quantum mechanics, from its birth to re-examination. In Ezawa H (ed) Physics of 20th century. Saiensu

Fisher ME (1983) Scaling, universality and renormalization group theory. Lecture Notes in Physics 186

Gaulin BD, Spooner S, Morii Y (1987) Kinetics of phase separation in $Mn_{0.67}$ $Cu_{0.33}$ Phys Rev Lett 59:668-671

Goldstein R, Zimmer M, Oono Y (1992) Modeling mesoscale dynamics of formation of δ'-phase in Al_3Li alloys. p255 in Chen H, Vasudevan VK (ed) Kinetics of ordering transformations in metals, a symposium held in honor of Professor Jerome B Cohen. TMS

Gowers T (ed) (2008) The Princeton companion to mathematics. Princeton

Graham A (1994) The *Hox* code out on a limb. Curr Biol 4:1135-1137

Hadamard J (1954) The psychology of invention in the mathematical field. Dover

Hemmer PC, Kac M, Uhlenbeck GE (1964) On the van der Waals theory of the liquid-vapor equilibrium, I-III. J Math Phys 4:216-228, 229-247, 5:60-74

Jones CKRT (2008) Is mathematics misapplied to the environment? Notices Amer Math Soc 55:481-484

Kicheva A, Pantazis P, Bollenbach T, Kalaidzidis Y, Bittig T, Jülicher F, Gonzálex-Gaitán M (2007) Kinetics of morphogen gradient formation.

---

[92] Recall *Noether's theorem* (A. E. Noether 1882-1935).

Science 315:521-525
Kmita M, Tarchini B, Zàkàny J, Logan M, Tabin CJ, Duboule D (2005) Early developmental arrest of mammalian limbs lacking *HoxA/HoxD* gene function. Nature 435:1113-1116
Kohn RV, Otto F (2002) Upper bounds on coarsening rates. Comm Math Phys 229:375-395
Kuris AM, et al (2008) Ecosystem energetic implications of parasite and free-living biomass in three estuaries. Nature 454:515-518
Lanczos C (1974) The Einstein decade (1905-1915). Paul Elek Ltd
Lebowitz JL, Marro J, Kalos KH (1982) Dynamical scaling of structure function in quenched binary alloys. Acta Metall 30:297-310
Lorenz K (1973) Die Rückseite des Spiegels. R Piper & Co Verlag
Lorenz K (1973) Die acht Todsünden der zivilisierten Menschheit. R Pier & Co
Maini PK, Baker RE, Chuong C-M (2006) The Turing model comes of molecular age. Science 314:1397-1398
McCartin BJ (1998) Seven deadly sins of numerical computation. Am Math Month December 1998, p929-941
Meinhardt H (2003) The algorithmic beauty of sea shells (third Edition). Springer
Nishiura Y (2002) Far-from-equilibrium dynamics (Translations of Mathematical Monographs vol 209, Am Math Soc (the original Japanese edition 1999)
Nizan P (1932) Les chiens de garde. Rieder
Newman SA (1996) Sticky fingers: *Hox* genes and cell adhesion invertebrate limb development, BioEssays 18:171-174
Newman SA, Comper WD (1990) 'Generic' physical mechanism of morphogenesis and pattern formation. Development 110:1-18
Ohta T, Kawasaki K (1986) Equilibrium morphology of binary copolymer melts. Macromolecules 19:2621-2632
Ohta T, Kawasaki K (1990) Comment on the free energy functional of block copolymer melts in the strong segregation limit. Macromolecules 23:2413
Ohta T, Kawasaki K, Sato A, Enomoto Y (1987) Domain growth in a quenched tricritical system. Phys Lett A 126:94-96.
Onuki A (2002) Phase transition dynamics. Cambridge University Press
Oono Y, Shinozaki A (1989) Cell dynamical systems, Forma 4:75-102
Oono Y, Shiwa Y (1987) Computationally efficient modeling of block copolymers and Benard pattern formation. Mod Phys Lett B 1:49-55
Oono Y, Puri S (1988) Study of phase separation dynamics by use of cell dynamical systems I. Modeling. Phys Rev A 38:434-53
Pickering MJ, Garrod S (2007) Do people use language production to make predictions during comprehension? Trends Cognitive Sci 11:105-110
Pilkey OH, Pilkey-Jarvis L (2007) Useless arithmetic—why environmental scientists can't predict the future. Columbia University Press
Popper K (1984) Evolutionary epistemology. In Pollard J(ed) Evolutionary Theory: paths into the future. Wiley
Qui T-C (1996) Self-organization of reactive polymer systems. Butsuri 56:590-593
San Martin L, Oono Y (1998) Physics-motivated numerical solvers for partial differential equations. Phys Rev E 57:4795-4810
Serre T, Oliva A, Poggio T (2007) A feedforward architecture accounts for rapid categorization. Proc Natl Acad Sci USA 104:6424-6429
Shinozaki A, Oono Y (1993) Spinodal decomposition in 3-space. Phys Rev 48: 2622-2654
Shnerb M, Loouzoun Y, Bettelheim E, Solomon S (2000) The importance of being discrete: life always wins on the surface. Proc Natl Acad Sci USA 97:

10322-10324
Simberloff D (2007) An angry indictment of mathematical modeling. Bioscience 57:884-885
Spooner S, In Family F, Landau DP (ed) Kinetics of aggregation and gelation. North-Holland
Takenaka M, Hashimoto T (1992) Scattering studies of self-assembling processes of polymer blends in spinodal decomposition. II. Temperature dependence. J Chem Phys 96:6177-6190
Togashi Y, Kaneko K (2005) Discreteness-induced stochastic steady state in reaction diffusion systems: self-consistent analysis and stochastic simulations. Physica D 205:87-99
Tomita H (1991) Preservation of isotropy at the mesoscopic stage of phase separation processes. Prog Theor Phys 85:47-56
Turcotte DL, Rundle JB (2002) Self-organized complexity in the physical, biological, and social sciences. Proc Natl Acad Sci USA 99 suppl 1:2463-2465
Waley A (1989) The analects of Confucius. Vintage
Wellik DL, Capecchi MR (2003) *Hox10* and *Hox11* genes are required to globally pattern the mammalian skeleton. Science 301:363-367
Wong N-C, Knobler CM (1978) Light scattering studies of phase separation in isobutyric acid + water mixtures. J Chem Phys 69:725-735
Wittgenstein L (2001) Philosophical investigations. Blackwell (original 1953)
Yabe T, Aoki T (1994) Shoemaker-Levi comet and Jupiter collision simulation. Butsuri 49:909-912
Yabe T, Xiao F, Utsumi T (2001) The constrained interpolation profile method for multiphase analysis. J Comp Phys 169:556-593
Yamaguchi M, Yoshimoto E, Kondo S (2007) Pattern regulation in the stripe of zebrafish suggests an underlying dynamic and autonomous mechanism. Proc Natl Acad Sci USA 104:4790-4793
Yeung C, Desai RC (1994) Pattern formation in laser-induced melting. Phys Rev E 49:2096-2114
Yokouchi Y, Sasaki H, Kuroiwa A (1991) Homeobox gene expression correlated with the bifurcation process of limb cartilage development, Nature 353:443-445

Chapter 5
Toward Complexity

The purpose of this book is to be a guide to the exploration of the world filled with nonlinearity through explaining conceptual analysis and phenomenology. In contrast to many 'nonlinear books,' however, such standard topics as chaos and fractals appeared only as vehicles to illustrate the points the author believes crucial.

The important feature of the nonlinear world is scale interference, because of which the events on the unknowable scales can affect our lives. The effects are often large, but still it is often the case that they do not show up haphazardly at arbitrary places. This is because there are (at least approximately) renormalizable structures in the broadest sense in the world. This is the secret of the world that allows us to gain its phenomenological understanding, a sort of general understanding not restricted to the individual understanding of particular systems. This is also the secret that allows the existence of intelligent beings as ourselves. As we can see from the difficulty of modeling systems that lack good phenomenology, the phenomenological viewpoint of the world exists as a basso continuo even when we try to make models of various phenomena.

There must be more phenomena no one has ever seriously studied than have already been studied by someone. Among them must be numerous 'simple' phenomena to which the methodology explained in Chapters 3 and 4 may be more or less straightforwardly applicable. However, there must be a lot of much more intractable phenomena: 'Complex phenomena' and 'complex systems' will become increasingly important targets of fundamental science. Nonlinearity is a decisive element of complex systems, because it 'pumps up' the unknown from the microscopic scales and fixes it on the meso or macroscopic scales as if new information were created.

In English something merely complicated is not described as *complex.* For example, a theater complex implies a collection of theaters under a certain plan and a ruin complex is a set of remains of what had been constructed with a certain intention. The word "complicated" is used to describe the situation that is intractable and is a considerable mess as can be seen in: "He died of complications of pneumonia" and must be contrasted with the word "complex."

In this last chapter, we recall that the concept 'complex' cannot be discussed without such concepts as meaning and value; consequently, the core of complex phenomena is inseparable from biology. This implies that the usual 'complex systems studies' are off the mark. An important characteristic feature of organisms as typical complex systems is that they are not spontaneously generated as stressed by Pasteur. The reason for this is that the 'tricks and contrivances' (which we will call fundamental conditions) coming from parent organisms are indispensable for their generation. Such systems are quite different from the systems physics has been studying seriously. What is the essence of this difference? What are the logical consequences of the difference? At present, we do not have any (mathematical) system to discuss complex systems precisely. Then, what kind of approaches are allowed to us? Where will they lead us?[1]

5.1 Meaning and value

An ambitious project to define 'complex systems' or 'complexity' is quite unlikely to be accomplished easily. As alluded in the preamble to this chapter, the so-called complex systems studies are not even on the target board. Therefore, it seems wise to start our exploration with a typical example as we always do.

We start with *a priori* acceptance of the proposition "Organisms are complex systems." The meanings of 'organism' and 'complex' are both unclear.[2] However, for 'organism' our starting point is that there are many systems that are unanimously recognized as organisms. The proposition "An organism or a cell is complex" will be accepted. The connotation of the word 'complex'

---

[1] Sections 5.1-3 are based on the first half of Y. Oono, "Complex systems study as biology," Int. J. Mod. Phys. B, **12**, 245 (1998). With the statement "In my view, meaning is intimately tied up with survival and natural selection," M. Mitchell's *Complexity: a guided tour* (Oxford University Press, 2009) is different from numerous garden variety complexity books. It gives a rudimentary molecular biology lesson as well. Still, regrettably, the main topic of the book is *pseudocomplexity*.

[2] ⟨⟨**Distinction between the quick and the dead**⟩⟩ "Many people have tried very hard to produce such criteria for separating the quick from the dead. Put briefly, they have all failed. This fact is most significant in itself, but perhaps of even greater significance is the further fact that we can somehow *know*, with certainty, that they have indeed failed." [R. Rosen, *Life Itself* (Columbia UP, 1991) p18] Such statements as "Life is a network of autocatalytic reactions," or "Life is a physical realization of recursive information system," sounds rather hollow.

has been discussed above, and to respect this connotation and our starting point are consistent.

If a portion of a DNA molecule codes histone 3,[3] then we should recognize that the base sequence is intricate and not simple (that is, in short, complex in the common sense). However, if we do not know the function of the base sequence, we would regard the sequence as a meaningless random sequence. Actually, when people did not understand the functions of non-protein-coding DNA sequences at all, such sequences were regarded as junk, although junk is not the same as trash. Even if we take the statistics of nucleotides, it is not easy to tell whether a given DNA sequence is (biologically) meaningful or not. The base arrangement of a coding sequence is not very redundant (not highly compressible), so although it is not as uncorrelated as a Bernoulli process on 4 symbols, still it is not very easy to distinguish it from a sample sequence of an appropriate Markov chain. That is, the reason why we say a sequence is complex in the usual sense of this word is not because the sequence itself has a special base arrangement pattern, but simply because it has a biological meaning.[4] We admit that the number sequence of π is intricate not because we sense something special from its number arrangement patten, but because we know its crucial roles in geometry and analysis. In short, we have a tendency to conclude that if a non-regular number or symbol sequence has a certain meaning, it is a complex sequence, but that if we do not recognize any meaning, it is random.[5]

So far we have used the word 'meaning,' but it is used to indicate 'significance' rather than merely implying 'semanticity.' Why do we feel that a base sequence coding a histone or a number sequence for π is 'significant'?

The assertion that something is abstractly and absolutely 'meaningful' is empty. 'Meaningful' implies 'meaningful to someone.' "An object is mean-

---

[3] ⟨⟨**Histone**⟩⟩ Histones are basic proteins making 1 to 1 complex (in weight) with DNA of eukaryotes and are necessary to fold DNA into compact conformations to form chromosomes; histones make nucleosomes that are building blocks of chromatins. Histone 3 is molecular-evolutionarily stable, so its biological importance is clear. C. Das, J. K. Tyler and M. E. A. Churchill, "The histone shuffle: histone chaperones in an energetic dance," Trends Biochem. Sci. **35**, 476 (2010) contains a good summary of histones.

[4] ⟨⟨**Correlations in base sequences**⟩⟩ The base sequence of DNA is not simply non-periodic, but with definite systematic changes corresponding, e.g., to starting and termination points of transcription or replication. There is a vague periodicity corresponding to nucleosome positioning. Furthermore, several different codes (protein code, splicing code, etc.) may be superposed on a single sequence.

[5] As we already know, π is highly compressible, so the reader would say it is not a good example. However, notice that it is a revealing example, even with its compressibility being taken into account. Cf. Discussion 5.1.

π is known to be statistically indistinguishable from a random sequence. An interesting related paper is D. H. Bailey and R. E. Crandall, "On the random character of fundamental constant expansions," Experimental Math. **10**, 175 (2001).

ingful to us" implies at least that it is meaningful for us to pay attention to. To pay attention to something, that is, to concentrate one's neural resources to something, is biologically costly. Therefore, recognizing an object as meaningful is equivalent to judging, albeit unconsciously, that it is worthy of some sort of investment. Therefore, in order for us to judge something to be 'meaningful,' as a prerequisite we must have a *value system.* For example, that we regard geometry or analysis meaningful implies that we value that at least someone studies it. We conclude that π is meaningful because it is indispensable to various mathematics branches whose value we recognize. As we have already seen, even basic factual judgment is based on value judgement; *value judgment* is always the foundation of any judgment.

What is 'value'? According to the *New Oxford American Dictionary* 'value' is: the regard that something is held to deserve; the importance or preciousness of something. Therefore, we may ask what 'good' means. It should be obvious that 'good' is not an absolute concept. That something is good implies that it is good for 'someone.' Then, what is the implication of 'good for someone'? This question makes sense, only if this 'someone' exists. The minimum requirement for "something to be good for someone" is that this something at least helps the continued existence of this 'someone.' Therefore, in the most fundamental sense, 'good' for someone or the value system of someone hinges upon "this someone to continue to exist (as an organism)." We know what is good in the fundamental sense of this word as long as we choose to live. In other words, we intrinsically have a 'value system' in the most fundamental sense of this word.[6]

The value system is needed for us to 'continue to exist as an organism,' because we must appropriately adapt to our environment. For bacteria, for example, the value system is 'hardwired' in the organism. Even for us the most part of this system is hardwired, so we do not consciously recognize it as such (i.e., we do not have any conscious room to choose; we are made alive). We must choose food, mates, etc., to exist as organisms. Conscious choice is virtually impossible in real time if we do not have an innate value system.[7] We call the system of bias or weight factors (employed when we

[6] "Humanistic ethics takes the position that *if man is alive he knows what is allowed.*" [E. Fromm, *Man for himself — an enquiry into the psychology of ethics—* (Routledge & Kagan Paul Ltd. 1950), Chapter 5] (the italics are in the original).

[7] ⟨⟨**Choice or decision by organisms**⟩⟩ When we speak of 'choice,' the reader might have a mental picture of selecting an item from itemized choices. If such a choice is required, evaluating each item is time consuming, so even if there is a device to assign bias (weight) on each item (i.e., the value system) computational difficulty cannot be overcome. However, this conclusion is due to misunderstanding the decision process in organisms. The decision process begins at quite an abstract level; for example, survival or reproduction. Then, a commitment cascade starts. There are many biases in this choice or decision process. For us human beings emotional bias has a decisive effect. Besides, it is highly likely that such a high-level choice is performed by almost completely hardwired devices (almost automated; our consciousness feels it only after the decision as if it decided the matter). Although what A. Damasio, *Descartes'*

decide) our value system in the most fundamental sense of this word.

As we have already seen at the end of Chapter 2, to some extent, and as can be seen from the above discussion, even if 'complexity' is not clearly defined, there is no room to doubt that it is a disparate concept from 'randomness.' At the least, 'complexity' is fundamentally different from such a concept as 'algorithmic randomness' that is unrelated to 'meaning.' Whether a sequence can be information-compressed or not is inessential to complexity (as the sequence we constructed in Section 2.14 illustrates this). After compression, a literary masterpiece is hardly distinguishable from a random sequence statistically. The practical success of information theory eloquently tells us that meaningless and meaningful writings have no difference as letter sequences. Intricate structures of a literary masterpiece are diametrically different from the ones detected statistically.

Discussion 5.1. π is not algorithmically random. If there is agreement about what 'π' is between the sender and the receiver of the message, to send a million digits of π starting from the millionth digit, we need not send the information about π except for its name. Even if the receiver does not know what π is, its algorithm can be appended to the message, so sharing the knowledge of π is not needed to compress the message. Now, let us try to send the structure of histone 3, or on a more grand scale, the genome of *Caenorhabditis elegans*. If there is agreement about *Caenorhabditis elegans* between the sender and the receiver, sending the name (plus the specification of the strain) is enough, essentially (ignoring individual differences). A tremendous information compression can be accomplished. However, if the

---

Error, emotion, reason, and the human brain (G. P. Putnam, 1994) discusses is the decision process at the very final stage where only a few decision items are explicitly and consciously presented, even at this stage emotional bias is decisive as is vividly described in this book. Notice that organisms basically evade the so-called frame problem

⟪**Choice and emotion**⟫ When we decide, we request advice from our ancestors through our body: D. T. Gilbert and T. D. Wilson, "Prospection: experiencing the future," Science **317**, 1351 (2007). We simulate the future using our prefrontal cortex and medial temporal lobe. The simulation result induces the emotional reaction to evaluate the future events. "When we preview the future and prefeel its consequences, we are soliciting advice from our ancestors."

If we mimic Lorenz' statement that the semicircular canals are the intuitive form of the three-dimensionality of the space, emotional reaction is the intuitive form of value judgment. *Emotionalism* claiming that ethical statements are the expressions of emotion captures correctly at least one important aspect of the truth about ethics.

Since a neural system cannot feel itself, to have an effective feedback, it must rely on something external to itself. It is highly plausible that the neurotransmitters (such as various nonapeptides like vasopressin and oxytocin) and other chemical concentration fields serve as the needed feedback device; thus, the chemical field interacting with the neural system is the essence of emotion. Since neurotransmitters and their analogues have a very long history, the concept of emotion must not be confined to human beings. For example, see M. Bateson, S. Desire, S. E. Gartside, and G. A. Wright, "Agitated honeybees exhibit pessimistic cognitive biases," Current Biol., **21**, 1070 (2011).

receiver has no knowledge of this nematode, what happens? This case is fundamentally different from the case of π. What is the lesson?[8] □

If we reflect on the usage of the word 'complex,' we must conclude that 'complexity' must be closely related, as many people must have felt, to meaning and value. We have already seen that these concepts are meaningful only in conjunction with organisms. Therefore, the core of 'complexity' we see in reality is a biological phenomenon. Many readers must feel this is an obvious conclusion. If one recognizes the connotation of the word 'complex,' it is insensitive for him to indiscriminatingly describe everything complicated and intractable as 'complex.' It must already be obvious to the reader that almost all the concepts regarded as keywords of complexity are irrelevant or even red herrings, but, to be sure, their irrelevancy is summarized in Note 5.1.

Note 5.1. What are complex systems not?
What are complex systems not, or what shouldn't they be? Books entitled 'complex systems'[9] use earthquakes, weather and climate, turbulence, market and society, biological systems, complex fluids, etc., as examples. The list includes both genuine and pseudocomplex systems. Chaos, fractal, dissipative structure, self-organizing system, and emergency are among the key words. As we already know, chaos is essentially related to randomness, and fractal is unrelated to meaning or value. The significance of chaos in the context of complexity is that it clearly shows us with simple examples the 'open-endedness' of the world we observe. After the remarkable success of the theory of critical phenomena people began to look at various phenomena through the lenses of self-similarity

[8] G. E. Moore said he could define a horse; we can vividly feel how he was indifferent to the actual world.

[9] **⟨⟨What do 'complex systems books' preach?⟩⟩** G. Nicolis and I. Prigogine's *Exploration of Complexity* (W. H. Freeman and Co., New York, 1989) may be used as an example. On p8, after pointing out that the fascination with biology is responsible for a somewhat diffuse identification of the idea of complexity with the phenomena of life, they write, "Surprisingly, this perennial idea will be the first to break down as we pursue an understanding of complexity. Since the 1960s a revolution in both mathematical and physical sciences has imposed a new attitude in the description of nature. Parallel development in the thermodynamic theory of irreversible phenomena, in the theory of dynamical systems, and in classical mechanics have converged to show in a compelling way that the gap between 'simple' and 'complex,' between 'disorder' and 'order,' is much narrower than previously thought." Immediately after this passage is a mention of chaos as an example of complex behavior. The lack of reflection on the word 'complex' is perfect. As a newer book, we could mention R. Badii and A. Politi, *Complexity: hierarchical structures and scaling in physics*, (Cambridge UP, 1997). The book (especially beyond p200) can be recommended as an introductory book for analysis techniques of chaos with, e.g., explanation of algorithmic randomness, the book is confined only to syntactically understandable things, and is irrelevant to the understanding of complexity itself. M. M. Waldrop, *Complexity: the emerging science at the edge of order and chaos* (Simon & Schuster, 1992) is a popular book, but discusses only complexity in the sense of 'complex systems studies' (i.e., *pseudocomplex systems*). M. Mitchell's *Complexity: a guided tour* (Oxford University Press, 2009) discusses complex systems studies, but the author does not ignore the genuinely complex systems, paying respect to biology. Unfortunately, however, the author seems to be mired in the conventional studies.

and related scaling laws uncritically. We must not forget that self-similarities and scaling laws purportedly found in biological systems are often due to incompleteness of data or inaccuracy of observations.[10]

Then, how about dissipative structures that seem to be a key for Prigogine and others to claim that complex systems are not special? Typical examples are the Turing pattern in reaction-diffusion systems and patterns observed in fluid dynamical systems. These patterns do not appear in equilibrium, and emerge only after a kind of phase transition when the extent of nonequilibrium is increased (for example, for the Benard pattern when the heat flux from the bottom of the fluid layer to the top is increased). Since biological systems are maintained in the states out of equilibrium, it is a natural guess that dissipative structures are important in biological systems, but we must not forget that dissipative structures are all very fragile structures. Even if a new structure emerges after something like a nonequilibrium phase transition, the important element is the microscopic structural change, so the transition is a virtual equilibrium phase transition; for example, when a system is maintained close to an equilibrium phase transition, a nonequilibrium perturbation, say, shear stress, could align molecules to order the system. Needless to say, if such a phenomenon is useful, biological systems would exploit it. Some patterns are based on dynamic concentration profiles of certain chemicals. For example, the pattern formation of the *Drosophila* embryo illustrated in Fig. 4.10 starts with the Bicoid protein gradient, which is established by the translation, transport (perhaps active) and degradation of this protein,[11] so the pattern is not at all static. However, as stressed in Chapter 4, it is hard to interpret the resultant pattern as an example of dissipative structure.

What is common to the concepts such as chaos, fractal, dissipative structure, etc.? Something apparently complicated or nontrivial emerges sponta-

---

[10] ⟨⟨**Scaling laws?**⟩⟩ "Log-log plot is a standard trick to hide error bars," the author's adviser Professor Tomo-o Oyama emphasized.

For example, see R. M. May, "Network structure and the biology of populations," Trends Evo. Eco., **21**, 394 (2006); In many purportedly power law examples (linear log-log plots) are better explained with different laws. As to the criticism of the physiological scaling laws see: D. S. Glazier, "Effects of metabolic level on the body size scaling of metabolic rate in birds and mammals," Proc. Roy. Soc. **275**, 1405 (2008), which finds significant deviations from the '3/4-power law.' T. Kolokotrones, V. Savage, E. J. Deeds and W. Fontana, "Curvature in metabolic scaling," Nature **464**, 753 (2010) demonstrates that the relationship between mass and metabolic rate has convex curvature on a logarithmic scale, and is therefore not a pure power law, even after accounting for body temperature. A. E. Sieg, M. P. O'Connor, J. N. McNair, B. W. Grant, S. J. Agosta, and A. E. Dunham, "Mammalian metabolic allometry: do intraspecific variation, phylogeny, and regression models matter?" Am. Nat., **174**, 720 (2009) points out the significance of intraspecific variation: the slope of the interspecific allometry is typically estimated without accounting for intraspecific variation in body mass or phylogenetic constraints on metabolism. "We find little support for a universal metabolic rate-body mass scaling relationship." Or, the scale-free nature of biological networks such as gene or protein interaction networks was advocated as if true, but it is due to the gross incompleteness of the data. See, for example, N. Pržlj, D. G. Corneil and I. Jurisica, "Modeling interactome: scale-free or geometric?" Bioinformatics **20**, 3508 (2005).

[11] Actually, it is more complicated: S. C. Little, G. Tkačik, T. B. Kneeland, E. F. Wieschaus, and T. Gregor, "The formation of the Bicoid morphogen gradient requires protein movement from anteriorly localized mRNA" PLoS Biology **9**, e1000596 (2011).

neously under simple rules, or without any of our special efforts or any particular preparations. It may not be easy to understand emergent patterns (as turbulence), but they are easily realized. We have only to set up a small number of right environmental parameters and to leave the systems alone. If something emerges spontaneously, it must be simply because it is in a certain sense simple.[12] Such systems are the antipode of truly complex systems (e.g., the reader of this book). We must concentrate our attention on the systems that cannot be constructed by self-organization alone.

Do earthquakes and weather/climate deserve to be called complex phenomena? For example, to forecast tomorrow's weather a tremendous amount of data is required. This is a computationally very complex problem, because as a typical nonlinear phenomenon scales from about a few tens of meters to the global scales are entangled. However, obviously, the atmospheric phenomena have nothing to do with meaning or value, so we must separate them from the true complex systems. Similarly, earthquakes can give interesting many-body problems, but should not be regarded as genuinely complex problems.

We must not forget that the so-called complex systems studies have political implications. The idea that without intervention of 'authority' something wonderful arises resonates very conveniently with the market fundamentalism and neoconservative philosophy. These ideas are characterized by neglect of initial conditions.[13] Neglect of history also comes from the same root. For example, 'progressive thinkers' tend to deny the special nature of the Western civilization, but history is not that simple.[14]

---

[12] ⟪**Self-organization and time scale**⟫ These days, the term 'self-organization' is used without paying any attention to the time scale. If the world is appropriate, even (the concept of) God can emerge according to Darwinism, so almost anything can emerge spontaneously. However, this is an abuse of the term 'self-organization' disregarding the time scale. We must not use the term 'self-organization' to describe the process whose time scale is much longer than the time scale of the system itself that emerges by the organization process; Crudely put, ontogeny and phylogeny should not be confused. In place of such an abused and loaded word like 'self-organization,' it may be wiser to use the term 'spontaneous organization' in order to describe the ontogeny of the system to disambiguate the discussion.

[13] ⟪**Equal opportunity vs equal outcome**⟫ Emphasizing only equal opportunity is just neglecting that we are complex systems under strong influence of auxiliary conditions (e.g., the initial condition). "There are people who say that only equal opportunity is necessary and that equal outcome is not. I believe this is totally wrong. I believe what is important is equal outcome, within reason, of course.

Why? Because we do not know what kind of human beings we will be born as before our births. The process is akin to rolling the dice over and over again, as it were. First, what will the country of birth be? Japan? The US? Where? Next, what disabilities would I have?, etc.

As a result, frankly, I believe that whether a person eventually becomes a big earner or not is greatly influenced by luck." Thus, the redistribution policy is a deferred insurance against the inevitable uncertainties surrounding our birth and infancy. (by Y. Iida, translated by Shoko Oono) `http://business.nikkeibp.co.jp/article/topics/20100924/216357/?P=3`

Paul Newman said anybody who is successful in life and who doesn't use the word 'luck' is a liar.

[14] ⟪**Western European history is the key to modern science**⟫ T. E. Huff, *Intellectual Curiosity and the Scientific Revolution: A Global Perspective* (Cambridge

5.2 Pasteur process

> I insist that words are totally absent from my mind when I really think ⋯.
> ⋯ the more complicated and difficult a question is, the more we distrust words, ⋯.[15]

We more or less understand what complex systems should not be, and we clearly notice that to call some system unrelated to organisms complex is often inappropriate; confusing complex and pseudocomplex systems hinders the correct appreciation of truly complex systems, leading to an oversimplified worldview and its dire consequences.

However, we have not seen any positive characteristic of complex things other than the existence of their relation to organisms. Therefore, to know more about the characteristic feature of complex systems, let us briefly look at the distinction between machines (artificial systems) and organisms (or things directly related to them).

As an example, let us compare computer language (e.g., C++) and natural language. A sentence in C++ can be mechanically (syntactically) analyzed and interpreted completely, so it is simpler than any natural language. In other words, it can be translated into computer actions algorithmically without paying any attention to its meaning. On the other hand, any attempts to formalize natural language (i.e., to analyze purely syntactically ignoring any meaning at all) have failed.[16] Since considerably grammatically degraded utterances can often be understood, it is almost certain that we do not use some algorithm to parse natural language. There is obviously a fundamental difference between computer language and natural language. Its origin must also be obvious: natural language has emerged based on natural intelligence as its prerequisite,[17] but artificial intelligence comes after artificial language. Artificial language if mechanically compiled can perform various computational

University Press, 2011) demonstrates that during the Renaissance of the twelfth and thirteenth centuries the extraordinary fusion of Greek philosophy, Roman law, and Christian theology gave Europe a new and powerful civilizational coherence. This was the basis of the modern science. That is why there was no science in the true sense of the word in other civilizations.

[15] J. Hadamard, *The Psychology of Invention in the Mathematical Field*, Ibid., p96.

[16] **⟪Introduction to generative linguistics⟫** The basic textbook of Chomsky's generative theory is N. Chomsky, *The Minimalist Program*, (MIT Press, 1995). S. Pinker, *The Language Instinct: how the mind creates language* (Harper Collins, New York, 1994) contains a popular account. S. Pinker, *Words and Rules* (Basic Books, 1999) illustrates how hard the purely syntactic approach is to understand natural languages.

[17] **⟪Intelligence required by linguistic capability⟫** However, if we take into account the phenomenon called 'linguistic idiot savant,' to handle a language does not seem to require the full power of natural intelligence (Notice that apparently meaningful sentences can be generated with a recursive mechanism as can be seen from: `http://www.elsewhere.org/pomo/` or A. C. Bulhak, "On the simulation of postmodernism and mental debility using recursive transition networks," (Monash

operations, but in contradistinction, natural language is terribly incomplete and cannot be used to support any operations in our brain.

In the first place, our language is likely to be very new. It is probably not older than genus *Homo*. It is likely that it is not even older than our own species *Homo sapiens*. Just as civilization is not a characteristic of our species, our language is not a characteristic of our species, either.[18] It is obvious that our fellow members of Hominoidea (apes including us) all have (logical) thinking capability. It is immediately clear, if we get acquainted with dogs and cats, that they really have high intelligence. Obviously, natural intelligence is far older than natural language (it is certain that our brain and our language evolved hand in hand, but this must only be a very recent phenomenon).

The relation between natural language and natural intelligence is parallel to what Pasteur taught us long ago: "Life comes only from life." The crucial point of Pasteur's assertion is that a *complex system* is only made by another complex system, or that if there is a complex system, there is another complex system that produced it (in other words, complex systems lack spontaneous-organizing capability and are not simply emergent[19]). Let us call a process by which a complex system is formed from another complex system a *Pasteur process*. The process through which a child becomes capable of speaking natural language is a good example. That is, grammar does not emerge in a self-organizing fashion as a result of general intellectual development, but is due to the unfolding of something already prepared phylogenetically. What Chomsky affirms is, in short, that in order to pro-

---

University Department of Computer Science, Technical Report 264 (1996)). Perhaps, natural intelligence consists of several semi-independent modules, and the part required by the linguistic capability might be one of them. Therefore, in contrast to the assertion of 'hopeful humanities' according to, e.g., Piaget, there need not be any correlation between the developments of logical capability and linguistic capability.

[18] ⟨⟨**Evolution of humans**⟩⟩ A nice introductory review may be M. Liangand R. Nielsen, "Q&A: Who is *H. sapiens* really, and how do we know?" BMC Biol., **9**, 20 (2011). Anatomically modern humans evolved in Africa recently (180 ka before present), but admixed with endemic archaic hominids—Neanderthals, Denisovans, or even *Homo erectus*—as they spread throughout the world. It is said that the cultural complex including language emerged about 110-120 ka before present (the Mesolithic Revolution; however, there are people who criticize this as groundless).

[19] ⟨⟨**Emergence**⟩⟩ The word 'emergence' is also used without paying attention to the time scale. The original meaning of 'emergence' is, "A property of the system as a whole that is not a simple sum of the properties of the parts," but if it is used disregarding the time scale, it can imply evolution as well; virtually anything is possible. When the emergent property of the universe is considered, the time scale must be of the cosmic scale. In contrast, when the emergence of a nervous system is discussed, it is the time scale of ontogeny, the developmental process; it means, e.g., the formation of the brain by a developmental process, so it is a process of a rather short-time scale. In this book, the word 'emergence' or 'emergent' is restricted to the short-time (i.e., ontogenetic) processes. As is emphasized repeatedly, time scales must be respected.

duce complexity within a short time some complexity is required previously. Chomsky's so-called poverty of stimuli indicates the essence of complex systems. This is closely related to phenomena physics has not studied seriously.

At present, the assertion of Pasteur (1822-1895) "Life comes only from life" is absolutely correct.[20] Therefore, before Darwin,[21] people thought this was always true. Consequently, if we go back along the chain of Pasteur processes (*Pasteur chain*), a very natural guess is that we must hit the primordial complex system—if personified, God. Of course, this is a wrong inference just as: since the sun rose not only today but also yesterday, this must have always been true. The achievement of Darwin (1809-1882) and Wallace (1823-1913) is to point out the (logical or theoretical) possibility that, after a sufficiently long time, complex systems can emerge from non-complex systems.[22] The process that can generate Pasteur chains and can subsequently increase (intuitively speaking) complexity[23] should be appropriately called the *Darwin process*.

In the following, it will be explained why grasping the relationship between the natural intelligence and our language is crucial for 'complex systems' studies. This is related to why this book emphasizes real examples as noted at several places. *The reader may skip the remaining part of this section to go to the next.*

For a digital computer to function, it requires a language, but we can think without language.[24] Before language is used, the important part of deep thoughts (especially in natural sciences) is over. Language is, so to speak, a

[20] ⟪**Spontaneous generation theory: why was this theory so long-lived?**⟫ Notice that the spontaneous generation theory was maintained largely by religious motivations. If the spontaneous generation theory were correct, the problem of the common ancestor could have been avoided. Furthermore, one could say that tapeworms and other 'filthy creatures' did not exist in the Garden of Eden.

[21] ⟪**Before Darwin**⟫ Correctly speaking, this is wrong. C. Darwin's grandfather and Lamarck (1744-1829) asserted evolution as a fact. As to Lamarck, S. J. Gould's *The structure of evolution theory* (Harvard University Press, 2002) Chapter 3 is good.

[22] ⟪**Pasteur and Darwin**⟫ Correctly speaking, historically Darwin is earlier than Pasteur. Pasteur was politically very conservative, and did not have a good feeling about 'dangerous ideas' such as Darwinism. See G. L. Geison, *The private science of Louis Pasteur* (Princeton University Press, 1995).

[23] This 'increase' must be accepted as a fact, if the reader studies the history of the earth.

[24] ⟪**Generative linguistics and natural intelligence**⟫ Natural language is said to have different levels such as the S-structure and the D-structure. If the actual thinking process is regarded as the operations in the D-structure, we must conclude that the thinking process is inseparable from language. However, since the D-structure is understood to be peculiar to human language, we must conclude that the crucial part of thinking has been over before the use of this structure. After the so-called Second Cognitive Revolution Chomsky tries to regulate the D-structure through establishing the principles, but if there are principles at all, they must be the principles of natural

mere I/O interface.[25] Digital computers cannot have 'thoughts' that cannot be expressed in their language. When we feel we are thinking with the aid of language, what is actually going on is that we are making efforts to encode what we have already thought out into language, or, in other words, that we are trying to come up with nice linguistic expressions.[26] The mathematics that is a record of observation looks much deeper than the mathematics that is a record of (conscious) linguistic thinking process.[27] Deep thinkers and sensitive poets must have been troubled by the limitation of language expressions, but this trouble is alien to computers.

What we can think and know is much richer than what we can unambiguously express in language. To recognize this limitation of language is fundamentally important in the study of 'complex systems.' It may well be the case that we recognize the complexity of the biological world to a considerable extent, but cannot completely express it in language.[28] This is the biggest reason why in this book discussions are all based on examples and why we try to look at the world phenomenologically starting with examples. However, our ancestors in the Mesolithic era were awakened to the potential of natural language; it almost overwhelmed them with awe. They became oblivious to its creator, the more powerful natural intelligence behind it,

---

intelligence prior to language. We must recognize a fundamental problem in that we can rarely spot the term 'natural intelligence' in the books of generative linguistics.

[25] ⟪**Language and thoughts**⟫ Many readers must feel that there is a reasonable amount of truth in the proposition that the thinking process using language is important. When we calculate consciously, we just feel this is true, and accurate numbers may not be handled without the aid of language [P. Pica, I. C. Lemer, V. Izard, and S. Dehaene, "Exact and approximate arithmetic in an amazonian indigenous group," Science **306**, 499 (2004)]. However, even in such cases who finally judges that the result is right or that the answer is it? The author feels the judge is more fundamental than language. We might conclude that those who cannot do mathematics without language (without linguistic help) are intrinsically the people who cannot do mathematics.

[26] There can be feedback from I/O. That is, pursuit of ease of expressibility could modify the thoughts.

[27] ⟪**The depth of observation records**⟫ "I believe that mathematical reality lies outside us, that our function is to discover or *observe* it, and that the theorems which we prove, and which we describe grandiloquently as our 'creation', are simply our notes of our observations." (G. H. Hardy, *A mathematician's apology* (Cambridge University Press, 1940) 22. "According to my observation, just as physics describes physical phenomena, mathematics describes the existing mathematical phenomena, so to understand mathematics it is crucial to grasp the intuitive picture of mathematics directly appealing to our senses." (K. Kodaira, *Introduction to Analysis* (Iwanami, 1976) Preface p1). See also *Kurt Gödel Collected Works* (Oxford University Press, 1995) vol. 3, Gödel *1951 and its introduction by G. Boolos.

[28] However, whether something cannot be expressed in language or not can be recognized only after a sizable amount of efforts to express it in language. We should not easily conclude that it cannot be expressed in words and neglect efforts as might be seen in some philosophy books. Therefore, it is unproductive to ignore things we cannot express at present unambiguously in words as unscientific.

and, even worse, became oblivious to their biological bodies supporting it. Probably, we have not yet recovered from this shock (remembered as Paradise Lost; we became even ashamed of our biological bodies). Consequently, intelligence has been unfortunately equated with linguistic ability. This serious error was even sanctified at the beginning of the New Testament book of John.[29] A portion of humanities may still be under the spell of this 'ancient curse.'[30]

5.3 Fundamental conditions

Complex systems are closely related to meaning and value, so the core part of the study of complex systems would be fundamental biology. To what extent is fundamental biology related to what physics, which is supposedly the fundamental science, has not been doing?

The Pasteur process teaches us that a complex system must not be self-organizing, although its parts may well be spontaneously organizing.

Take a ribosome, the translator of a mRNA into a protein chain, as an example. This molecule consists of numerous RNA's and proteins. To reconstitute it from these parts *in vitro*, even under an ideal condition without any protein and other cell components which are not the parts of a ribosome, two stages controlled externally are needed. Furthermore, the process requires

---

[29] ⟨⟨***Logos***⟩⟩ Even though the language of the New Testament is Greek, its philosophy is not Greek philosophy, so we must look at the original Hebrew word *davar* translated as *logos*. According to T. Izutsu (*Consciousness and essence* (Iwanami, 1991)), "In Hebrew *davar* clearly has two meanings: thing and word. That is, in the depth of consciousness, those who have Hebrew as the mother tongue word and thing are intrinsically one." The beginning of John merely reflects this.

To recognize the inherent difficulty in such a mode of thinking can be an important contribution to human culture.

[30] ⟨⟨**Is knowledge of the humanities deep?**⟩⟩ It is sometimes said that knowledge due to the humanities is deeper than that due to natural sciences, but is this really true? What does 'deep' mean?

We must clearly recognize the narrow-mindedness of the idea that regards natural sciences as a local form of intellectual activity: a particular form of intellectual activity born in the 17th century Europe that tries to understand the world in terms of 'rational and analytical logos.' The question of what science is does not have (just as such questions as what rigor is) any definitive answer, but at least what is important in its core are the humility that constantly makes us reflect on whether we really know and whether our methodology and logic are correct, and the resultant skepticism and the mindset to pursue impartiality with respect to metaphysics. Also, the skepticism applied to itself is crucial: to cut the chain of skepticism off at appropriate positions and 'to experiment.' (Isn't mutation that provides the substance for evolution skepticism itself on the status quo?) This must be cosmically universal beyond us human beings wherever there are intelligence and conscience. [In this fundamental respect, there cannot be masculine science or feminine (or feminism) science.]

unrealistically high temperature for cells.[31] The necessity of high temperature *in vitro* suggests that the rate-determining steps are to resolve wrong conformations of, e.g., RNA's. In the actual cell, the process requires a lot of chaperones (see footnote 111 in this chapter). It is not a process to mix all the components at once. The process *in vivo* has to overcome problems due to the presence of numerous other molecules in contradistinction to the *in vitro* reconstitution.[32] Even if self-organization (spontaneous construction) is possible as a centriole, its spontaneity is suppressed by the cell to regulate the totality of the cell.[33] We see that complex systems are made of parts with closely regulated spontaneity. That is, the key to the organization of a complex system lies in when, where, and how the potential spontaneity of the parts should be exploited. In other words, the key elements lie in the factors that cannot be spontaneously determined. If an organism has a self-organizing property (of not only each parts but also of the whole), then the organism would never die (or can be resurrected relatively easily). When we consider complexity, "Memento mori" shows us an important research direction. Those who do not die have never lived; those who can be apparently resurrected have never lived. We must not forget that even the cell wall of bacteria does not spontaneously constitute itself.

To distinguish complex systems from not complex systems, let us compare

---

[31] ⟪**Reconstitution of ribosomes**⟫ All the r-proteins and r-RNAs are mixed at once, but the temperature and magnesium ion concentration must be changed in two stages. See K. H. Nierhaus, "The assembly of prokaryotic ribosomes," Biochimie **73**, 739 (1991).

[32] ⟪**Ribosome assembly *in vivo***⟫ The assembly *in vivo* is a much more complicated process. In *Saccharomyces* more than 180 assembly factors and 100 small nucleolar ribonucleoproteins (snoRNPs) associate with pre-rRNA to catalyze ribosome assembly [cf. J. P. Staley, and J. L. Woolford Jr, "Assembly of ribosomes and spliceosomes: complex ribonucleoprotein machines," Comm. Cell Biol., **21**, 109 (2009)]. After being pre-assembled in the nucleus and exported to the cytoplasm, ribosomes undergo final maturation. This step involves the release of trans-acting shuttling factors, transport factors, incorporation of the remaining ribosomal proteins, and final rRNA processing steps [cf. V. G. Panse and A. W. Johnson, "Maturation of eukaryotic ribosomes: acquisition of functionality," Trends Biochem Sci., **35**, 260 (2010)]. Throughout the process of molecular machines, assembly chaperones are crucial [cf. A. Chari and U. Fischer, "Cellular strategies for the assembly of molecular machines," Trends Biochem. Sci., **35**, 676 (2010)]. The assembly in *Escherichia coli* is not simple, either. Ribosome biogenesis is assisted by assembly factors such as helicases, GTPases, and other proteins [A. E. Bunner, A. H. Beck and J. R. Williamson, "Kinetic cooperativity in *Escherichia coli* 30S ribosomal subunit reconstitution reveals additional complexity in the assembly landscape," Proc. Natl. Acad. Sci., **107**, 5417 (2010); cf. M. T. Sykes and J. R. Williamson, "A complex assembly landscape of ribosomal subunit," Ann. Rev. Biophys., **38**, 197 (2009)].

[33] A. Rodrigues-Martins, M. Riparbelli, G. Callaini, D. M. Glover, and M. Bettencourt-Dias, "Revisiting the role of the mother centriole in centriole biogenesis," Science, **316**, 1046 (2007). We could say that cancer is a failure in controlling various spontaneities, that are normally suppressed to construct a complex system.

these systems from the viewpoint of the 'thoroughgoing' reductionist.[34] As soon as we learn elementary quantum mechanics, it is easy for us to write down the equation of motion for a collection of particles not moving too fast (compared with light).[35] We have only to write down the Schrödinger equation (see Note 1.1) for a collection of particles interacting with the Coulomb forces. Thanks to the superposition principle (Chapter 1), this is 'easy.' The equation for a droplet of salt water containing a small salt crystal and the equation for a droplet of water containing a single cell of *Escherichia coli* are not much different. They take the following form:

$$i\hbar\frac{\partial\psi}{\partial t} = \left[-\sum_i \frac{\hbar^2}{2m_i}\frac{\partial^2}{\partial \boldsymbol{r}_i^2} + \sum_{i>j}\frac{q_i q_j}{4\pi\varepsilon_0|\boldsymbol{r}_i - \boldsymbol{r}_j|}\right]\psi, \tag{5.1}$$

where m_i is the mass of the i-th particle (nucleus or electron), q_i its charge, and $\boldsymbol{r}_i$ its position vector (in the three-dimensional space). The sum is over all the particles. The wave function ψ is a function of the position vectors $\{\boldsymbol{r}_i\}$ of all the particles and time t. We can 'write down' such an equation, if we perform elemental analysis.

No one can expect that we will be able to solve the equation numerically in the future, but we can do 'analogue computation.' That is, we have only to perform actual experiments (Fig. 5.1). Such an experiment is an ultramodern

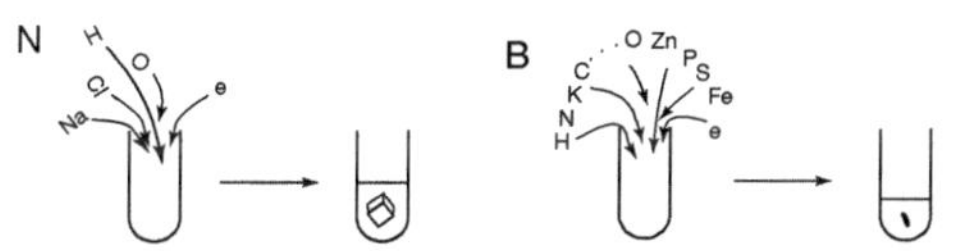

Fig. 5.1 Comparison of quantum mechanical computations for simple systems and complex systems. In a Dewar jar are appropriate numbers of various nuclei and electrons to maintain electrical neutrality. The boundary condition may be a homogeneous Dirichlet condition. With a reasonable initial energy almost immediately a salt crystal would be formed in N, but in B nothing interesting would happen even after a year.

version of Pasteur's famous experiments. We will find a fundamental difference between the salt-water system N and the system B containing a cell of *E. coli.* What is the key difference? Let us perform experiment B, isolating

[34] There seem to be several kinds of reductionism, but here, we adopt the ontological reductionism that everything we observe daily is made of atoms and molecules obeying quantum mechanics (and electrodynamics). The following part was taken from Y. Oono, "Bio-physics manifesto—for the future of physics and biology," in *Physics of Self-organization Systems* (edited by S. Ishiwata and Y. Matsunaga, World Scientific, 2008) p3.

[35] For a system around room temperatures, the condition is satisfied if we do not have nuclei with too large charges (i.e., $Z < 100$).

the whole system from the external world (i.e., under a homogeneous Dirichlet condition); in a Dewar jar we put specified number of nuclei and electrons and adjust the total energy appropriately, and then leave the system alone.[36] Unless the experimenter is extremely lucky, she will never get the cell (will never observe the cell with a "reasonably accessible" probability) within a year. Since it is unthinkable that a sonicated cell can return to its original normal state within a year, it is not at all surprising that (5.1) cannot produce a cell of *E. coli*. However, in contradistinction, a similar experiment for the salt-water system N must produce a floating crystal almost immediately. If we are shown such a system, it is very natural that we wish to conclude that the equation of motion (the Schrödinger equation) is everything for the understanding of the system.

Discussion 5.2. Among anti-reductionists there are people who claim that the impossibility of numerical simulation of a real macroscopic system with a digital computer is evidence that reductionism is untenable. How does the reader counter? □

What is the fundamental difference between systems N and B? The equations are partial differential equations, so we need initial and boundary conditions, the so-called auxiliary conditions. In the present case, the boundary condition is not so important, as we have already seen. For the salt-water case N, almost any initial condition with the right energy should give a thermodynamically natural state before long; this is the reason why equilibrium statistical mechanics works even though its theoretical framework is crude. In contrast, if we wish to produce a cell of *E. coli* within, say, a day, an exquisitely fine-tuned initial condition must be prepared. For the salt-water system N, to know the equation of motion is a much harder problem than to specify a 'correct' initial condition. In contrast, in the *E. coli* case B, to know an appropriate initial condition is much harder than to know the equation of motion.[37]

The reader might think that a very complex initial condition could be imposed on system N as well to make a minuscule salt tablet inscribed with a masterpiece literature (or the genome sequence of a certain strain of *E. coli*) or to make a detailed 3D sculpture of an *E. coli* cell microstructure. However, such nanoscopically sculpted structures floating in saturated aqueous salt solution would erode with thermal fluctuations very quickly before we could observe them on our time scale (say, one second).

Now, we know a distinction between a truly complex system and a not so

[36] The reader might say an organism cannot exist in an isolated system, but it must be possible to confine a live cell of *E. coli* for a minute in an isolated vessel. The equation of motion of such a system is (5.1), so under a homogeneous Dirichlet condition, (5.1) must have a solution describing such a system.

[37] Even though the initial condition is nontrivial, the reader may say that fundamental laws of physics are still crucial. Of course, they are. If the digestive canal of Shakespeare did not function well, no masterpiece could have ever been produced.

complex system: complex systems require a lot of elaborate auxiliary conditions[38] and can retain the information in them. If a system is worth imposing a set of initial conditions with a large amount of information, the system must be a sort of memory device, because imposed information must be retained against ubiquitous noise. For example, take a 2D nonconserved CDS model (4.6) and impose the monochromatic photocopy of *A Girl with a Pearl Earring*. The masterpiece would soon disappear. A CDS model (with a reasonable diffusion constant) has almost no capability to retain patterns, so valuable intricate initial condition cannot be exploited to make another masterpiece.

Thus, to make a complex system we need a lot of auxiliary conditions and materials that can retain/utilize the former. As the reader has noticed, 'complexity' (discussed in this book) is a feature of an ensemble of microstates characterized by the properties observable on our (or at least mesoscopic) space-time scale (i.e., the equivalence class of mechanical microscopic states due to some at least mesoscopically observable characteristics; especially in the case of eukaryotes). Therefore, the auxiliary conditions must contain a lot of information as to structures and materials. The system must have a certain size which is big enough to have 'collective variables' that resist noise with the aid of the law of large numbers. Such variables are formed when the system symmetry is broken. We have already glanced at a mechanism to break the system symmetry through collective interactions of microscopic entities. This is essentially related to equilibrium phase transitions (see Fig. 1.2). Symmetry breaking could occur purely without microscopic structural changes as is known, e.g., in the so-called Turing mechanisms found in reaction-diffusion systems. However, the resultant patterns are very often fragile without supporting microscopic structural changes. Thus, complex systems need special substances that respond microscopically collectively to the external influences (by, e.g., changing the molecular structure allosterically, making molecular aggregates as can often be found in surface receptors, etc.).[39] In short, complex systems need collective behaviors of microscopic entities to produce mesoscale variables. This is the very reason why self-organization is often relevant to

---

[38] **⟨⟨Extensive auxiliary conditions⟩⟩** However, we cannot simply say any system requiring a large amount of auxiliary conditions is complex. It is often said that the weather system is the most complicated system, but in this case the required amount of conditions is proportional to the system size. If the amount of required conditions increases extensively, we cannot say the system is really complex. However, here, we are not directly interested in defining 'complexity,' but in a simpler problem: finding features that are shared by all genuinely complex systems.

[39] Allosteric conformation changes and folding of proteins and the ordinary phase transitions are not fundamentally different. Of course, a single or a few protein molecules cannot make really macroscopic systems, so the mathematical singularity we see in thermodynamic phase transitions cannot be found, but conceptually they can be understood as phase transitions in finite (small) systems. For example, the analogy between the first-order phase transition and globular protein folding was clearly recognized by N. Gō long ago. Read his memoir: "Physics and biology of protein folding," Prog. Theor. Phys. Supp. **170**, 198 (2007).

complex systems; self-organization is important not because it can spontaneously create structures and order, but because collective variables emerge due to symmetry breaking whose actual values may be specified to retain the information in the initial condition.[40] Notice that even a simple on-off behavior cannot be spontaneously specified; the on or the off state is selected by a reason not in the microscopic and materials-scientific structure of the system.

In summary:
(i) A complex system must have a sort of stability to retain the information in auxiliary conditions.
(ii) This stability requires space-time slow varying variables which are robust against noises. Compare the time scales of atoms and molecules (0.1 fs is the order of time steps in molecular dynamic simulations) and of the cell reaction to stress (very rapid reaction may be on the time scale of order 0.1 s).
(iii) These slow variables are mainly produced by collective interactions of microscopic entities that are often (materialistically) special, requiring specific interactions.

Let us repeat the crucial point. Self-organization is important because it produces collective variables whose actual values are *not* prespecified.[41] A complex system is something like a painting created with numerous meso- and macro-variables prepared often by self-organization or at least by collective behaviors of microscopic entities to withstand noise (ultimately) with the aid of the law of large numbers. To this end a complex system needs a special set of materials that sustain collective behaviors/symmetry breaking and a set of auxiliary conditions that specify resultant collective variables.

---

[40] Since there is apparently no difference between the Schrödinger equations for the salt water and for *E. coli*, the reader may well question the meaningfulness of the argument here. The point is, in the case of the Schrödinger equation for *E. coli*, there is a way to impose the initial condition that can persist against 'noise.' That is, the initial condition is required to construct memories (and materials supporting them) as well. This is the reason, as the reader will see shortly, why the fundamental conditions must specify the materials.

[41] **⟨⟨Typical misunderstanding of the significance of self-organization⟩⟩** There is a well-accepted misunderstanding exemplified by the following quotation from S. Levine, *Fragile Dominion—Complexity and the Commons* (Perseus Publishing, 1999) p12: "Self-organizing systems have been the fascination of scientists from a diversity of disciplines because the concept of *self-organization* provides a unifying principle that allows us to provide order to an otherwise overwhelming array of diverse phenomena and structures. By self-organization I mean simply that not all the details, or "instructions" are specified in the *development* of a complex system." The author of this quoted book, who stresses that not all the details need be specified, does not recognize why self-organization is crucial for complex systems. The stress must be placed at the emergence of parameters (variables) that must be (or can be) specified meso- or macroscopically *extrinsically*. The significance of self-organization is in the production of variables that must be specified. That is, the significance of self-organization to complex systems is in preparation of canvasses and paints and not in actually creating paintings.

These sets cannot be invented easily, so our existence requires our parents.

Let us look back on what physics has been doing. The kernel of the Newtonian Revolution, which was the springboard of modern science, was the recognition that a phenomenon can be dissected into fundamental laws (e.g., equations of motion) and (contingent) auxiliary conditions:

$$\text{phenomenon} = \text{auxiliary conditions} + \text{fundamental laws}.$$

Physicists have traditionally been studying systems for which the fundamental laws are much more nontrivial than auxiliary conditions (like system N). There are a lot of such phenomena, and to understand them only slightly allowed us to do many practically useful things. However, the comparison above clearly tells us that many systems in the world are not like them.

If we pursue further understanding of the equation of motion = fundamental laws, we would proceed to the direction of complete reductionism[42] (Fig. 5.2). This is the path the so-called fundamental physics followed after Newton. This has made physics a discipline to study matter, field, and sym-

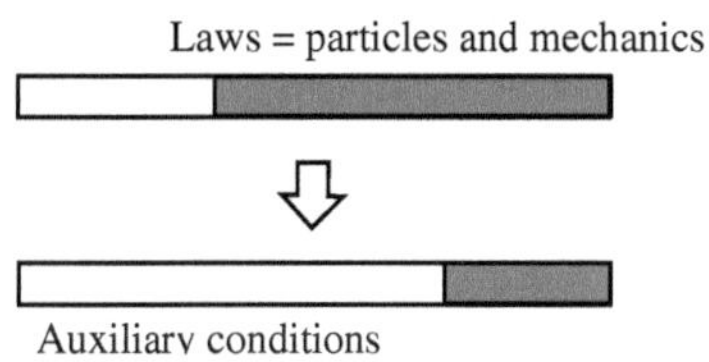

Fig. 5.2 The history of physics is the history of the pursuit of universal fundamental laws. It is a history of removing contingent chaff called auxiliary conditions such as initial-boundary conditions.

metry breaking. Condensed matter physics is no exception; it is the study of collective variables formed by symmetry breaking.[43] We should reflect on whether this really takes us to a deeper understanding of the world. Weinberg asked why the world consisted of three fields of quarks, light, and electrons,[44] and said that the answer to this question led us to the fundamental understanding of the world. However, we might say that the fundamental laws become this simple merely because we have restricted our questions. The question "what are inessential details?" must intrinsically be a deep question, but physics has automatically regarded everything in the auxiliary conditions inessential. Therefore, these conditions have been called *auxiliary* conditions.

---

42 Here, the word 'reductionism' is used in its usual sense: molecules can be dissected into atoms, atoms into elementary particles, etc. Building blocks are further dissected into smaller building blocks and "to understand the laws governing smaller blocks amounts to understanding the world more deeply."

43 See, e.g., P. W. Anderson, *Concepts in Solids: Lectures on the Theory of Solids* (World Scientific, 1998).

44 S. Weinberg, *Dream of a Final Theory, The Scientist's Search for the Ultimate Laws of Nature* (Vintage, 1992) p25.

We may analogize the ultimate fundamental laws with the universal Turing machine (Section 2.10). It can compute any computable function (Section 2.9), but to understand a function, understanding the universal Turing machine is of almost no use, because almost everything is written on the tape as the input. There are many interesting questions such as "what is a computable function?", "how much is the computational cost?", etc., but these questions are not answerable even if we understand universal Turing machines.

Discussion 5.3. Read L. Wolfenstein, 'Lessons from Kepler and the theory of everything," Proc. Nat. Acad. Sci. **100**, 5001 (2003) and summarize the comparison between the 'emergentist' and the reductionist. □

If there is another revolution in science, isn't it from the analysis of what has not had any light shed on it after the Newtonian Revolution? The author does not say that the current direction of fundamental studies has reached a dead end. However, phenomena that cannot be understood with the currently accepted fundamental laws will not become understandable through further 'purification' of fundamental laws. Our frontier, then, must be the earth left intact after the frontier of physics in the ordinary sense has zipped through like wildfire. *Complex systems* are the systems for which auxiliary conditions are crucial. Therefore, to cultivate this earth is the study of complex systems.[45]

Auxiliary conditions are our main research objects, but not all the auxiliary conditions are important. For example, auxiliary conditions that specify the values of collective variables are the ones we must pay attention to, so such important auxiliary conditions will be called *fundamental conditions*. We wish to understand the world in terms of fundamental laws and fundamental conditions.

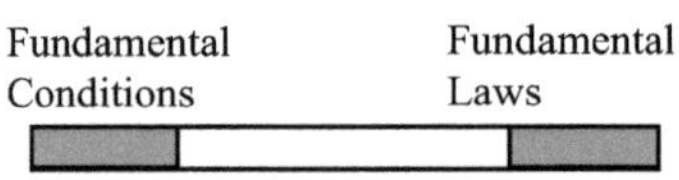

Fig. 5.3 Among auxiliary conditions that are squeezed out from the fundamental laws is an important core. Reversing the conventional point of view, we may even say that the fundamental laws are the auxiliary elements required to utilize (or unfold) fundamental conditions.

As is noted above, collective behaviors must be sustained often by special materials, so an auxiliary condition is often accompanied by a particular material system. Therefore, specification of required materials (or supplying

[45] The following paper may point in the same direction: G. J. Vermeij, "Historical contingency and the purported uniqueness of evolutionary innovations," Proc. Nat. Acad. Sci. **103**, 1804 (2006). The paper says, "Details of initial conditions, evolutionary pathways, phenotypes, and timing are contingent, but important ecological, functional, and directional aspects of the history of life are replicable and predictable." (perhaps, a bit exaggerated)

them) must also be included in the auxiliary conditions, so at least some of them must be counted among fundamental conditions.

5.4 What do 'fundamental conditions' imply?

Complex systems (within the time scale of individual systems) do not emerge spontaneously.[46] This implies that a complex system requires a special set of prerequisite conditions for it to be generated within a short time.[47] We have decided to call the important core of the conditions 'fundamental conditions.' The reader who has at least browsed through this book up to this point would realize that our tentative goal must be the conceptual analysis and phenomenology of fundamental conditions. The preceding chapters are preparatory steps.

As the reader correctly guesses, ideally, the book should be a guidebook to the land of complexity (or the guidebook to the land of organisms for theoretical physicists), but the guide himself has not yet reached there. What he could do best is to give a rough and incomplete description of the land seen from a distance.

Since complex systems 'die' and are not resurrected, the fundamental conditions cannot be constructed within a short time.

Fundamental conditions may be classified into two major categories, structural (SF) and instructional (IF).[48] We must first take note that in this world any information must be stored in or carried by a certain structure or field (the principle of no-ghost). Consequently, even fundamental conditions must be stored in objects made of materials (and their combinations). Furthermore, to execute instructions (e.g., to realize phase transitions/bifurcations) special materialistic systems are desirable. Thus, one category of fundamental conditions is structural fundamental conditions. For organisms their spatial structure is quite important even for Bacteria. If information is abstract (as described by a symbol sequence), it can never specify a structure in 3-space. As has been remarked repeatedly, we need an adaptor to connect these two different categories; if we wish to dispense with adaptors, then structures themselves must be specified by structures. For example, if we wish to specify the orientation of the crystal lattice of a forming crystal, the lattice direction must be specified physically. This information may be provided by a crystal

---

[46] As already stated, the ontogeny of an individual member and the genesis and evolution (phylogeny) of the ensemble (a group, a taxonomic unit, etc.) must be distinguished. The evolution process is a very long (time-consuming) massively parallel computation process.

[47] i.e., a much shorter time scale than that of evolution (phylogeny)

[48] There must be materials-related fundamental conditions as noted above, but they will not be emphasized or be mixed with SF in this book.

nucleus with its actual lattice orientation. Examples of SF in organisms include the cell wall of bacteria, some cellular organelles as centrioles that can function as nuclei of the same structure, etc.

The other category is instructional fundamental conditions. Think of a key to turn on a switch. It must be made of a certain material and must have a definite structure, since there is no ghost in the world. However, the particular structure/material itself is dissociated from its function; we can replace the lock and the corresponding key on the same switch with a different set without affecting the function of the switch. The fundamental conditions dissociable from the structure of their carriers are IF. For example, the information as to the condition when to transcribe a portion of DNA is regulated by transcription factors and numerous cofactors,[49] but (just as in the case of the key and the lock) their structures are used as a sort of symbols. Numerous small RNAs[50] are also examples. In the following, 'structural fundamental (SF) condition' may be abbreviated as 'structural condition,' and 'instructional fundamental (IF) condition' as 'instructional condition.'

Structural conditions cannot construct a large structure directly by themselves. The basic usage (functional mode) of these conditions is to make copies of the structures (as templates, stamps, etc., do), so (unless some repetition is utilized as crystals) basically they can make systems of the same size as the carrier of SF. Therefore, if only molecular motions are used to move and deform structures, no macroscopic systems can be constructed.

A complex system is made of many parts, so without spatially localizing them no complex system is formed. Therefore, if they are not united as a single macromolecule or a rather stable molecular complex, in order to utilize the structural condition effectively, there must be a structure that can demarcate its own components from the rest at least to some extent. The demarcated space must be of the size that its components can move around and interact with each other with the aid of Brownian motion. This must essentially be the significance of the cell. One of the two pillars of modern biology is *cell theory*: all organisms are made of cell(s) and their products (the other pillar is Darwinism). The universal significance of the cell may be understood by the fact that any system that requires structural conditions must be a spatially confined system. We may say that the cell is the largest domain where structural conditions can be utilized effectively.

---

49 ⟪**Transcription factors**⟫ Near the portion of a coding DNA is a sequence called the promoter that determines whether the part is transcribed to RNA or not. This decision is made by various molecular factors binding to the promoter and nearby (sometimes not so nearby) sequences called enhancers. The major attaching molecular factors are transcription factors and their cofactors.

50 ⟪**small RNAs**⟫ In this century it has gradually been understood that many kinds of small (20-30 base) RNAs play significant regulatory roles. For example, animal microRNAs often enhance degradation of mRNAs through binding to its 3' end. H. Grosshans and W. Filipowicz, "The expanding world of small RNAs," Nature **451**, 414 (2008) is a good introduction.

To make a larger system, one can make the cell (or the unit) larger, or prepare many of them. To make the unit larger, we must organize molecular motions in terms of nonequilibrium means such as molecular motors. Furthermore, we need instructional conditions that specify where particular structural conditions are used (as the delivery code of proteins). Thus, it is natural to classify complex systems into two large classes according to the extent of the use of IF. This fundamental distinction corresponds to the distinction between prokaryotes and eukaryotes.[51] This dichotomy need not correspond to phylogenetic taxonomic conclusions.

Even eukaryotic cells cannot be indefinitely large, so in order to make larger systems, multiple cells must be used. Although there is no proof that this is the only way to make a larger system, multicellularity must be the most natural way to make larger systems. Multicellular systems must use instructional conditions to organize the system as a whole, because no structural conditions can be impose beyond the scale of cells. In other words, megaeukaryotes like us are the 'fruits' of instructional conditions. The structure beyond the scale of cells is not inherited from its predecessor as structural conditions. When a structure cannot be communicated directly,[52] there is no other way but to communicate the set of instructional conditions that specify the symmetry-lowering modes of the parts (notice that differentiation, e.g., tissue specification, is a symmetry-breaking process). Note that on larger scales (spontaneous) symmetry-breaking processes become more crucial than on smaller scales as tissue specifications exemplify. Therefore, instructional conditions are the key to complexity. In the usual organisms, this is called the genetic control of the developmental process. The non-coding part of the genome plays a crucial role. Thus, the fact pointed out by Mattick et al. (Fig. 5.4[53]) seems to be a logical consequence of multicellular organisms, and

---

[51] ⟪**Eukaryota and the rest**⟫ The residents of the biological world can be divided into Eukaryota and the rest, Archaea and Bacteria, collectively called prokaryotes. Eukaryota have cells with nuclei, but prokaryotes have no nuclei (but see the footnote 102 about the PVC phyla). The representative size of the eukaryal cells is 10 μm, and that of the prokaryal cells is about one order as small. The total length of eukaryotic DNA is of order 1m, but that for prokaryotic DNA is the order of 1mm.

[52] ⟪**Speed and structural conditions**⟫ If one wishes to maximize reproductive success, reducing instructional conditions must be effective. Bacteria seem to have evolved in this direction; they are the ultimate Darwinian machines. P. Forterre says, "I suspect that Archaea and Eukarya are the only two lineages that survive the extraordinary success of bacteria." (in reviewer's report 1 to R. E. Valas and P. E. Bourne, "The origin of a derived superkingdom: how a gram-positive bacterium crossed the desert to become an archaeon," Biology Direct **6**, 16 (2011))

[53] The size of the DNA (the so-called C-number of the genome) varies by one order easily within a taxonomic group (see, e.g., Fig. 1 A of B. S. Gaut and J. Ross-Ibarra, "Selection on major components of angiosperm genomes," Science **320**, 484 (2008)). However, the C-number expansion is largely due to polyploidy, so the 'noncoding ratio' does not seem to change very much as pointed out in the original paper quoted here. If one wishes to avoid this potential C-number problem completely, one could choose the smallest C-number organism as the representative of each taxonomic

complexity.[54]

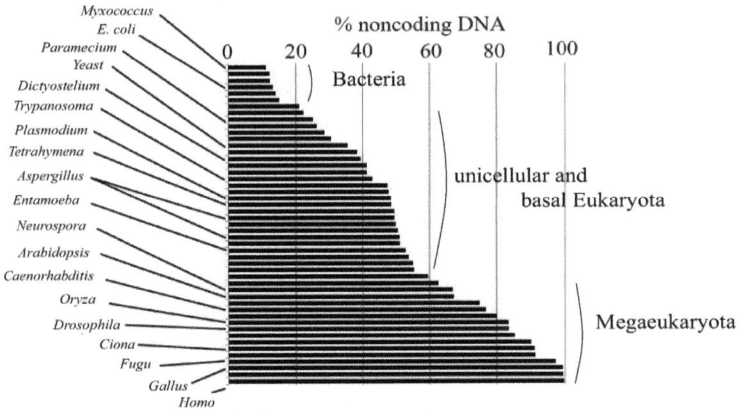

Fig. 5.4 The relation between the common-sense complexity of organisms (i.e., the so-called higher/lower organisms) and the relative amount of the non-coding DNA in the genome. Based on Figure 1 of R. J. Taft, M. Pheasant, and J. S. Mattick, "The relationship between non-protein-coding DNA and eukaryotic complexity," BioEssays **29**, 288 (2007) [with kind permission of Professor Mattick]. The common recognition or the notion of higher/lower organisms is almost justified. Furthermore, the figure might hint at the possibility of natural quantification of 'complexity.'

That the developmental process consists of a cascade of symmetry-breaking processes is what Waddington (1905-1975) affirmed long ago. His figure (Fig. 5.5 left) is an expression of the mathematical essence of the system that requires fundamental conditions;[55] which branching valleys should be chosen is instructed by the fundamental conditions inherited from the parent system(s). Incidentally, Figure 5.5 right also expresses a deep idea about the relation between genes and the mechanism that determines the cascade conditions.[56]

---

group. Such organisms as *Fugu*, *Arabidopsis*, etc., are included in this figure. Besides, the organisms whose genomes have so far been read are mostly near the small end of the C-number scale in their own taxonomic groups.

[54] Mattick's philosophy may be found in J. S. Mattick, "Deconstructing the dogma, a new view of the evolution and genetic programming of complex organisms," Ann. N.Y. Acad. Sci. **1178**, 29 (2009). An update of the paper cited in Fig. 5.4 is the following review article: J. S. Mattick, "The central role of RNA in human development and cognition," FEBS Letters **585**, 1600 (2011).

[55] C. H. Waddington, *The Strategy of the Gene* (Allen and Unwin, London, 1957).

[56] The figure may suggest rampant pleiotropy and polygenic traits (i.e., many-gene effect, in short). According to our general discussion in the last section of this chapter, organisms must have somehow avoided complicated many-gene effects, a hard combinatorial question. To ask the question as to how they accomplish this must be a better research strategy than to ask how genes entangle to produce many-gene effects.

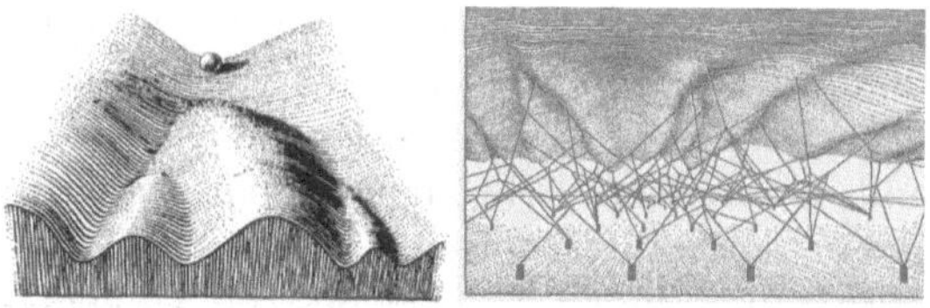

Fig. 5.5 Left: the *epigenetic landscape*, a metaphor of the developmental process. The developmental process is illustrated as the very cascade of symmetry-breaking processes. Right: how this landscape is supported by the genome: pegs are genes and the ropes their effects. This metaphor also exhibits a deep idea. We wish to be thinkers like Waddington.

Most people must believe that organisms are the outcome of evolution, so they must believe that the Darwin process is indispensable for the existence of the biological systems. Can we demonstrate this belief? Such a question must be of almost no interest to biologists: obviously, the Darwin process is definitely required. However, fundamental scientists (who are especially mathophiles) would think it would be nice if this conclusion could be deduced from a set of natural premises. The premises are that formation (ontogeny) of a complex system requires a lot of prerequisite information (i.e., FC), but the needed fundamental conditions cannot be formed *de novo* within a short time, and that there is omnipresent noise that destroys any information sooner or later. The latter is a logical consequence of the nonlinear nature of the world; due to scale interference the world of organisms is not closed informationally (Chapter 1). Thus, fundamental conditions decay inevitably due to noise. This means that the continued existence of a complex system is threatened during the geological time span. If, still, complex systems continue to exist in the world after a very long time, there must be a certain means to restore or repair corrupted fundamental conditions.

Is it possible to repair a damaged set of fundamental conditions within a short time? If there remains an intact set, this may be possible, but, if there is no way to tell whether another set is intact or not, we cannot use it to restore the damaged set. One might still expect that the majority of the sets is intact, so the majority rule could indicate the intact set of conditions. In this case, however, we must accept that the God point of view to survey the whole world at once is prohibited to any real system. Consequently, the law of large numbers cannot be fully exploited, and sooner or later, the noise will win; that is, eventually we will mistake a noise-damaged set for an uncorrupted set.

This implies that the comparison method eventually fails to restore the original fundamental condition set. Besides, we must take the comparison costs seriously,[57] since, after all, the 'purpose' of the fundamental conditions is to make complex systems quickly. Since the role of the fundamental condi-

[57] The reader might recall the *word problem for semigroups*, which is undecidable.

tion is to generate a complex system in a short time, the condition need not be exactly maintained. If the condition can produce a complex system, that is enough. Possibly-noise-corrupted FC are used to construct complex systems and then defective systems would be eliminated through comparison. After comparison, if we do not repair but destroy the selected systems, it is just competition for survival (even if there is a judge who decides the winners).

In short, selection due to competition must be one of the keys to rescue adequate fundamental conditions,[58] and is the secret to 'preserve' complex systems through time. Thus, the functioning of Darwinism must be deducible from the continuous persistence of complex systems and the existence of noise. Even if complex systems might have been constructed by 'Intelligent Design,' as long as the design exists within this real world (i.e., it is stored in the configuration of matter and field), it cannot be preserved without Darwinism (if one wishes to avoid God's incessant intervention = lawlessness of the world).

Darwinism, though absolutely necessary, is not sufficient for the existence of complex systems. Darwinism does not guarantee the complexification in the common sense of this word. This is obvious from Spiegelman's experiment:[59] if the world is Paradise, those who invest all of their resources and capabilities to reproduction win. Thus, whether complex systems emerge or not (even whether FC can be preserved or not) depends on the nature of the world. The same applies to the genesis of intelligence. If there is no lawfulness at all in the world (i.e., it is information theoretically incompressible, see Sections 2.8 and 2.11), intelligence is not only useless but even harmful, because it is a waste of resources. Only in a world that allows both the scale filled with noise and the scale with some lawfulness (i.e., a world diverse enough but one that allows its phenomenological understanding) can complex systems emerge and be maintained. Generally speaking, it is a very interesting question to what degree of complexity is possible in a given world.[60]

---

[58] Cf., "The true basis of his (= the individualist's) argument is that nobody can know *who* knows best and that the only way by which we can find out is through a social process in which everybody is allowed to try and see what he can do." [F. A. Hayek, *Individualism and Economic Order* (University of Chicago Press, 1948) I. Individualism: True and False p15] Notice that this assertion coupled with the idea in footnote 13 is not contradictory to socialism.

[59] ⟨⟨**Spiegelman's experiment**⟩⟩ S. Spiegelman, "An approach to the experimental analysis of precellular evolution," Quart. Rev. Biophys. **4**, 213 (1971); D. L. Kacian, D. R. Mills, F. R. Kramer, and S. Spiegelman, "A replicating RNA molecule suitable for a detailed analysis of extracellular evolution and replication," Proc. Natl. Acad. Sci. **69**, 3038 (1972). A 4500 base RNA obtained from the Qβ virus was replicated in the solution containing RNAase from the same virus (Qβ replicase), free nucleotides, and electrolytes. After a while a small portion of this solution was introduced into another test tube with fresh solution. After 74 such generations the original strand 'evolved' into a strand consisting of only 218 bases.

[60] Just as functions and dynamical systems definable on it can characterize a manifold

5.5 How can we approach complex systems?

We have reached a conclusion that the need for a set of fundamental conditions for their formation is a crucial characteristic feature of complex systems. We have also seen a possibility to deduce the two pillars of modern biology through considering fundamental conditions. However, our imagination and creative power are limited, so deductive studies alone will lead us sooner or later to a dead end. We must look at the reality. To study actual examples of genuinely complex systems closely must be the task we must first undertake. We have started with a proposition that organisms are complex systems; it is hard to find clear examples of genuinely complex systems unrelated to organisms. Then, in order to study genuinely complex systems the fundamental scientist must make efforts toward construction of phenomenology called the '*Integrative Natural History*.'[61]

Integrative natural history naturally consists of inductive and deductive aspects. We wish to pursue as general an understanding of the world as possible, but according to the consideration in Chapter 3, this is unrealizable unless we can phenomenologically summarize empirical facts. Consequently, the inductive aspects include:

(A) To survey facts in order to establish a phenomenological summary of the world of organisms.[62]

The eventual outcome might be that there is no nice phenomenology.[63] Then,

---

to some extent, possible complex systems could characterize a world in which they are maintained, although we need a more explicit definition of 'complexity.'

[61] ⟪**Natural history**⟫ The original meaning of *natural history* is a story about nature. What is meant in this book is to weave a consistent phenomenology encompassing all scales from molecules to ecosystems and disciplines from mathematics to fieldworks.

[62] ⟪**Lesson from linguistics**⟫ Another reason why the author stresses phenomenology is a lesson from linguistics. Generative linguistics neglected collection and field observation of various languages 'in the wild' to waste 50 precious years. See J. Ebert, "Tongue tied," Nature **438**, 148 (2005): Small and endangered languages often display rare characteristics that help linguists understand the limits and versatility of language. Because of Chomsky they have been looking for universals, but quirky characteristics of little-known languages often challenged the language universals. "Every time we find another language, another universal bites the dust." (Doug Whalen, Haskins Laboratories)."

This seems to tell us that the best contribution we can make now for the future of biology is to concentrate our energy and resources on describing biodiversity and various ecological systems rather than 'wasting our energy' on molecular biology and biophysics.

[63] ⟪**Comparison with history**⟫ Even for history "It may be true that only its general structure can be predicted as to the future. However, the general structure is the only thing that we can understand truly about the past and the present. Therefore, if you wish to look at your own era well, you should look it from a distance. How distant? It is very easy. You have only to look from distance as far as you cannot

it simply means that we cannot understand organisms from the standpoint cultivated by fundamental physics.[64]

Surveying what is already known is not enough.

(B) To develop methods and tools to explore the world of organisms.

These two are the pillars of the inductive side of integrative natural history. However, inductive studies are very often futile unless one knows what one must pursue.[65] According to our consideration up to this point, the core of the phenomenology of complex systems must be the study of fundamental conditions.[66] Thus, how to look at what aspects of fundamental conditions dictates the direction of research. Questions such as what kinds of fundamental conditions there are or how they are utilized (how they specify complex systems) are fundamental ones that easily come to our mind. For the organisms we know, these questions are mainly dealt with by molecular and cell biology. Those who pursue phenomenology cannot ignore these disciplines, and must know their details to some extent. This is because organisms are not simple meso- or macroscopic systems, but have aspects as 'extended microscopic systems.'[67]

Fundamental conditions are selected to make the whole organism, so their meaning cannot be deciphered without referring to the whole organism. Molecules have been selected under the light of natural history. Therefore, we must know some details of biology at and above the organism level. That is why we must have some familiarity with classical biology or natural history.

---

see the nose of Cleopatra." [José Ortega y Gasset, *The Revolt of the Masses* (Norton & Co., New York 1932) p62]). [Incidentally, this is the essence of renormalization group theory.] "It is nonsense to preach that history is alien to generalization. It is on generalization that history grows." [E. H. Carr, '*What is history?* (Vintage, 1967), III History, science and ethics].

However, the author is afraid that the phenomenology of history is not very clear.

[64] It is a sensible opinion that biology should be studied as biologists study it, but the pursuit of universal understanding cannot be totally futile. Even to recognize that the pursuit is actually totally futile is considerable progress in emancipating us from a wrong viewpoint which looks apparently not obviously futile.

[65] **⟪Simpson says, "Facts are elusive..."⟫** "In reality, gathering facts, without a formulated reason for doing so and a pretty good idea as to what the facts may mean, is a sterile occupation and has not been the method of any important scientific advance. Indeed facts are elusive and you usually have to know what you are looking for before you can find one." [G. G. Simpson *The meaning of evolution— a study of history of life and of its significance for man—* (Yale University Press, 1967) p272].

[66] Already it was written that fundamental conditions are conditions to specify the destinations (sectors) after symmetry is broken, but there may be other important conditions, so here we avoid precise characterization of the set of fundamental conditions, and understand it as a subset of auxiliary conditions that cannot be specified with considerable leeway.

[67] Although this book does not pay much attention to the materials aspect of fundamental conditions as already noted, the collective modes of microscopic entities strongly depend on microscopic details, so understanding of SFC and IFC requires knowledge about the materials aspect of the fundamental conditions.

We cannot ignore the history and the diversity of the biological world. We need Integrative Natural History as the phenomenology synthesizing facts of the (organismic) world on all scales from microscopic to macroscopic.

As already noted, a considerable portion of our core program may be stated as: Perform the conceptual analysis of fundamental conditions while phenomenologically summarizing relevant facts. Fundamental conditions consist of structural and instructional conditions, and the organisms genuinely utilizing these conditions in a synergetic fashion are eukaryotes. Therefore, the author believes that the ultimate research target must be eukaryotes, and the study of complex systems is, in large part, eukaryote physics. However, this opinion seems to be becoming increasingly a minority opinion in the era of the renaissance of microbiology. For example, Gould declares that complexity is not worthy of discussion or, even worse, says that paying attention to complexity is a point of view corrupted by anthropocentrism.[68] Also this viewpoint neglects flourishing Archaea and Bacteria. A similar opinion is eloquently stated by Woese.[69] However, notice that the considerable difference in the amount of DNA entails a fundamentally different method of regulating genes.[70] Furthermore, as can be exemplified by the significance of numerous ncRNA we must clearly recognize qualitative differences between the complexity of eukaryotes and that of prokaryotes. Therefore, although the point of view advocated by Gould and others sounds very sensible and wise, in

---

68 ⟨⟨**Gould's criticism of complexity study in biology**⟩⟩ Gould asserts that the point of view advocated by E. Szathmáry and J. Maynard-Smith, "The major evolutionary transitions," Nature **374**, 227 (1995) is a point of view poisoned by anthropocentrism. "The sequence of bacterium, jellyfish, trilobite, eurypterid, ⋯ does express 'the temporal history of the most complex structure in a rough and rather anthropomorphic perspective.' " [S. J. Gould, *The Structure of Evolution Theory* (Harvard University Press, 2002) p897]; "I fail to find any rationale beyond anthropocentric hope and social tradition for viewing such a sequence as a fundamental signal, or an expression of the main weights and tendencies in life's history." [*ibid.*, p1321]

69 ⟨⟨**Carl Woese's biology**⟩⟩ C. R. Woese, "Default taxonomy: Ernst Mayr's view of the microbial world," Proc. Nat. Acad. Sci. **95**, 11043 (1998): its last part reads, "Dr. Mayr's biology reflects the last billion years of evolution; mine, the first three billion. His biology is centered on multicellular organisms and their evolutions; mine on the universal ancestor and its immediate descendants. His is the biology of visual experience, of direct observation. Mine cannot be directly seen or touched; it is the biology of molecules, of genes and their inferred histories. Evolution for Dr. Mayr's is an 'affair of phenotypes'. For me, evolution is primarily the evolutionary process, not its outcomes. The science of biology is very different from these two perspectives, and its future even more so."

This may suggest that study of Bacteria can be within the conventional physics, and may lead to physics-inspired biology. In contrast, we might wish to see biology-inspired physics, but what we really wish to do is to understand natural history.

70 K. Stuhl, "Fundamentally different logic of gene regulation in eukaryotes and prokaryotes," Cell **98**, 1 (1999). We may conclude that a qualitative change occurs in the regulation mode according to the quantitative difference in the amount of those being regulated.

reality it is not. To understand this distinction is a fundamental question of the study of complex systems. The idea that the common ancestor of the extant organisms is rather close to Eukaryota (without mitochondria)[71] is not outright nonsense. The day will come when we will be able to declare that the core of fundamental biology is the study of Eukaryota.[72] See Section 5.7 for further discussion on this point.

5.6 Is there a 'theory of biological systems'?

We discussed briefly how the inductive side of the general approach to complex systems should be. The paragon of phenomenology, thermodynamics, demands that phenomenology must transcend a mere summary of phenomena. We discussed intuitively the relation between the fundamental conditions and the two pillars of modern biology, but the discussion is still a mere verbal discourse as in the humanities. The theory of complex systems or of biological systems must provide such a discussion with a mathematical framework. Can there be such a (deductive) framework?[73]

What is a good mathematical theory? It seems natural to demand of a good mathematical subfield at least the following elements:
(0) It has its own sufficiently general but definite research objects.
(1) It has its own intrinsic classic problems and remarkable facts that shake our 'common sense.'
(2) It has its own concepts and methodology/tools.
(3) It has simple but nontrivial examples to test tools and concepts of its own.
(4) It can contribute to other subfields of mathematics.

Let us look at examples that are accepted by everyone as good subfields of mathematics: equilibrium statistical mechanics and theory of dynamical systems.

For the theory of dynamical systems, the items listed above are as follows (no technical terms are explained, but are mostly looked up on the Web or can be found in the already quoted references in Chapter 2):
(0) The objects are one-parameter semigroups on manifolds.

[71] However, do not mix this up with the extant Eukaryota without mitochondria such as *Giardia*.

[72] Industrially, prokaryotes are tractable and eukaryotes are hard to handle, so the study of prokaryotes will continue to be the core of technological or industrial biology, but this is of no concern of ours.

[73] The reader may wonder whether 'abductive studies' of biological systems are conceivable. Although the author already stated that it is not easy, K. Kaneko, *Life: An Introduction to Complex Systems Biology* (Springer, 2010) may be counted as an example.

(1) Chaos observed by Poincaré in the three-body problem, ergodic hypothesis conceived by Boltzmann, and random motions observed by Cartwright and Littlewood in periodically perturbed nonlinear oscillators. These are all (apparently) stochastic phenomena found in deterministic systems. Their discovery overturned the common-sense understanding of dynamical determinacy.
(2) Ergodicity, chaos, strange attractors, etc., are special concepts. Ergodic theorems, thermodynamic formalism, Artin-Mazur's zeta function, etc., are tools developed in the field.
(3) We can easily think of examples such as restricted three-body systems, Bernoulli shifts, Smale's horseshoe, Arnold's cat, etc.
(4) Ergodic theory has had a big impact on mathematics in general. If we adopt a viewpoint that to be able to contribute to number theory is a sign of maturity, the theory of dynamical systems is definitely a mature and respectable subfield of mathematics.

For statistical mechanics,
(0) The objects are many-body systems allowing thermodynamics.
(1) The existence of phase transitions and symmetry breaking that are qualitative changes observed only in the thermodynamic limit; we must not forget that there were people who suspected that statistical mechanics could not explain phase transitions.
(2) Examples are symmetry breaking, phase transitions, Gibbs measure, etc.
(3) For example, the Ising model, the Hubbard model, augmented van der Waals fluid (the Kac model), etc. These models are realistic enough to be called, deservingly, minimal models (see Chapter 4), but still in-depth mathematical analyses are feasible.
(4) Gibbs measure is important in probability. Computational complexity problems are studied by mapping them to problems of spin glasses and other random systems. The thermodynamic formalism for dynamical systems is a rehash of the theory of Gibbs measures. The empirical relation between the Riemann zeta function and random matrices suggests that statistical mechanics has a deep connection to number theory.

How about the 'theory of biological systems'?
(0) We know numerous examples of 'organisms.'
(1) What is the most important and classical observation? That life comes only from life is a crucial fact. That every organism dies and cannot be resurrected may be such a fact as well.
(2) What are the new mathematical concepts and tools peculiar to the subfield? The Darwin process seems to come at the top of the list according to the accepted view, but as we have already considered, self-reproducible fundamental conditions must be a more fundamental concept.
(3) In biology, when we say models, we tend to think about *E. coli*, *Drosophila melanogaster*, *Arabidopsis thaliana*, etc., but these are not models in the sense we use this word here. The Ising model is not an actual magnet. Is there a

model of organisms corresponding to the Ising model? None at all.[74] We wish to have mathematical models that embody self-reproducible fundamental conditions.
(4) Therefore, we do not know whether we would contribute to other branches of mathematics.

The theory of biological systems discussed above and existing mathematical biology are distinct. This can be understood easily, if the reader recalls the difference between mathematical physics and (applied) mathematics for physics (physical mathematics). We are aiming at a theory corresponding to mathematical physics, and not mathematical biology as a branch of applied mathematics. The existing mainstream mathematical biology is applied

---

[74] ⟨⟨**Self-reproducing machines**⟩⟩ The reader might point out that self-reproducing machines are models of organisms, but their relevance to biology is not simple. Many models have been devised since von Neumann, but, in short, they are as trivial as something like a 2D lattice of gas burners: igniting at one point causes burning of its nearest neighbor burners and eventually all the burners would be burning (alive). The existing models merely demonstrate that worlds can be constructed in which self-reproducing machines can live happily; if we are free to invent a world, we can make a very simple self-reproducing machine in it [J. A. Reggia, S. L. Armentrout, H.-H. Chou and Y. Peng, "Simple systems that exhibit self-directed replication," Science **259**, 1282 (1993); the model was further simplified. However, none seem to be stable against random perturbations, since they are all fine-tuned.] These studies with simplifications are thus almost irrelevant to biology.
⟨⟨**von Neumann's agenda**⟩⟩ If, however, we read von Neumann's original article [J. von Neumann, "The general and logical theory of automata," in Vol. V of *John von Neumann Collected Works* (A. H. Taub general editor, Pergamon Press, 1961-3)], we learn, from its last part entitled "The concept of complication: self-reproduction," that his motivation was to understand the qualitative difference between complex and non-complex systems (von Neumann uses 'complication' instead of 'complexity'), or to formulate the rigorous concept of "complication." The very last part of the article is worth quoting: All these are very crude steps in the direction of a systematic theory of automata. They represent, in addition, only one particular direction. This is, as I indicated before, the direction towards forming a rigorous concept of what constitutes "complication." They illustrate that "complication" at its lower levels is probably degenerative, that is, that every automaton that can produce other automata will only be able to produce less complicated ones. There is, however, a certain minimum level where this degenerative characteristic ceases to be universal. At this point automata which can reproduce themselves, or even construct higher entities, become possible. This fact, that complication, as well as organization, below a certain minimum level is degenerative, and beyond that level can become self-supporting and even increasing, will clearly play an important role in any future theory of the subject. [End of quote]
The existence of very simple self-reproducing automata implies that von Neumann's original idea that self-reproducibility could be the threshold of 'complexity' is not very significant. Rather, it is now clear that each world has a self-reproducibility threshold, and that there are many worlds for which the systems even beyond the thresholds are very simple.
No one constructed a self-reproducing machine in a given world which is natural. It is certain that a self-reproducing machine would emerge on the huge lattice of 2D life game, but we can only guess its existence. In this case the self-reproducing system would likely be too gigantic to be studied as a model.

mathematics to solve problems posed by biological facts and observations, utilizing tools developed by mathematicians. We should not forget that this has stimulated some new developments of bifurcation theory and a certain aspect of (partial) differential equations in conjunction with pattern formation, for example. Also (the popularization of) catastrophe theory is closely related to biology.[75] However, can existing mathematical biology close in on the mathematical essence of being alive? The author regards the contribution of mathematical biology to the understanding of biology as qualitatively different from, e.g., the contribution of symbolic dynamics to the understanding of stochasticity of dynamical systems or the contribution of the Lee-Yang theory to the understanding of phase transitions. It should be repeated that neither mathematical statistical mechanics nor the theory of dynamical systems is applied mathematics.

Incidentally, what we can say only with the aid of computers is not a general proposition, and almost all general propositions must be provable without computers. Therefore, just as in the cases of dynamical systems and statistical mechanics, computers will not play fundamental roles in the solution of fundamental and conceptual problems, although we should gratefully use computers to aid our intuition.

There has already been a program called Artificial Life (AL) that purportedly pursues general understanding of life.[76] From our point of view based on phenomenology, AL tends to neglect reality. Its salient features may be the following three points: (a) putting too much confidence in computers and computer models, (b) excessive reliance on dynamical-system-theoretical tools, and (c) neglect of (spatial) structures. (c) is probably a logical consequence of (a) and (b).

There were attempts to construct a general theory of organisms/biology. A famous example may be von Bertalanffy's general systems theory. We could count cybernetics among the examples as well. The author feels these examples are rather hollow. This is likely due to their neglect of reality, the actual phenomenon, simply because they did not observe biological phenomena closely. If we look at the biological world frankly, the Pasteur chain is a key fact. A system with life (organism) must be a system that requires

---

[75] ⟨⟨**Thom's delicate feeling**⟩⟩ According to A. Jackson, *Comme appelé du néant*— as if summoned from the void: the life of Alexandre Grothendieck," Notices Am. Math. Soc. **51**, 1038 (2004), René Thom wrote, in *Fields Medalists' Lectures* [M. Atiyah and D. Iagolnitzer, eds. (World Scientific, second edition, 2003)] as follows: "His (= Grothendieck's) seminar attracted the whole of Parisian mathematics, whereas I had nothing new to offer. That made me leave the strictly mathematical world and tackle more general notions, like morphogenesis, a subject which interested me more and led me towards a very general form of 'philosophical' biology."

[76] The following paper conveniently summarizes important problems from the AL point of view: M. A. Bedau, J. S. McCaskill, N. H. Packard, S. Rasmussen, C. Adami, D. G. Green, T. Ikegami, K. Kaneko and T. S. Ray, "Open problems in artificial life," Artificial Life **6**, 363 (2000).

FC. Under omnipresent and all-destroying noise, we need Darwinism and metabolism to maintain FC. Thus, life must be capable of (or be forced to be capable of) evolution and must be self-maintaining.

'Self-maintaining' might remind the readers of the word '*autopoietic.*'[77] An autopoietic system produces, based on an external flow of molecules and energy, its components which, in turn, continues to maintain the organized bounded structure that gives rise to these components. This statement itself does not necessarily imply that an autopoietic system is regarded as self-organizing without any aid of preexisting and not 'producible' devices and contrivances, but it is customary to forget about such aids, so naturally autopoietic systems are compared with dissipative structures.

It must be clear to the reader that if we wish to avoid the same mistakes, we must look at the world of biology closely to begin with. As discussed repeatedly up to this point in this book, we should try to make a phenomenological summary of biological systems, paying special attention to various aspects of fundamental conditions.

5.7 How do fundamental conditions evolve?

To conclude this book, observed facts closely related to the phenomenological characterization of the world of organisms will be discussed in this and

---

[77] ⟪**Problems with autopoiesis**⟫ The Wikipedia article 'autopoiesis' emphasizes the cellular internal structures, "based on an external flow of molecules and energy, *produce* the components which, in turn, continue to maintain the organized bounded structure that gives rise to these components. An autopoietic system is to be contrasted with an allopoietic system, such as a car factory." However, this article goes into the following wrong statement: "More generally, *autopoiesis* resembles the dynamics of a non-equilibrium system; that is, organized states (sometimes also called dissipative structures)..." "an autopoietic system is autonomous and operationally closed, in the sense that there are sufficient processes within it to maintain the whole."

If comparable to dissipative structures, there is no surprise that uninteresting counterexamples performing autopoiesis exist (e.g., self-reproducing micelles). L. Damiano and P. L. Luisi, "Towards an autopoietic redefinition of life," Origin Life **40**, 145 (2010) recognizes this and says that such counterexamples lack something essential to be alive and point out that what is missing is the adaptative interaction with the environment, which is described by the theory of autopoiesis as a 'cognitive' interaction. Therefore, they propose that an adaptive autopoietic system is a living system.

If some phenomenon is observable at all, it must be stable under small perturbations. How does a (locally) stable system respond to external perturbations? Of course, in an adaptive way, if 'adaptive way' is understood (as usual) in a way to maintain the system integrity. Thus, the revision suggested above is still insufficient.

Generally speaking, in short, all the definitions proposed for life are satisfied by a flame of fire. A flame is also adaptive: a small disturbance cannot extinguish it.

subsequent sections. Inevitably, the discussion is brisk and incomplete.

The central piece should be the phenomenology of fundamental conditions. Sections 5.7-5.9 discuss the historical aspect, and the last section will discuss more general observations that amount to the lessons we should learn. For complex systems evolution through the Darwin process is required. It is the evolution of fundamental conditions. How do the fundamental conditions evolve?[78] Let us look at three important aspects of the general pattern of change (history) from our point of view: Does history repeat itself? Does history proceed uniformly on the average? How generally do organisms complexify? This section discusses the first two questions.

Does history repeat itself? The famous showdown on the Burgess Shale between Gould and Conway Morris concerns this question.[79] If the history of life is replayed, Conway Morris asserts, "I believe that a creature with intelligence and self-awareness on a level with our own would surely have evolved." However, as we will see in Section 5.8, the trigger of complexification hinges upon, first of all, the preparation of the stage required by complexification. Then, the genuine complexification step may take place on this stage through higher-order integration of existing elements (i.e., if we use the phase transition analogy, this process is a nucleation-like process). Consequently, the history cannot easily retrace itself. Conway Morris seems to underestimate the importance of the very nature of organisms as complex systems. In contrast, Gould's understanding of this key point is accurate.[80]

Whether changes occur gradually or in fits and starts (as punctuated equilibrium) was also a point of dispute. Gould affirmed the latter, and as its cause he proposed developmental constraints, asserting that organisms should not

---

78 ⟪**We do not discuss the origin of life**⟫ Here, we consciously avoid the origin problems (e.g., the origin of life on the earth). They are intellectually highly interesting problems and thinking about numerous possibilities is fun and also not useless from the applied science point of view. However, we avoid them, because required empirical evidences are largely unavailable. We should not forget that judging whether various chemical and isotopic signatures are abiotic or biotic is often contentious.

79 ⟪**Burgess showdown**⟫ S. Conway Morris and S. J. Gould, "Showdown on the Burgess Shale," Natural History Magazine, **107** (10), 48 (1998). Gould's *Wonderful Life* (W. N. Norton, 1989) and Conway Morris' *The Crucible of Creation* (Oxford University Press, 1998) led to this showdown. The whole text can be downloaded from `http://www.stephenjaygould.org/library/naturalhistory_cambrian.html`.

80 ⟪**Improbability of intelligent organisms**⟫ We may say that the genesis of life is fairly certain on a planet like the earth, but Gould asserts that whether the history actually takes the route to organisms with consciousness of our level is highly contingent: "Consciousness at our level of language and conceptual abstraction has evolved but once on earth in a small lineage of primates (some 200 species), within a small lineage of mammals (some 4,000 species, while the more successful beetles now number more than half a million), within a phylum that prevailed by contingent good fortune from the Burgess draw."

be considered according to panselectionism (because the mechanism to develop complex systems cannot be modified so easily as adaptationists believe). Most evolutionary biologists frowned on his assertion.[81] It is very likely to be true that developmental mechanisms cannot cause any constraints in the long run. However, the constraint problems should not be considered without paying attention to time scales. It must be impossible for a wolf to give birth to a chihuahua whatever the cause of constraints is. This may sound like a stupid comment, but this trivial statement is enough to demonstrate that the nonexistence of constraints cannot be logically asserted. We must accept that developmental and other constraints do exist generally; it is only that we cannot develop a general argument without paying attention to time scales.

In more general terms, the issue may be understood about the memory effect, since developmental constraints may be interpreted as the effect of the past evolution results. Biological systems are complex systems, so their very existence relies on fundamental conditions. Thus, dependence on history must be there. That is, constraints should have affected history which may be developmental constraints, ecological or geographic constraints, etc. However, as we just asserted, we must respect the time scale.

A conspicuous fact one can easily recognize by analyzing the history of life on the earth is that the atmospheric oxygen partial pressure has strong effects on the emergence of Eukaryota, multicellular organisms, etc. It is impossible for organisms to predict when the atmospheric composition would become favorable, so the above observation implies that, for example, attempts to make multicellular organisms occur sufficiently often, but fail because the oxygen pressure is not enough. However, once the oxygen pressure exceeds a certain threshold, the attempt turns out to be successful. In other words, there are no constraints or restrictions on the organism side to create multicellular organisms. However, we must be conscious about the time scale of our argument. The relation between evolution and oxygen pressure teaches us that on the time scale around ten to a hundred million years, various organismal constraints such as developmental constraints are virtually non-existent.

However, on the time scale of one million years, which is relevant to the coevolution of mammalian herbivores and their carnivorous predators, it is highly plausible that relevant innovations can occur only sporadically. For example, to be able to run very fast is not so easy, so even if running fast is advantageous to herbivores the running speed cannot be improved easily. Consequently, if we look at the fossil records on the mentioned time scale,

---

[81] ⟪**Criticism of punctuated equilibrium**⟫ "By the nineties, most evolutionary biologists had simply stopped paying attention to punctuated equilibrium. At best, the theory looked like a moving target, veering now at breakneck speed toward Darwinism. At worst, its chief advocate seemed muddled, a mixture of radical rhetoric and malleable ideas. Punctuated equilibrium was down, if not out." (H. A. Orr, "The descent of Gould," (New Yorker, 2002-09-30)). Then, on the developmental constraints, Orr wrote, "Look at the corn on the table and a chihuahua on the sofa."

(significant) evolution is only rare.[82] That is, at present, in Africa zebras and lions are *not* being improved gradually to be able to run faster. Rather, they are running as fast as they can at the (constant) limit (boundary) of the capabilities granted to them.

Adaptation processes are often visualized as mountain climbing in the *adaptive landscape* (Fig. 5.6A). An important defect of this picture is, math-

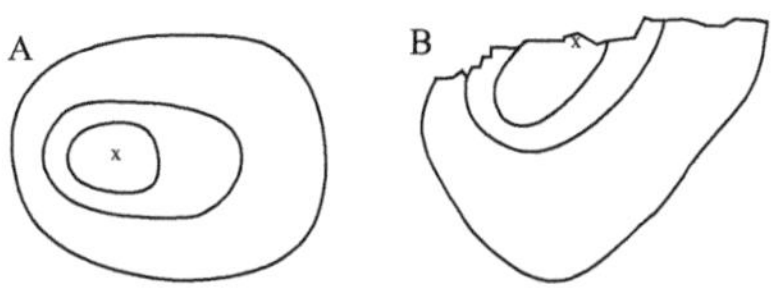

Fig. 5.6 The contour maps of two landscapes: The adaptive landscape is not like a gentle mound A as suggested originally by Wright and in many textbooks, but is better analogized with a mountain with a cliff as B. The highest point (denoted by x) is on the edge or very close to it. The cliff represents the viability limit. Needless to say, this is an idealization; the cliff means a steep slope, and the crucial point is that the shape of the hill is extremely asymmetric around its top.

ematically put, that it forgets that the maximum (or supremum) value of a continuous function is not restricted to an extreme value. It is a better picture that zebras and lions are at the highest point on the boundaries due to various constraints of the sets of their allowed phenotypes than on the top of a round hill. It seems natural that most optimal states of organisms are not realized as extrema, but realized as the highest point at the domain boundary (actually, close to it with a 'safety margin' for various noises). Significant evolutions are likely due to the expansion of the search space (the domain of the fitness function).[83]

Even though Gould denounced paying attention to *complexification* as a form of anthropocentrism, still complexification is a major trend of eukaryotic evolution. As was noted by Mattick and coworkers, an observed complexification tendency is not a mere figment due to the anthropocentric bias. Then, the reader might wish to strengthen Gould's objection by saying that Eukaryota is but a tiny portion of the biological world, and prokaryotes are the

[82] **⟪The limit of adaptation realized by evolution⟫** See R. T. Bakker, "The deer flees, the wolf pursues: incongruence in predator-prey coevolution," in *Coevolution* edited by D. J. Futuyma and M. Slatkin (Sinauer, 1983). Bakker writes as follows: "Most species are adequate, sometimes barely adequate, compromises of heritage and adaptation to local conditions, compromises that survive only until a better compromise appears." He concludes his paper as follows: "The fossil record demands more emphasis on explaining why populations do not evolve, even when considerations of functional morphology argue they should. Perhaps, as a complement to our Society of the Study of Evolution, we need a Society for the study of the Prevention of Evolution, to explore explanation of the apparent rarity of major adaptive change."

[83] This could be realized by stabilizing the peak with the aid of chaperones. Chaperones could widen the peak, and the adaptive landscape could become like table mountains (mesas).

center of (21st century) biology. The author totally disagrees with this point of view as will be discussed in Section 5.9; prokaryotes are streamlined organisms. Our last common ancestors must have been much more complicated than the existing prokaryotes. Consequently, if we wish to understand complex systems, the main target must be Eukaryota. Let us try to summarize the complexification history phenomenologically in the next section.

5.8 How do systems complexify?

A survey of the history of organisms may be summarized as the conceptual diagram in Fig. 5.7. Complexification requires (numerous) symmetry-lowering (breaking) processes and the coexistence of a resultant variety of states. The

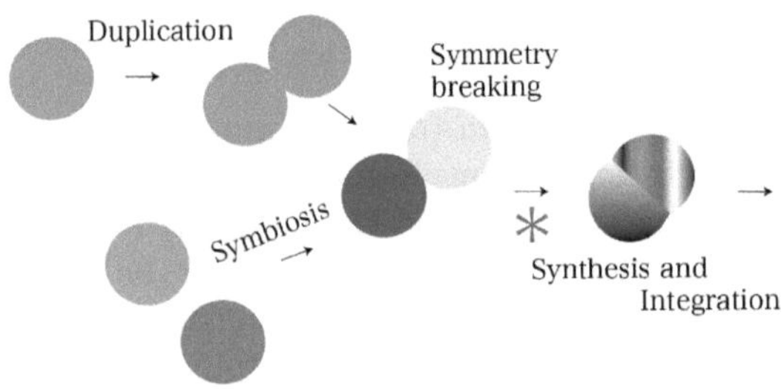

Fig. 5.7 The 'unit process' of complexification consists of two steps. The first step before the asterisk consists of a stage-widening process by duplication/juxtaposition + the subsequent symmetry-lowering (breaking) process. One method to expand the stage is duplication/multiplication in the broad sense: to juxtapose a copy with the original. Duplications of genomes and genes, multicellularization, formation of colonies, and groups are examples. The parts juxtaposed or coexistent after duplication are initially more or less the same, but symmetry-lowering processes such as specialization/subfunctionalization and differentiation often ensue. The other method for the first step is to import something from outside (e.g., endosymbiosis) or to cooperate with something external (symbiosis in general). In this case, the joined systems are already different from the beginning, but as can be exemplified by endosymbiosis often the lack of symmetry is further enhanced, and neither side becomes possible to continue to exist (live) without the other. The step before the asterisk stressed by S. Ohno is, so to speak, the step to prepare the stage (scene) and the actors for a complexification drama. The second process after the asterisk is the integration process. Now, the actors (elements) prepared in the preliminary step before the asterisk are organized into a higher organization to stage an intricate play of complexification, so to speak. This synthesis step has not been characterized clearly, but this step is *the* actual complexification step. Examples may include genesis of Bilateria, complexification of body plans of arthropods and vertebrates, emergence of eusociality, emergence of language, and genesis of civilization. Synthesis in the system prepared by symbiosis may not exist.

expansion of the stage where these symmetry breakings occur is a prerequisite to accommodate the resultant subsystems with different states. To expand the stage, there are two principal methods: *symbiosis* and *duplication*. The former expands the stage by an element imported from outside, and the latter uses a copy of (the part of) the system to expand the stage. The importance of duplication was stressed by S. Ohno in the context of gene duplication.[84] *Multicellularization* is also duplication and its significance is beyond any dispute.[85] Furthermore, Bilateria (the names of many taxonomic groups appearing henceforth are explained briefly in Note 5.2 at the end of this section) is likely to have been formed based on the repetition of a basic unit such as a segment. The importance of symbiosis is duly stressed by Levine these days.[86] Except for mass extinction, major events in the history of organisms may have always been related to symbiosis. Eukaryota came to the foreground thanks to the symbiotic mitochondria. It is generally accepted that the landfall of plants is due to their symbiosis with fungi. The key to the landfall of animals may have been a symbiotic relationship with gut microbes. The first land Eukaryota may well have been lichens.

Symbioses and duplications and subsequent symmetry lowering, that is, the processes up to the asterisk in Fig. 5.7, do not accomplish complexification that makes really qualitatively distinct level of organization possible. For example, formation of gene families (e.g., chemosensory receptor genes, heat shock factors, etc.) is a good example of the first step, but elaboration of the original function of the ancestral gene of the family is basically the outcome. The second step that integrates the parts formed by various symmetry-lowering processes is the truly crucial step for complexification. The endosymbionts already mentioned such as mitochondria and plastids may look highly integrated but it is better to be interpreted as the ultimate form of symmetry breaking and not the result of integration to realize complex systems at the higher level of complexification. That is, the cause of the symmetry breaking (the interaction between the symbiont and the host) reinforces the mutual interaction (with positive feedback) between the same participants. Really novel integration is not the outcome of the reinforcement of the same driving force causing the symmetry breaking.

---

[84] ⟪**Duplication of genes**⟫ S. Ohno was the first to point out its significance: *Evolution by Gene Duplication* (Springer 1970). A recent review is J. S. Taylor and J. Raes. "Duplication and divergence: the evolution of new genes and old ideas," Ann. Rev. Genetics **38**, 615 (2004). The following paper asserts that all of our genes are the results of duplication: R. J. Britten, "Almost all human genes resulted from ancient duplication," Proc. Nat. Acad. Sci. **103**, 19027 (2006).

[85] ⟪**Prokaryotes never complexify**⟫ Multicellularization processes are known to have occurred for both Bacteria and Archaea several times and systems like biofilms are known, but it is highly unlikely that they can complexify, because *adaptive sweeps* annihilate any historical accumulation. Furthermore, there are examples of endosymbiosis for prokaryotes, but this is quite unlikely to happen generally, because prokaryotes are usually specialized to maximize their reproduction rates.

[86] S. A. Levine, "Fundamental questions in biology," PLoS Biol. **4**, e300 (2006).

Generally speaking, it seems unlikely that symbiosis can cause a qualitatively different level of complexity. After a high level of symmetry breaking between the participants, there must be a reason why no qualitatively novel features emerge. If the symmetry-breaking process proceeds extensively in symbiosis, one party is reduced to the state of a rather extreme parasite (or excessively specialized), so no extra degree of freedom remains that can be used to make novel relations with the partner. In contrast, in the case of duplication, as can easily be seen for multicellularization, even if differentiation proceeds, the members do not degenerate excessively, so further complexification due to novel interactions is possible.

That the synthesis/integration step, the second step of complexification, is the key to realize higher-level complexity is clear if we review the history of Bilateria. It is well known that Choanoflagellata has many signal transduction elements and intercellular communication tools.[87] These molecules are of course used by Choanoflagellata (perhaps to communicate with an extracellular matrix), but Porifera are the first to use them fully to make a multicellular regime. *Hox* genes are famous for their role in determining the body plan of Bilateria like us, but some of them can already be found in Cnidaria, which are regarded as 'primitive' compared with Bilateria (they are inferred to be lost from Porifera[88]). Thus we may guess that the common ancestor of Cnidaria and Bilateria must have a considerable number of *Hox* genes.[89] In this case as well the basic elements existed beforehand, and Bilateria emerged due to revolutionary changes in the usage of the preexisting elements. Subsequently, just as in the evolution of gene families, the set of the genes used in a novel fashion would be further (partially) duplicated and elaboration of the novel features may well ensue.

There are many similar examples. It is natural to consider our natural language as an example. The elements that enable us to use language are mostly already found in higher Primates.[90] The integration step is the rate-determining step of evolution. This is clear from the fact that the 'modern language' is very likely to be much younger than *Homo sapiens* as noted before. The genesis of societies and civilizations may also be understood similarly. In these cases, integration steps were intertwined with drastic changes

---

[87] Here, the genome analysis of Choanoflagellata is cited: N. King et al., "The genome of the choanoflagellate *Monosiga brevicollis* and the origin of metazoans," Nature **451**, 783 (2008).

[88] K. J. Peterson and E. A. Sperling, "Poriferan ANTP genes: primitively simple or secondarily reduced?," Evol. Dev. **9**, 405 (2007).

[89] J. F. Ryan, M. E. Mazza, K. Pang, D. Q. Matus, A. D. Baxevanis, M. Q. Martindale and J. R. Finnerty, "Pre-Bilaterian origins of the Hox cluster and the Hox code: evidence from the sea anemone, *Nematostella vectensis*," PLoS ONE **2**, e153 (2007).

[90] An example very recently demonstrated is: L. Heimbauer, M. J. Beran, M. J. Owren, "A chimpanzee recognizes synthetic speech with significantly reduced acoustic cues to phonetic content," Curr. Biol. **21**, 1210 (2011).

in, e.g., environments.[91]

Let us look at the integration step more closely in the next section.

Note 5.2. ABC of (animal) taxonomy
We Vertebrata together with Cephalochordata (e.g., *Amphioxus*) and Urochordata (e.g., *Ciona*) belong to Phylum Chordata. Chordata belongs to Deuterostomia with Xenoturbellida, Acoelomorpha, Echinodermata (sea urchin, sea star, etc.) and Hemichordata. Deuterostomia is one of the three major groups of Bilateria (which contains most invertebrates). The remaining two groups are Ecdysozoa [Arthropoda (crabs, insects, spiders, etc.), Nematoda, etc.] and Lophotrochozoa [Mollusca (clams, squids), Annelida (polychaetes, earthworms), Brachiopoda, etc.]. Bilateria and Cnidaria (sea anemones, jellyfish, etc.) are together called Eumetazoa. This with Porifera (sponges), etc., is Metazoa (the author (at least at present) does not wish to include Placozoa, Ctenophora, etc., in Eumetazoa). Metazoa + Choanoflagellata (e.g., *Monosiga*) is close to Fungi and together make up Kingdom Opisthokonta of Domain Eukaryota. Opisthokonta together with Kingdom Amoebozoa make up a superkingdom Aplastida, which bisects Eukaryota (the rest are called Plantae = Plastida and contain ordinary plants, brown algae, etc.).[92] Finally, three domains, Bacteria, Archaea and Eukaryota make up the world of (ordinary ribosomal) organisms. However, the author is not sure whether Bacteria and Archaea are natural monophyletic groups.

5.9 Integration step and its logical consequence

The second real step to complexification (the real step to modify FC in order to realize innovative complexification) is the innovative use or new combination of already existing elements prepared by the previous steps of complexification. For the result of combination to realize something meaningful, a 'sufficient' collection of 'parts' is required beforehand. That is, *exaptation* is required. Instead, can't it be possible to prepare needed elements as they become necessary? That is impossible, because organisms do not have any foresight, and desirable mutations cannot be prepared at the critical moment of their necessity due to the very nature of the Darwin process.

The members of a taxonomic group often individually accumulate duplicated parts (e.g., genes), but these members are more or less the same without big organizational differences until the integration step enables the organisms to utilize integratively the duplicated (and often differentiated) parts. The

---

[91] For example, see the case of the Egyptian civilization: S. Kröpelin et al., "Climate-driven ecosystem succession in the Sahara: the past 6000 years," Science **320**, 765 (2007).

[92] ⟨⟨**Six kingdoms of Eukaryota**⟩⟩ See, for example, P. J. Keeling et al., "The tree of eukaryotes," Trends Ecol. Evol., **20**, 670 (2005). This explains the five kingdom scheme combining Opisthokonta and Amoebozoa separated in the text.

evolution direction of individual members due to duplications may be different (e.g., by creating different gene families), but their 'magnitudes of complexity' are largely the same.[93]

However, even within the same high-level taxonomic group if a group assumes inactive life styles (sessile, parasitic or interstitial life styles), the situation is completely different. In Chordata, if we compare *Hox* genes of subphylum Vertebrata to which we belong and of Cephalochordata, nothing is lost in the phylogenetic line leading to us, but our sister subphylum Urochordata lost some of them. It is natural to guess that this is a consequence of the sessile life style.[94] This observation teaches us the nature of the bottleneck in the complexification processes.

As was stressed above, the necessary ingredients must be ready when the new integrated system is developed. It was also stressed that the system cannot prepare needed ingredients when it 'realizes' their necessity. Furthermore, it is highly unlikely (due to the stochastic nature of the Darwin process) that the elements emerge together at once. That is, elements (or the FC required for them) emerge one by one at different times in the prehistory of the system. If an element is not in use, then ubiquitous noise will degrade it sooner or later.[95] Therefore, during the waiting period before the asterisk in Fig. 5.7,

---

[93] **⟨⟨Equality of capabilities⟩⟩** When we compare the natural intelligence of dogs and that of human beings, we tend to believe there is a tremendous difference. However, if we take into account the very high processing ability of olfactory information by dogs, it is more reasonable to think that the difference in the natural intelligences should be understood as difference in directions but not in magnitudes. When we study the intelligence of a creature, it is highly questionable to use the test that measures the capability we value. We must be extremely retarded if measured by the intelligence test for ordinary dogs. To the zeroth order the intelligence of all the mammals should be approximately 'the same.' Since we share many key neurotransmitters (e.g., nonapeptides, neuroamines) even with Protostomia, we should share emotional reactions with (at least 'higher) protostomes such as cephalopods and some insects (as already quoted in footnote 7). Since we know emotion is the basis of natural intelligence, we should not underestimate the natural intelligence of, e.g., mice. Our future generation will be appalled by what we are doing to lab mice, now. The journal *Nature* often reports as if it were a big news that a kind of bird can use tools, but use of tools is possible for almost all 'higher' organisms (including insects), if it is advantageous ecologically; if they do not use tools, it is simply that tools are not needed and/or are not advantageous from the cost-performance point of view.

[94] **⟨⟨Consequence of sessile life⟩⟩** Likely loss of *Hox* from Porifera must also be an example. The common ancestor of Cnidaria and Bilateria must be much closer to us than to Cnidaria, and must have had a considerable amount of *Hox* genes, but they were lost by sessile Anthozoa. In short, the planula (= creeping larvae of Cnidaria) theory is, if not literally correct, captures a crucial truth. Notice the parallelism between the relationship between planula and sea anemones and that between Cephalochordata and Urochordata. We should not regard Porifera and Choanoflagellata as close to our primitive ancestors, because both are essentially sessile.

[95] Remember that higher-order organizations ('higher' means farther away from molecules) tend to be lost more easily. Recall how easily cultural elements would

all the existing elements must be in use. As noted already the integration step is something like a nucleation process, so it may not occur for a long time even if all the ingredients are there. Even during the waiting period other elements may be added, so the system before the complexification step may look rather baroque at least in retrospect (that is, other elements that would not be integrated into the higher-level complex system may abound).

The Synthesis/integration step may look like a discovery of new patterns as schematically illustrated in Fig. 5.8B. Here, the original 'baroque' pattern must be prepared by the previous steps of complexification, and then must

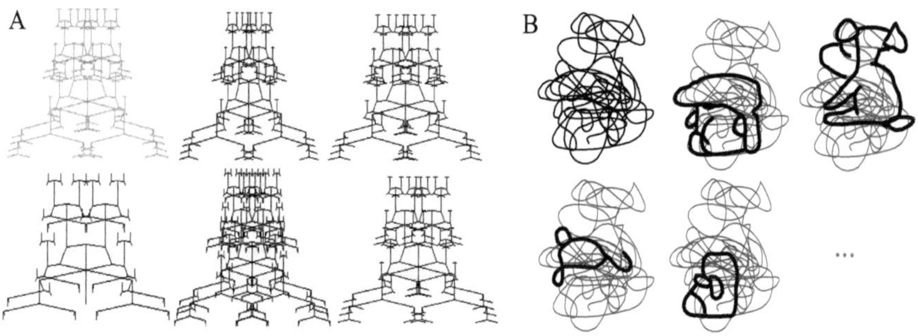

Fig. 5.8 **A** is the famous Morph of Dawkins (This figure is from `http://www.rennard.org/alife/english/biomgb.html`). In this case step by step a 'good' pattern is selected, and mutation is added to it before the next selection step; here the pattern in the upper left corner is the ancestor, and the other five are the results of one-step mutations. This is just the "descent with modification." The complexification due to this process is only gradual and it is unlikely to explain major complexification steps in the history. **B** is the mode of innovative complexification. Some primitive pattern is formed (top left); this must have been prepared by the previous complexification steps, and the 'pattern' must be in use to be maintained for a long time before the next 'discovery' occurs. In this illustration at some point in the history the 'usefulness' or 'advantage' of a pattern is 'realized' serendipitously and selected ('serendaptation'?) because reinforcing and streamlining the newly 'realized' pattern is advantageous. Here, a continuous line figure is used to illustrate the idea, but we can replace the figure with one consisting of discrete elements. We could easily conceive a 'morph'-assisted serendaptation game for fun. [B is from the Kyushu U seminar in June, 2009]

be maintained for a sufficiently long time, because serendipitous discovery requires happy chance (with the right environment).[96] Just as crystalliza-

---

be lost. Thus, the author is afraid that our modern mathematics is rather vulnerable to neglect. Also 'positive' devolution was mentioned in footnote 3 of Chapter 3.

[96] ⟪**Ramsey-type theorems and evolution?**⟫ J. A. Poulos points out in his *Irreligion* (Hill and Wang, 2008) (esp., p104) that a *Ramsey-type theorem* might be behind Kaufman's 'order for free' [A. Kaufman, *The origins of order* (Oxford University Press, 1993)]. A typical Ramsey-type theorem asserts that if a set or a geometrical object (say, a graph) is large enough, surely a subset (substructure) satisfying a certain prescribed property is found in it. A watered-down version of the

tion in a (mildly) supersaturated solution, the initiation step of integration may not begin easily. If the 'lifestyle' does not require the maintenance of the original pattern in B, no new patterns would be discovered. Also, a very rigorous selection process may be against the preservation of the baroque system. Thus, the so-called constructive neutral evolution especially in small population groups as recently stressed by Lynch may well be crucial for the second step.[97]

We could summarize that the first step of complexification prepares a big stage and many actresses and actors and that the second step gives an interesting scenario to play. Adaptation to sessile life and filter feeding makes the stage narrower and causes the loss of many players before any interesting play begins. This is why loss of motility is often fatal for the realization of

---

original Ramsey's theorem is: suppose we color every edge of a complete graph blue or red. if the graph is large enough (i.e., the number of vertices n is large), then surely there is a blue or red complete subgraph with m vertices [if $m = 3$, $n = 6$ is enough; Even for $m = 5$ minimum n is not known ($43 \le n \le 49$ is known)]. Perhaps the most famous theorem of this type other than the original is *van der Waerden's theorem*. Its watered-down version reads: if a set of consecutive numbers is large enough, then any randomly constructed subset or its complement surely contains an arithmetic progression of any prescribed length (say, 1250). Poulos quotes Diaconis's dictum: if it's big enough, "almost any damn thing will happen." Notice that the theorems are not probabilistic theorems.

All the Ramsey-type theorems are about precise statement; for example, in van der Waerden's theorem precise arithmetic progression is required. If only approximate fulfillment of the requirement is required, the needed size of the set or the object could be much smaller. We need only this type of approximate Ramsey-type theorems.

Note that Ramsey-type theorems are all non-constructive, so actually finding the object guaranteed to exist is generally not easy. Thus, finding the pattern (i.e., $*$) becomes an evolutionary bottleneck.

97 **⟪Crucial neutrality for complexity evolution⟫** M. W. Gray, "Irremediable complexity?" Science **330**, 920 (2010) has a discussion on the molecular 'apparently gratuitous' complexity (e.g., spliceosomes) of organisms. One possible cause is the 'constructive neutral evolution' [P. S. Covello and M. W. Gray, "On the evolution of RNA editing," Trends Genet., **9**, 265 (1993); A. Stoltzfus, "On the possibility of constructive neutral evolution," J. Mol. Evol., **49**, 169 (1999)].

Lynch [e.g., M. Lynch, "Evolution of the mutation rate," Trends Genet., **26**, 345 (2010); A. Fernández and M. Lynch, "Non-adaptive origins of interactome complexity," Nature, **474**, 502 (2011)] invoked fixation of neutral or slightly deleterious features as a general and unavoidable source of complexity in taxa with small populations. Neutrally fixed complexity could be neutrally 'unfixed' (through random reversion), but ratchet-like (one-directional) tendencies might prevent this.

At the organismal level, Maynard Smith and Szathmary [*The Major Transitions in Evolution* (Oxford Univ. Press, Oxford, 1995)] proposed that a ratchet mechanism called contingent irreversibility might render previously independent evolutionary units interdependent for 'accidental reasons that have little to do with the selective forces that led to the evolution of the higher entity in the first place.' An example is the mutual loss of autonomy by the symbionts that became mitochondria or plastids and the cells that harbored them.

further complexification steps.[98]

What is the most general conclusion we can draw from the general property of the second step? When a new (higher-level) taxonomic group (say, class) is founded, it is often due to (whole genome) duplication as we know for vertebrates and angiosperms. Thus, the first step often creates (higher-order) taxonomic groups. According to the general discussion above, complexification would be facilitated by the preservation of features obtained by the first step. Therefore, this strongly suggests that the possibility to complexify is most widely open to the organisms that most preserve the common ancestor's features of the taxonomic group.

This applies to Chordata, as already noted. We Vertebrata share a lot of features with Cephalochordata, but the closer group Urochordata lost many key features of Cephalochordata. Loss of motility is fatal; it is true that Urochordata contains pelagic groups, but they never recover the active movement of the chordate ancestor. Once going through a sessile lifestyle the group does not seem to recover an active lifestyle. The same logic applies to Deuterostomia as a whole. We can infer this from the *Hox* gene cluster structure.[99] We preserve the ancestral gene cluster structure of the common ancestor of Deuterostomia. It is expected that among Bilateria we keep this feature. That is, we preserve most features of the common ancestor of Bilateria. Interestingly, the gene structures of *Nematostella* (Anthozoa in Cnidaria) have lots of commonalities with us in contrast to Protostomia. Thus, we conserve the ancestral features of Eumetazoa most.

A next natural question is our position in Eukaryota. It is probably reasonable to divide Eukaryota into two major groups Plastida and Aplastida.[100] Aplastida is ancestral, and we are there. They are mostly motile. The primitive form of Eukaryota is unclear; even whether it had an undulipodium or not is unclear, but it is certain that it was an active form. Plastida filter feed

---

98 ⟨⟨**Importance of being able to move**⟩⟩ For complexification of organisms it seems very important that organisms be able to move by themselves. T. Ikegami [M. M. Hanczyc and T. Ikegami, "Chemical basis for minimal cognition," Artificial Life **16**, 233 (2010)] stresses that to improve the cognitive capability of machines, it is important that the machine can move around. This observation seems to apply much more generally in the context of complexification.

99 G. Fritzsch et al., "PCR survey of *Xenoturbella bocki* Hox genes," J. Exper. Zool., **310**, 278 (2008) is a useful paper.

100 Traditionally, the Unikonta-Bikonta dichotomy was proposed, but recent findings indicate that this implies Amoebozoa are paraphyletic. Thus, the presence or absence of plastid endosymbiosis seems a much better dichotomy. A. J. Roger, and A. G. B. Simpson, "Revisiting the root of the eukaryote tree," Curr. Biol., **19**, R165 (2009) is a good introductory review. This divides Eukaryota into Opisthokonta+Amoebozoa and Plantae. Cf., H. Nozaki et al., "Phylogenetic positions of Glaucophyta, green plants (Archaeplastida) and Haptophyta (Chromalveolata) as deduced from slowly evolving nuclear genes," Mol. Phyl. Evo., **53**, 872 (2009). Incidentally, it is interesting to note that the O+A vs P dichotomy is quite parallel to our naive dichotomy of organisms into animals and plants.

on photons, so to speak, and are likely to lose many ureukaryote features.

If we extrapolate what we have observed so far, what can we conclude reasonably? It is the most natural guess that we preserve most features of the last common ancestor of Eukaryota. However, a much more interesting conclusion inferred 'by linear extrapolation' is that since we are among Eukaryota, the common ancestor of eukaryotes and prokaryotes must have been more Eukaryota-like. Prokaryotes are a specialized dead end from the complexity point of view and should not be regarded as primitive. This may sound heretical, but even if we believe the usual branching order of Bacteria, Archaea and Eukaryota, the phylogenetic tree does not tell us anything about the nature of its trunk.[101] Supposedly early-branching Bacteria have much more cellular internal structure (including the nucleus-like structure and the endomembrane system).[102]

Generally speaking, it is an outstanding prejudice to believe that the precursor of something is simpler than itself. The second integrative step of complexification may well be a process making a more streamlined and organized system from the precursor with a less organized state having many more parts than required later (i.e., more complicated). Fig. 5.8B illustrates this; the not-used curves in the figure may be erased by 'disuse' after 'integration.'

Then, we should learn one more lesson. When the origin of life is studied, almost all attempts study how the needed minimal elements emerge (RNA first, protein first, etc.). However, it is much more natural to understand the emergence of life as the second step of complexification. Many bits and pieces may have formed, and then as if some vague image emerges from a fog, a loosely connected complex system emerges as 'life' with numerous redundant and superfluous or even parasitic parts. Such a primitive baroque system gradually streamlines and refines. The ultimately refined results are Bacteria,[103] the zenith of the Darwin process in our world.

---

[101] C. G. Kurland, L. J. Collins and D. Penny, "Genomics and the irreducible nature of eukaryote cells," Science, **312**, 1011 (2006). A unicellular raptor with a larger, more complex cell structure than that of present-day prokaryotes is envisioned as the host of the ancestral endosymbiont. This is in accordance with the general consideration of complex system evolution.

[102] ⟨⟨**PVC superphylum**⟩⟩ P. Forterre and S. Gribaldo, "Bacteria with a eukaryotic touch: a glimpse of ancient evolution?" Proc. Natl. Acad. Sci., **107**, 12739 (2010); J. A. Fuerst and E. Sagulenko, "Beyond the bacterium: planctomycetes challenge our concepts of microbial structure and function," Nature Rev. Microbiol., **9**, 403 (2010). However, these features could be simply due to convergence.

[103] A paper pointing in the same general direction just appeared: M. Wang, C. G. Kurland, and G. Caetano-Anollés, "Reductive evolution of proteomes and protein structures," Proc. Natl. Acad. Sci. **108**, 11954 (2011).

5.10 'Lessons' we learn from complex systems

If we wish to aim at Integrative Natural History, we must observe real phenomena closely. However, the author's attempt to make a phenomenological summary is only less than a few years old, so no phenomenological system is in order yet. In any case, to attempt such an enterprise single-handedly is crazy. This is the main reason why we discuss only some preliminary inductive aspects and primitive observations of Integrative Natural History. The chapter now concludes with two qualitative features of the world of organisms and the discussion of their significance. They are still mere stories, but they could affect the direction of research. Furthermore, they could mean something even when we handle more general problems of social and political significance. We should realize that a wrong characterization of 'complex systems' and 'complexity' can have (actually, the author believes, have already had) dire consequences.

(I) Don't rely on foresight.
One may not be praised for not having any foresight, but living well without foresight is rather important. For example, a molecular machine (e.g, processive molecular motor) often waits patiently for the realization of a desirable state by thermal fluctuations. Once a desirable state is reached, the molecule pays free energy and secures the state, For example, ATP or GTP is hydrolyzed and then inorganic phosphate is discarded and the state is preserved. If the effect of thermal fluctuations is decisive, to aim at the destination is not a wise strategy. To try to control fluctuations is reckless. Following the wind and not missing good opportunities is much wiser. For example, a ribosome captures the correct state when a correct amino acid (correct amino acylated tRNA) arrives through hydrolyzing GTP and then throwing away a phosphate ion just as throwing away the key after closing the door. Nonequilibrium is used not to drive something, but to secure a convenient state when it is realized. Consequently, the cell is almost in the thermal equilibrium condition as the state of matter, so the equilibrium theory of matter is almost enough to understand the materials making the cell.[104]

The other method to do without foresight is, just as we human beings have been doing, to make full use of technology to make the future predictable, that is, to regulate or control the future. This is to drive a system in the direction we wish, so no foresight is needed, but nonequilibrium processes are maximally used, and we cannot avoid dissipation. Organisms avoid such waste. However, we tend to regard molecular systems as if they are engineered systems (we call them 'molecular machines'), so, for example, the

[104] Needless to say, globally the system is in nonequilibrium. The situation is similar to the pattern formation in the reaction-diffusion system. In this case the system is not away from equilibrium microscopically, so we do not need any special theory of nonequilibrium state of matter.

protein folding process tends to be regarded as a process driven by a free energy difference. However, what actually happens in the foldable proteins is that impasses and traps are 'marked,' so to speak, and the folding trajectories are prevented from being trapped. No right passages are marked; only bad ways are 'plugged' beforehand. Thus, it is a nonsensical question whether there is a unique folding path or not, generally speaking. Some proteins may well have a "well-trodden" path or two, but such proteins must be exceptional fast folders. We should recall that the evolution process is also to preserve the state when something good appears.[105]

Foresight requires insight, so it is not possible for everybody, and there is no guarantee of success to begin with. Therefore to rely on foresight is not advantageous for a system that must last stably for a long time. This is quite parallel to the fact that the standard strategy for safety is foolproof and to avoid reliance on human quality.

(II) Don't try to solve difficult problems.
Biological systems never solve difficult problems. Generally speaking, there seem to be three ways to cope with potentially difficult problems:
(i) To use only very special cases or special solutions.
(ii) To change the problem completely with the aid of sophisticated machinery.
(iii) To destroy the problem through cutting the Gordian knot.[106]

A good example of (i) is the way biology copes with the so-called *Levinthal paradox* of protein folding: since there are so many different conformations for a given protein molecule, it is impossible to find the correct 'natural conformation' in real time. Despite this possible paradox small (the so-called single domain) proteins can be folded very quickly in less than 1 msec.[107] The necessary energy difference seems to be only about a few k_BT.[108] Then, has Biology found a general way to fold peptides quickly (to minimize its free

---

[105] "Humiliating to human pride as it may be, we must recognize that the advance and even the preservation of civilization are dependent upon a maximum of opportunity for accidents to happen." [F. A. Hayek, *The Constitution of Liberty* (U Chicago Press, 1960) Part I Chapter 2, The Creative Power of a Free Civilization, p29]

[106] ⟪**Legend of the Gordian Knot**⟫ According to Plutarch's *Lives*, Alexander "$\cdots$ conquered Phrygians, at whose chief city, Gordium, $\cdots$ he saw the famous chariot fastened with cords made of the rind of the cornel-tree, which whosoever should untie, the inhabitants had a tradition, that for him was reserved the empire of the world. Most authors tell the story that Alexander finding himself unable to untie the knot, the ends of which were secretly twisted round and folded up within it, cut it asunder with his sward." (Plutarch, *The lives of the noble Grecians and Romans* (translated by J. Dryden, revised by A. H. Clough) p813 (Modern Library, New York, 1977)

[107] This time scale can be much less than 1msec, say less than 1 μsec. See V. Daggett and A. Fersht, "The present view of the mechanism of protein folding," Nature Rev. Mol. Cell Biol. **4**, 497 (2003).

[108] R. Zwanzig, A. Szabo and B. Bagchi, "Levinthal's paradox," Proc. Nat. Acad. Sci. **86**, 20 (1992).

energy)? The actual picture seems to be as follows. For small proteins (single domain proteins consisting of less than ~300 amino acid residues), Biology chose only well-designed peptide chains.[109]

Larger proteins are often constructed from concatenating foldable smaller proteins. The resultant multidomain proteins, if made by tandem repeats, have evolved to minimize the problems of interdomain interference causing misfolding.[110] Generally, larger proteins are hard to fold. Organisms leave them to folding catalysts called chaperones.[111] This is an example of (ii).

---

[109] It is believed that they are made of the repetition of some basic oligopeptides.

[110] ⟪**Multidomain proteins**⟫ See J. H. Han, S. Batey, A. A. Nickson, S. A. Teichmann and J. Clarke, "The folding and evolution of multidomain proteins," Nature Rev. Mol. Cell Biol. **8**, 319 (2007); More than 70% of eukaryotic proteins are composed of multiple domains. Tandem repeats (like titin; 20% of the proteins in a genome are tandem repeats) present an interesting case; evolutionary mechanisms to avoid misfolding might include: sequence divergence of adjacent domains, kinetic partitioning and the conservation of gatekeeper residues.

[111] ⟪**Chaperones**⟫ In a cell many kinds of chaperones make a sort of chaperone atmosphere, and its homeostasis (chaperone homeostasis) is crucial for maintaining cell health. When a peptide is folded, there are chaperones working at earlier stages of folding, chaperones acting at the completion stage of folding, etc. For a single peptide to be folded, a few tens to a hundred chaperones take part in [J. C. Young, V. R. Agashe, K. Siegers and F. U. Hartl, "Pathways of chaperone-mediated protein folding in the cytosol," Nature Rev. Mol. Cell Biol. **5**, 781 (2007)]. As soon as a polypeptide chain emerges from the ribosome upon translation of mRNA many chaperones called trigger factors insulate it from other macromolecules (even a spontaneously foldable protein is not allowed to fold; as we stress in complex systems self-organizing capability is alway under tight regulation [R. S. Ullers, D. Ang, E. Schwager, C. Georgopoulos, and P. Genevaux, "Trigger factor can antagonize both SecB and DnaK/DnaJ chaperone functions in *Escherichia coli*," Proc. Natl. Acad. Sci. **104**, 3101 (2007)].

Chaperones renature denatured proteins and resolve undesirable aggregates and interactions among proteins, repairing damages due to stresses such as heat shock (Hsp = heat-shock proteins are famous). Elimination of aggregates can be achieved by solubilization of aggregates and either refolding of the liberated polypeptides or their proteolysis [K. Liberek, A. Lewandowska and S. Ziętkiewicz "Chaperones in control of protein disaggregation," EMBO J. **27**, 328 (2008)].

Chaperones usually work with numerous other non-chaperone proteins, and their interaction network is really complicated. F. U. Hartl, A. Bracher and M. Hayer-Hartl, "Molecular chaperones in protein folding and proteostasis," Nature **475**, 324 (2011) is a good review.

The function to repair damages due to stress can easily be extended to repairing or neutralizing/alleviating the results of mutations and diseases (including tumors). Indeed, it is recognized that Hsp90 masks effects of mutations, and can increase genetic diversity behind a given phenotype [S. L. Rutherford and S. Lindquist, "Hsp90 as a capacitor for morphological evolution," Nature **396**, 336 (1998); a more recent paper is T. A. Sangster, N. Salathi, H. N. Lee, E. Watanabe, K. Schellenber, K. Morneau, H. Wang, S. Undurraga, C. Queitsch and S. Lindquist, "HSP90-buffered genetic variation is common in *Arabidopsis thaliana*," Proc. Natl. Acad. Sci. **105**, 2969 (2008)].

We must not forget that a cell is a very crowded system.[112] Therefore, even spontaneously foldable proteins must be 'protected' from interferences. Thus, realistically, we must regard it as a rule that *de novo* protein folding is usually orchestrated by chaperones. For example, in *Escherichia coli* the chaperones, trigger factor, DnaK (Hsp70), and GroEL (Hsp60), are the keys. The ribosome-bound trigger factor is the first chaperone to interact cotranslationally with nascent polypeptides. Spontaneously foldable chains must often be controlled its folding to be secreted, for example.[113]

Even chaperones cannot fold arbitrarily large peptides. Thus, Biology has not solved the folding problem in general, but use strategy (i) for smaller proteins, and then follow (ii) using chaperones to extend the set of foldable special proteins. Statistical mechanical study of general polypeptide folding is almost irrelevant to what we are interested in in Biology. (ii) becomes possible when a special device is invented.[114]

Another example of (ii) is the vesicle and membrane shapes. For example, in ER (= endoplasmic reticulum) or Golgi, lipid membranes are folded in a very complicated manner. They could pose differential geometrically difficult problems, if elasticity and molecular compositions are to be used to shape them (e.g., budding). Statistical physicists studied shape changes of vesicles and colloidal particles, but in reality clathrin and other membrane coating

---

[112] ⟪**Crowded cell**⟫ About 40% of the cell space is occupied by macromolecules. The protein concentration is about 300 g/ℓ. About half of water molecules are bound to proteins and cannot move very freely. See, for example, R. J. Ellis, "Macromolecular crowding: obvious but underappreciated," Trends Biochem. Sci. **26**, 597 (2001), which is a good starting point; also see J. J. Spitzer and B. Poolman, "Electrochemical structure of the crowded cytoplasm," Trends Biochem. Sci. **30**, 536 (2005); The cytoplasms are a 'bustling and well-organized metropolitan city,' and the cellular metabolism is vectorial throughout the cytoplasm because of channeling.

[113] ⟪**Trigger factor**⟫ Trigger factors oppose other chaperones to prevent the proteins from forming their natural states prematurely: R. S. Ullers, D. Ang, E. Schwager, C. Georgopoulos, and P. Genevaux, "Trigger Factor can antagonize both SecB and DnaK/DnaJ chaperone functions in *Escherichia coli*," Proc. Natl. Acad. Sci. **104**, 3101 (2007).

[114] ⟪**Are proteins really in their equilibrium states?**⟫ In many cases, we should not simply assume that proteins *in vivo* are in their equilibrium states. There is no experimental proof in any case except perhaps for very simple proteins. Some may not fold without interaction partners (prions may be considered such examples) [K. Sugase, H. J. Dyson and P. E. Wright, "Mechanism of coupled folding and binding of an intrinsically disordered protein," Nature **447**, 1021 (2007)], for example, or some may fold into different forms according to their synthesis modes (e.g., due to codon usage bias). Also we should not forget that proteins that can adopt more than one native folded conformation may be more common than previously thought [A. G. Murzin, "Metamorphic proteins," Science **320**, 1725 (2008)]. Furthermore, not naturally folded states may be biologically important [F. Bemporad, J. Gsponer, H. I. Hopearuoho, G. Plakoutsi, G. Stati, M. Stefani, N. Taddei, M. Vendruscolo and F. Chiti, "Biological function in a non-native partially folded state of a protein," EMBO J. **27**, 1525 (2008)].

proteins are used that bind stoichiometrically to the surface.[115]

A typical example of (iii) is a topological problem related to DNA. When the prokaryotes DNA was recognized as a ring polymer, naturally, some mathematicians were delighted, because replication causes topological problems. Thus, applications of knot theory and other topological sophisticated theories were expected to be useful. However, virtually, there is no problem, because topoisomerase II cuts the Gordian knot.[116]

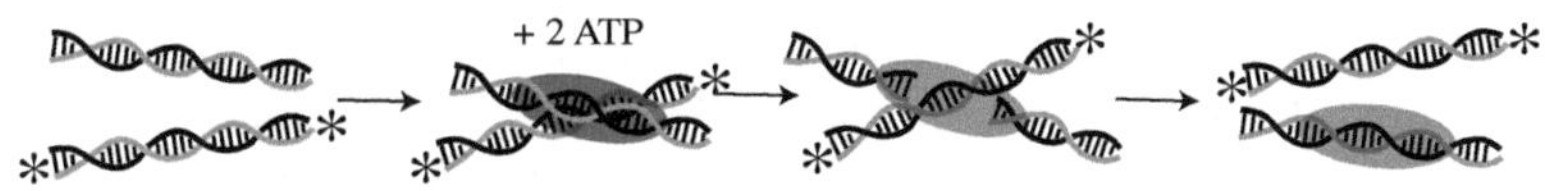

Fig. 5.9 Topoisomerase II. The gray oval denotes the topoisomerase II. Using two ATP molecules, the molecule binds at the crossing point of two DNA double strands. Cutting one of them to move the other through the cut, the topoisomerase reconnects the severed double strand. We may almost totally forget about the existence of topological invariants.

We may summarize that when mathematically difficult problems could occur, they are never solved generally. We have seen examples of combinatorial geometrical, topological, and differential geometrical difficulties. We should add a purely combinatorial problem as well: combinatorial regulation of genes. Naturally, we can expect the combinatorial explosion that we encounter in many NP complete problems, but biological systems should not have been plagued by such difficulties. That is, biological systems must be free from combinatorial difficulties. It is not because they have found how to solve the problems, but organisms design systems that are intrinsically free from this difficulty.[117] For example, ontogeny of eyes and that of ears utilize (apparently) the same module, but these modules are made of different genes (due to duplication) to avoid entanglement. In conjunction with decision making, it already has been stated that there is no frame problem

---

[115] For example, see G. K. Voeltz, W. A. Prinz, Y. Shibata, J. M. Rist, and T. A. Rapoport, "A class of membrane proteins shaping the tubular endoplasmic reticulum," Cell **124**, 573 (2006): Proteins that may be called structure-molding proteins divide and stabilize greatly curved membranes of ER.

[116] How topoisomerase II does its job is not our concern (largely irrelevant to our project; however, it is highly interesting): R. D. Smiley, T. R. L. Collins, G. G. Hammes, and T.-S. Hsieh, "Single-molecule measurements of the opening and closing of the DNA gate by eukaryotic topoisomerase II," Proc. Natl. Acad. Sci. **104**, 4840 (2007).

[117] As an example, see G. C. Conant and K. H. Wolfe, "Functional partitioning of yeast co-expression networks after genome duplication," PLoS Biol. **4**, e109 (2006).

Also, we must not forget the possibility that even though there is a combinatorial explosion, organisms may be indifferent. We have already mentioned pleiotropy-multigenic trait problems. Many qualitative traits are likely to be multigenic, but in most cases organisms do not care what combination is used to realize the traits.

for organisms.

A lesson we can obtain is: When we consider biological problems that have potential mathematical difficulty, it is meaningless to consider general solutions for the problems; it is productive to think how to avoid such problems.

The strategy of organisms to avoid possible or potential difficulties may be understood as the ultimate form of foresight by organisms that ironically lack the usual sense of foresight. If we look at contemporary ethics, the central problems seem to be how to deal with ethical aporias.[118] A lesson we should learn from complex systems study is that the most fundamental problems of ethics, environmental policies, etc., are to make efforts to eliminate the necessity to cope with difficult problems and choices beforehand. For example, for environmental problems, the population issue is an important key.[119] In ordinary ecological systems the population (of at least vertebrates) is maintained at a level that is considerably lower than the carrying capacity of the systems.[120] The possible difficulty is avoided beforehand.

It has already been pointed out that so-called complex systems studies had a strong political connotation. As noted before, the so-called complex systems studies that stress self-organization and emergent properties seem to have been supportive of market fundamentalism and misguided deregulations. Thus, even outside pure science, to study genuine complex systems has a significance for dispelling various delusions of self-organization. We have touched upon difficulties of the large-scale 'quantitative models' of environment, and expressed apprehensions about those who advocate 'changing the world.' Those who claim that the complex world emerges spontaneously must not forget that the process is a very long and tortuous time process. If our complex society is comfortable at all, it is because many people obey numerous tacit traditions and rules. To realize such a society requires a tremendously long time.[121] A good example is democracy, which cannot easily be transplanted; modernization of a society necessitates a long cultural tradition

[118] Examples may be found in Box 1 of E. Dupoux and P. Jacob, "Universal moral grammar: a critical appraisal," Trends Cognitive Sci. **11**, 373 (2007).

[119] D. Lamb, P. D. Erskine, and J. A. Parrotta, "Restoration of degraded tropical forest landscapes," Science **310**, 1628 (2005).

[120] J. E. C. Flux, "Evidence of self-limitation in wild vertebrate populations," Oikos **92**, 555 (2001); The population densities attained by wild species is about 10% of the carrying capacity, but their feral domesticated forms go beyond this limit and turn into pests, so wild species must have mechanisms to restrict their population. This is likely to be their social behavior.

[121] "Indeed, the great lesson which the individualist philosophy teaches us on this score is that, while it may not be difficult to destroy the spontaneous formations which are the indispensable bases of a free civilization, it may be beyond our power deliberately to reconstruct such a civilization once these foundations are destroyed." [F A Hayek, *Individualism and Economic Order* (U Chicago Press, 1948) I. Individualism: True and False p25]

as a civil society.[122] A practical significance of studying complex systems is in understanding the following ancient dictum by Zhuangzi: "Accomplishment of good things lies in long time."

References

Anderson, PW (1998) Concepts in solids: lectures on the theory of solids. World Scientific

Badii R, Politi A (1997) Complexity: hierarchical structures and scaling in physics. Cambridge University Press

Bailey DH, Crandall RE (2001) On the random character of fundamental constant expansions. Experimental Math 10:175-190

Bakker RT (1983) The deer flees, the wolf pursues: incongruence in predator-prey coevolution: In Futuyma DJ, Slatkin M (ed) Coevolution. Sinauer

Bateson M, Desire S, Gartside SE, Wright GA (2011) Agitated honeybees exhibit pessimistic cognitive biases. Curr Biol 21:1070-1073

Bedau MA, McCaskill JS, Packard NH, Rasmussen S, Adami C, Green DG, Ikegami T, Kaneko K, Ray TS (2000) Open problems in artificial life. Artificial Life 6:363-376

Bemporad F, Gsponer J, Hopearuoho HI, Plakoutsi G, Stati G, Stefani M, Taddei N, Vendruscolo M, Chiti F (2008) Biological function in a non-native partially folded state of a protein. EMBO J 27:1525-1535

Britten RJ (2006) Almost all human genes resulted from ancient duplication. Proc Natl Acad Sci USA 103:19027-19032

Bulhak AC (1996) On the simulation of postmodernism and mental debility using recursive transition networks. Monash University Department of Computer Science, Technical Report 264

Bunner AE, Beck AH, Williamson JR (2010) Kinetic cooperativity in *Escherichia coli* 30S ribosomal subunit reconstitution reveals additional complexity in the assembly landscape. Proc Natl Acad Sci USA 107:5417-5422

Carr EH (1967) What is history? Vintage

Chari A, Fischer U (2010) Cellular strategies for the assembly of molecular machines. Trends Biochem Sci 35:676-683

Chomsky N (1995) The minimalist program. MIT Press

Conant GC, Wolfe KH (2006) Functional partitioning of yeast co-expression networks after genome duplication. PLoS Biol 4:e109

Conway Morris S, Gould SJ (1998) Showdown on the Burgess Shale. Natural History Magazine 107(10):48-55

Covello PS, Gray MW (1993) On the evolution of RNA editing. Trends Genet 9:265-268

Daggett V, Fersht A (2003) The present view of the mechanism of protein folding. Nat Rev Mol Cell Biol 4:497-502

Damasio A (1994) Descartes' error, emotion, reason, and the human brain. G P Putnam

Damiano L, Luisi PL (2010) Towards an autopoietic redefinition of life. Origin Life 40:145-149

Das C, Tyler JK, Churchill MEA (2010) The histone shuffle: histone chaperones in an energetic dance. Trends Biochem Sci 35:476-489

Dupoux E, Jacob P (2007) Universal moral grammar: a critical appraisal.

[122] See, for example, the case of Japan is discussed in E. Ikegami, *Bonds of Civility: aesthetic networks and political origin of Japanese culture* (Cambridge University Press, 2005). Japan is a non-western society, but it has a long tradition of civil cultural tradition (such as haiku circles, gardening clubs, etc.).

Trends Cognitive Sci 11:373-378
Ebert J (2005) Tongue tied. Nature 438:148-149
Ellis RJ (2001) Macromolecular crowding: obvious but underappreciated. Trends Biochem Sci 26:597-604
Fernández A, Lynch M (2011) Non-adaptive origins of interactome complexity. Nature 474:502-506
Flux JFC (2001) Evidence of self-limitation in wild vertebrate populations. Oikos 92:555-557
Forterre P, Gribaldo S (2010) Bacteria with a eukaryotic touch: A glimpse of ancient evolution? Proc Natl Acad Sci USA 107:12739-12740
Fritzsch G, et al (2008) PCR survey of *Xenoturbella bocki* Hox genes. J Exper Zool 310:278-284
Fromm E (1950) Man for himself — an enquiry into the psychology of ethics—. Routledge & Kagan Paul Ltd.
Fuerst JA, Sagulenko E (2010) Beyond the bacterium: planctomycetes challenge our concepts of microbial structure and function. Nat Rev Microbiol 9:403-413
Gaut BS, Ross-Ibarra J (2008) Selection on major components of angiosperm genomes. Science 320:484-486
Geison GL (1995) The private science of Louis Pasteur. Princeton University Press
Gilbert DT, Wilson TD (2007) Prospection: experiencing the future. Science 317:1351-1354
Glazier DS (2008) Effects of metabolic level on the body size scaling of metabolic rate in birds and mammals. Proc Roy Soc Lond, Ser B: Biol Sci 275: 1405-1410
Gō N (2007) Physics and biology of protein folding. Prog Theor Phys Supp 170:198-213
Gödel K (1995) Collected Works vol 3 (ed. Fefferman S, Dawaon JW Jr., Goldfarb W, Parsons C, Soloway RN). Oxford University Press
Gould SJ (2002) The structure of evolution theory. Harvard University Press
Gray MW (2010) Irremediable complexity? Science 330:920-921
Grosshans H, Filipowicz W (2008) The expanding world of small RNAs. Nature 451:414-416
Hadamard J (1954) The Psychology of invention in the mathematical field. Dover
Han JH, Batey S, Nickson AA, Teichmann SA, Clarke J (2007) The folding and evolution of multidomain proteins. Nature Rev Mol Cell Biol 8:319-330
Hanczyc MM, Ikegami T (2010) Chemical basis for minimal cognition. Artificial Life 16:233-243
Hardy GH (1940) A mathematician's apology. Cambridge University Press
Hartl FU, Bracher A, Hayer-Hartl H (2011) Molecular chaperones in protein folding and proteostasis. Nature 475:324-332
Hayek FA (1948) Individualism and economic order, University of Chicago Press
Hayek FA (1960) The constitution of liberty. University of Chicago Press
Heimbauer L, Beran MJ, Owren MJ (2011) A chimpanzee recognizes synthetic speech with significantly reduced acoustic cues to phonetic content. Curr Biol 21:1210-1214
Huff TE (2011) Intellectual curiosity and the scientific revolution: a global perspective. Cambridge University Press
Ikegami E (2005) Bonds of civility: aesthetic networks and political origin of Japanese culture. Cambridge University Press
Izutsu T (1991) Consciousness and essence. Iwanami
Jackson A (2004) Comme appelé du néant — as if summoned from the void:

the life of Alexandre Grothendieck. Notices Amer Math Soc 51:1038-1056
Kacian DL, Mills DR, Kramer FR, Spiegelman S (1972) A replicating RNA molecule suitable for a detailed analysis of extracellular evolution and replication. Proc Natl Acad Sci USA 69:3038-3042
Kaneko K (2010) Life: an introduction to complex systems biology. Springer
Keeling PJ, et al (2005) The tree of eukaryotes. Trends Ecol Evol 20:670-676
King N, et al (2008) The genome of the choanoflagellate *Monosiga brevicollis* and the origin of metazoans. Nature 451:783-788
Kodaira K (1976) Introduction to analysis. Iwanami
Kolokotrones T, Savage V, Deeds EJ, Fontana F (2010) Curvature in metabolic scaling. Nature 464:753-756
Kröpelin S, et al (2007) Climate-driven ecosystem succession in the Sahara: the past 6000 years. Science 320:765-768
Kurland CG, Collins LJ, Penny D (2006) Genomics and the irreducible nature of eukaryote cells. Science 312:1011-1014
Lamb D, Erskine PD, Parrotta JA (2005) Restoration of degraded tropical forest landscapes. Science 310:1628-1632
Levine SA (1999) Fragile dominion—complexity and the commons. Perseus Publishing
Levine SA (2006) Fundamental questions in biology. PLoS Biol 4:e300
Liang M, Nielsen R (2011) Q&A: Who is *H sapiens* really, and how do we know? BMC Biol 9:20
Liberek K, Lewandowska A, Ziętkiewicz S (2008) Chaperones in control of protein disaggregation. EMBO J 27:328-335
Little SC, et al (2011) The formation of the bicoid morphogen gradient requires protein movement from anteriorly localized mRNA. PLoS Biology 9:e1000596
Lynch M (2010) Evolution of the mutation rate. Trends Genet 26:345-352
Mattick JS (2009) Deconstructing the dogma, a new view of the evolution and genetic programming of complex organisms. Ann N.Y Acad Sci 1178:29-46
Mattick JS (2011) The central role of RNA in human development and cognition. FEBS Letters 585:1600-1616
May RM (2006) Network structure and the biology of populations. Trends Evo Eco 21:394-399
Mitchell M (2009) Complexity: a guided tour. Oxford University Press
Murzin AG (1989) Metamorphic proteins. Science 320:1725-1726
Nicolis G, Prigogine I (1989) Exploration of complexity. W H Freeman
Nierhaus KH (1991) The assembly of prokaryotic ribosomes. Biochimie 73:739-755
Nozaki H, et al (2009) Phylogenetic positions of Glaucophyta, green plants (Archaeplastida) and Haptophyta (Chromalveolata) as deduced from slowly evolving nuclear genes. Mol Phyl Evol 53:872-880
Ohno S (1970) Evolution by gene duplication. Springer
Oono Y (1998) Complex systems study as biology. Int J Mod Phys B 12:245-256
Oono Y (2008) Bio-physics manifesto—for the future of physics and biology. In: Ishiwata S, Matsunaga Y (ed) Physics of self-organization systems. World Scientific
Orr HA (2010) The descent of Gould. New Yorker, 2002-09-30
Ortega y Gasset J (1932) The revolt of the masses. Norton & Co.
Panse VG, Johnson AW (2010) Maturation of eukaryotic ribosomes: acquisition of functionality. Trends Biochem Sci 35:260-266
Peterson KJ, Sperling EA (2007) Poriferan ANTP genes: primitively simple or secondarily reduced?. Evol Dev 9:405-408
Pica P, Lemer IC, Izard V, Dehaene S (2004) Exact and approximate arithmetic in an amazonian indigenous group. Science 306:499-503

Pinker S (1994) The Language instinct: how the mind creates language. Harper Collins
Pinker S (1999) Words and rules. Basic Books
Poulos JA (2008) Irreligion. Hill and Wang
Pržlj N, Corneil DG, Jurisica I (2005) Modeling interactome: scale-free or geometric? Bioinformatics 20:3508-3515
Reggia JA, Armentrout SL, Chou H-H, Peng Y (1993) Simple systems that exhibit self-directed replication. Science 259:1282-1287
Rodrigues-Martins A, Riparbelli M, Callaini G, Glover DM, Bettencourt-Dias M (2007) Revisiting the role of the mother centriole in centriole biogenesis. Science, 316:1046-1050
Roger AJ, Simpson AGB (2009) Revisiting the root of the eukaryote tree. Curr Biol 19:R165-R167
Rosen, R (1991) Life itself. Columbia University Press
Rutherford SL, Lindquist S (1998) Hsp90 as a capacitor for morphological evolution. Nature 396:336-342
Ryan JF, Mazza ME, Pang K, Matus DQ, Baxevanis AD, Martindale MQ, Finnerty JR (2007) Pre-bilaterian origins of the Hox cluster and the Hox code: evidence from the sea anemone, *Nematostella vectensis*. PLoS ONE 2:e153
Sangster TA, Salathi N, Lee HN, Watanabe E, Schellenber K, Morneau K, Wang H, Undurraga S, Queitsch C, Lindquist S (2008) HSP90-buffered genetic variation is common in *Arabidopsis thaliana*. Proc Natl Acad Sci USA 105:2969-2974
Sieg AE, O'Connor MP, McNair JN, Grant BW, Agosta SJ, Dunham AE (2009) Mammalian metabolic allometry: do intraspecific variation, phylogeny, and regression models matter?" Am Nat 174:720-733
Simpson GG (1967) The meaning of evolution— a study of history of life and of its significance for man—. Yale University Press
Smiley RD, Collins TRL, Hammes GG, Hsieh T-S (2007) Single-molecule measurements of the opening and closing of the DNA gate by eukaryotic topoisomerase II. Proc Natl Acad Sci USA 104:4840-4845
Spiegelman S (1971) An approach to the experimental analysis of precellular evolution. Q Rev Biophys 4:213-253
Spitzer JJ, Poolman B (2005) Electrochemical structure of the crowded cytoplasm. Trends Biochem Sci 30:536-541
Staley JP, Woolford Jr JL (2009) Assembly of ribosomes and spliceosomes: complex ribonucleoprotein machines. Comm Cell Biol 21:109-118
Stearns SC (2007) Are we stalled part way through a major evolutionary transition from individual to group? Evolution 61:2275-2280
Stoltzfus A (1999) On the possibility of constructive neutral evolution. J Mol Evol 49:169-181
Stuhl K (1999) Fundamentally different logic of gene regulation in eukaryotes and prokaryotes. Cell 98:1-4
Sugase K, Dyson HJ, Wright PE (2007) Mechanism of coupled folding and binding of an intrinsically disordered protein. Nature 447:1021-1025
Sykes MT, Williamson JR (2009) A complex assembly landscape of ribosomal subunit. Annu Rev Biophys 38:197-215
Szathmáry E, Maynard-Smith J (1995) The major evolutionary transitions. Nature 374:227-232
Taft RJ, Pheasant M, Mattick JS (2007) The relationship between non-protein-coding DNA and eukaryotic complexity. BioEssays 29:288-299
Taylor JS, Raes J (2004) Duplication and divergence: the evolution of new genes and old ideas. Annu Rev Genet 38:615-634
Ullers RS, Ang D, Schwager E, Georgopoulos C, Genevaux P (2007) Trigger

factor can antagonize both SecB and DnaK/DnaJ chaperone functions in *Escherichia coli.* Proc Natl Acad Sci USA 104:3101-3106

Vermeij GJ (2006) Historical contingency and the purported uniqueness of evolutionary innovations. Proc Natl Acad Sci USA 103:1804-1809

Voeltz GK, Prinz WA, Shibata Y, Rist JM, Rapoport TA (2006) A class of membrane proteins shaping the tubular endoplasmic reticulum. Cell 124:573-586

von Neumann J (1961-3) The general and logical theory of automata. In: Taub AH (ed) John von Neumann Collected Works Vol. V. Pergamon Press

Waddington CH (1957) The strategy of the gene. Allen and Unwin

Waldrop MM (1992) Complexity: the emerging science at the edge of order and chaos. Simon & Schuster

Wang M, Kurland CG, Caetano-Anollés G (2011) Reductive evolution of proteomes and protein structures. Proc Natl Acad Sci USA 108:11954-11958

Weinberg S (1992) Dream of a final theory, the scientist's search for the ultimate laws of nature. Vintage

Woese CR (1998) Default taxonomy: Ernst Mayr's view of the microbial world. Proc Natl Acad Sci USA 95:11043-11046

Wolfenstein L (2003) Lessons from Kepler and the theory of everything. Proc Natl Acad Sci USA 100:5001-5003

Young JC, Agashe VR, Siegers K, Hartl FU (2007) Pathways of chaperone-mediated protein folding in the cytosol. Nat Rev Mol Cell Biol 5:781-791

Zwanzig R, Szabo A, Bagchi B (1992) Levinthal's paradox. Proc Natl Acad Sci USA 86:20-22.

Index

MIX
Papier aus verantwortungsvollen Quellen
Paper from responsible sources
FSC® C105338
FSC
www.fsc.org

If you have any concerns about our products, you can contact us on
ProductSafety@springernature.com

In case Publisher is established outside the EU, the EU authorized representative is:
Springer Nature Customer Service Center GmbH
Europaplatz 3, 69115 Heidelberg, Germany

Printed by Libri Plureos GmbH
in Hamburg, Germany